HARR

Topics in Applied Physics
Volume 102

Available online at
SpringerLink.com

Topics in Applied Physics is part of the SpringerLink service. For all customers with standing orders for Topics in Applied Physics we offer the full text in electronic form via SpringerLink free of charge. Please contact your librarian who can receive a password for free access to the full articles by registration at:

springerlink.com → Orders

If you do not have a standing order you can nevertheless browse through the table of contents of the volumes and the abstracts of each article at:

springerlink.com → Browse Publications

Topics in Applied Physics

Topics in Applied Physics is a well-established series of review books, each of which presents a comprehensive survey of a selected topic within the broad area of applied physics. Edited and written by leading research scientists in the field concerned, each volume contains review contributions covering the various aspects of the topic. Together these provide an overview of the state of the art in the respective field, extending from an introduction to the subject right up to the frontiers of contemporary research.

Topics in Applied Physics is addressed to all scientists at universities and in industry who wish to obtain an overview and to keep abreast of advances in applied physics. The series also provides easy but comprehensive access to the fields for newcomers starting research.

Contributions are specially commissioned. The Managing Editors are open to any suggestions for topics coming from the community of applied physicists no matter what the field and encourage prospective editors to approach them with ideas.

Managing Editors

Dr. Claus E. Ascheron
Springer-Verlag GmbH
Tiergartenstr. 17
69121 Heidelberg
Germany
Email: claus.ascheron@springer.com

Dr. Hans J. Koelsch
Springer-Verlag New York, LLC
233, Spring Street
New York, NY 10013
USA
Email: hans.koelsch@springer.com

Assistant Editor

Adelheid H. Duhm
Springer-Verlag GmbH
Tiergartenstr. 17
69121 Heidelberg
Germany
Email: adelheid.duhm@springer.com

Hiroshi Imai   Masahito Hayashi (Eds.)

Quantum Computation and Information

From Theory to Experiment

With 49 Figures

Springer

Hiroshi Imai
Graduate School of Information, Science and Technology
The University of Tokyo
7-3-1 Hongo, Bunkyo-ku
Tokyo, 113-8656 Japan

and

ERATO Quantum Computation and Information Project
Japan Science and Technology Agency
201 Daini Hongo White Bldg
5-28-3, Hongo, Bunkyo-ku
Tokyo 113-0033, Japan
imai@is.s.u-tokyo.ac.jp

Masahito Hayashi
ERATO Quantum Computation and Information Project
Japan Science and Technology Agency
201 Daini Hongo White Bldg
5-28-3, Hongo, Bunkyo-ku
Tokyo 113-0033, Japan
masahito@qci.jst.go.jp

Library of Congress Control Number: 2006923435

Physics and Astronomy Classification Scheme (PACS):
03.67.Lx, 03.67.-a, 03.67.Dd, 03.67.Mn, 42.50.-p

ISSN print edition: 0303-4216
ISSN electronic edition: 1437-0859
ISBN-10 3-540-33132-8 Springer Berlin Heidelberg New York
ISBN-13 978-3-540-33132-2 Springer Berlin Heidelberg New York

This work is subject to copyright. All rights are reserved, whether the whole or part of the material is concerned, specifically the rights of translation, reprinting, reuse of illustrations, recitation, broadcasting, reproduction on microfilm or in any other way, and storage in data banks. Duplication of this publication or parts thereof is permitted only under the provisions of the German Copyright Law of September 9, 1965, in its current version, and permission for use must always be obtained from Springer. Violations are liable for prosecution under the German Copyright Law.

Springer is a part of Springer Science+Business Media

springer.com

© Springer-Verlag Berlin Heidelberg 2006
Printed in Germany

The use of general descriptive names, registered names, trademarks, etc. in this publication does not imply, even in the absence of a specific statement, that such names are exempt from the relevant protective laws and regulations and therefore free for general use.

Typesetting: DA-TEX · Gerd Blumenstein · www.da-tex.de
Production: LE-TEX Jelonek, Schmidt & Vöckler GbR, Leipzig
Cover design: *design & production* GmbH, Heidelberg

Printed on acid-free paper   57/3100/YL   5 4 3 2 1 0

Preface

Once the quantum effect had been regarded as an obstacle to suitable information processing in existing information systems. Recently, it has been discovered that quantum effect is, to the contrary, very useful as a resource used in information processesing. This research field is called quantum information and is rapidly growing as a new paradigm for information systems. For example, we can factorize a large number quickly by Shor's algorithm on a quantum computer once a quantum computer is available, and we can communicate securely without any assumption for computation complexity by using quantum key distribution. These quantum information protocols cannot be realized without quantum effects.

In the research of existing information processes, it is possible to study hardware and software separately because their roles are clearly divided. However, such separation between them becomes an obstacle for the whole research on quantum information. Toward the development of quantum algorithms and protocols, it is necessary to understand the mathematical description of quantum phenomena. The realization of quantum information systems requires development of quantum devices, for which we need to understand the theoretical scheme of quantum information science. Therefore, we need collaboration over the existing framework. To promote such collaboration, we bring this book as a collection of overviews of selected topics in quantum information.

This book organized as follows. We explain the power of quantum computation in Part I. Currently, only Shor's factorization algorithm and Grover's search algorithm are known to be faster on a quantum computer than on a classical computer. The ability of a quantum computer cannot be cleared up only by discovery of these algorithms. Now, many researchers are attempting to developing better quantum devices to build a quantum computer. However, the research for the power of quantum computers is as important as the research for quantum devices. Part I reviews the so-called identification problem of an unknown function f using a quantum computer, where the function f is often called an oracle. In fact, many problems in computer science are formulated in this form. For instance, Grover's search problem is also in this form. Part I discusses the superiority of a quantum computer over a classical computer for this type of problems. In particular, the Chapter by Ambainis et al. treats the case of no error in the computation process, and the Chapter by Iwama et al. covers the case where some errors happen in the specific points. By reviewing these topics, Part I signifies the importance of building a quantum computer.

Part II focuses on the bounds of the power of several quantum information processes and quantum entanglement, which is an important resource for quantum information protocols. The Chapter by Hayashi deals with theoretical issues on the identification of the density matrix of a quantum system. Since the perfect cloning of a quantum state is impossible and any measurement demolishes quantum states, precise identification requires a better measurement extracting much information from the quantum system. Hence, the selection of measurement is an important issue of this topic. On the other hand, an approximate cloning is possible. The Chapter by Fan discusses the bound of the performance of the quantum approximate cloning. Through the Chapters by Hiroshima et al. and Matsumoto, we give an overview of the research on quantum entanglement. The Chapter by Hiroshima et al. reviews approaches toward quantum entanglement from various viewpoints. In particular, entanglement is closely related to the problem of sending a quantum state via a noisy quantum channel. Such a relation is also discussed. The Chapter by Matsumoto focuses on the additivity problem, the hottest topic in quantum entanglement. This problem is essentially linked to the problem on sending classical information via a noisy quantum channel. We highlight this connection. Note that the problem on sending quantum state is different from the problem on sending classical information.

Part III treats secure quantum information processes. Shor's factorization algorithm makes the RSA public-key cryptosystem insecure once one builds a quantum computer. Hence, we have to prepare alternative cryptosystems as a countermeasure for realization of quantum computer. One idea is the development of public-key cryptosystem that is secure even for quantum computers. Another is an information-theoretically secure cryptographic system whose security does not depend on the assumption for computational complexity. The Chapter by Kawachi et al. highlights the former type of cryptosystems by discussing the concept of one-way functions, which is a basic concept for public key cryptosystems. Based on this concept, the quantum public-key cryptosystem is explained. This cryptosystem well works on the assumption that all component parts (eavesdropper, channel, sender, and receiver) are quantum. The Chapter by Wang treats an information-theoretically secure protocol that distributes a secret key via a quantum channel, which is called quantum key distribution. Perfect single photons and noiseless quantum channels are necessary for the realization of the initial protocol proposed by Bennett and Brassard. Hence, we need to consider the protocol of sending imperfect single-photons via noisy and lossy quantum channels. The security of the above realistic protocol is the main topic of this chapter. Secure protocols are not limited in cryptography. Steganography is known as a protocol that keeps the secret of the existence of the communication. The Chapter by Natori shows that quantum steganography exceeds classical steganography.

Finally, Part IV reports the research activities of realization of quantum information systems. This part contains experiments concerning quantum key distribution, a part of Shor's factorization algorithm, and generation of entangled states. First, we review 150 km transmission quantum key distribution and quantum key distribution with a real optical fiber of commercial use for 14 days. Next, we see how to realize quantum computation with 1024 qubits of a part of Shor's factorization algorithm. High-quality generation of entangled states is also discussed in this part.

This book is organized so that each chapter can be read independently. We recommend that the reader begins with the chapter of interest and then expand the range of this interest. We hope that the reader of this book would get interested in a wide research area of quantum information science.

In fact, the contents of this book mainly consist of research results obtained by the ERATO Quantum Computation and Information (QCI) Project. This project started in October 2000 by gathering interdiscipnary researchers from various research fields as one of Exploratory Research for Advanced Technology (ERATO) programs of Japan Science and Technology Agency (JST). This project finished in September 2005, and continued another program, Solution-Oriented Research for Science and Technology (SORST) of JST. Each chapter of this book is written by the researchers and a visiting researcher of this project.

We would like to express our gratitude to Mr. Hideo Ohgata, Mr. Jun-ichi Hoshi, Mr. Satoshi Asada, and Mr. Takanori Kamei, Department of Research Project in JST, for their kind management. We are also thankful to all the contributors for their interesting research manuscripts. We are also grateful to all the researchers of ERATO Quantum Computation and Information Project and their collaborators. Moreover, we are particularly indebted to our administrative and supporting staff, Mr. Michiyuki Amaike, Ms. Emi Bandai, Ms. Miho Inagaki, Ms. Chie Matsumoto, Ms. Minako Ooyama, Ms. Takako Sakuragi, Ms. Hiroko Takeshima, and Mr. Nobuyoshi Umezawa for their kind support. Finally, we wish to thank Dr. Claus E. Ascheron of Springer-Verlag for his excellent management for the publication of this book and his encourangement.

Hongo, Tokyo,   *Hiroshi Imai*
January 2006   *Masahito Hayashi*

Contents

Part I Quantum Computation

Quantum Identification of Boolean Oracles
Andris Ambainis, Kazuo Iwama, Akinori Kawachi, Rudy Raymond,
Shigeru Yamashita 3
1 Introduction 3
2 Formalization 5
3 General Upper Bounds 8
4 Relation With Learning Theory 11
5 Tight Upper Bounds for Small M 12
6 Classical Lower and Upper Bounds 14
7 Concluding Remarks 15
References 16
Index 18

Query Complexity of Quantum Biased Oracles
Kazuo Iwama, Rudy Raymond, Shigeru Yamashita 19
1 Introduction 19
2 Goldreich–Levin Problem and Biased Oracles 21
   2.1 The Model of Quantum Biased Oracles 26
3 Upper Bounds of the Query Complexity of Biased Oracles With
   Special Conditions 27
   3.1 Basic Tools for Quantum Computation 27
   3.2 Quantum Biased Oracles With the Same Bias Factor 28
   3.3 Quantum Biased Oracles With Resettable Condition 33
4 Lower Bounds of the Query Complexity of Biased Oracles 34
5 Concluding Remarks 39
References 40
Index 42

Part II Quantum Information

Quantum Statistical Inference
Masahito Hayashi 45

1	Introduction	45
2	Quantum State Estimation	47
	2.1 State Estimation in Pure State Family	47
	2.2 State Estimation for Covariant Pure States Family	48
	2.3 State Estimation in Gaussian States Family	49
	2.4 State Estimation in Nonregular Family	51
	2.5 Estimation of Eigenvalue of Density Matrix in Qubit System	52
3	Estimation of SU(2) Action With Entanglement	52
	3.1 One-Parameter Case	53
	3.2 Three-Parameter Case	53
4	Hypothesis Testing and Discrimination	54
	4.1 Hypothesis Testing of Entangled State	54
	4.2 Distinguishability and Indistinguishability by LOCC	55
	4.3 Application of Quantum Hypothesis Testing	55
5	Experimental Application of Quantum Statistical Inference	56
	5.1 State Estimation in the Two-Qubit System	57
	5.2 Testing of Entangled State in the SPDC System	57
6	Analysis on Quantum Measurement	58
	6.1 Quantum Measurement With Negligible State Demolition	58
	6.2 Quantum Universal Compression	58
References		59
Index		61

Quantum Cloning Machines
Heng Fan 63

1	Introduction	63
2	Bužek and Hillery Universal Quantum Cloning Machine	63
3	N to M UQCM (Gisin and Massar)	65
4	Universal Quantum Cloning Machine for General d-Dimensional System, Werner Cloning Machine	65
5	A UQCM for d-Dimensional Quantum State Proposed by Fan et al.	66
6	Further Results About the UQCM	67
	6.1 UQCM for 2-Level System	68
	6.2 UQCM for d-Level System	71
7	UQCM Realized in Real Physical Systems	74
8	UQCM for Identical Mixed States	79
	8.1 A 2 to 3 Universal Quantum Cloning for Mixed States	79
	8.2 General 2 to $M(M>2)$ UQCM	81
9	Phase-Covariant Quantum Cloning Machine	81
10	Transformation	82
11	Hilbert–Schmidt Norm	84
12	Bures Fidelity	88
13	Quantum Cloning for $x-y$ Equatorial Qubits	90
14	Quantum Cloning Networks for Equatorial Qubits	91
15	Separability of Copied Qubits and Quantum Triplicators	93

	15.1 Separability 93
	15.2 Optimal Quantum Triplicators 94
16	Optimal 1 to M Phase-Covariant Quantum Cloning Machines..... 96
17	Some Known Results About Phase-Covariant Quantum Cloning Machine for Qubits and Qutrits 98
18	Phase-Covariant Cloning of Qudits 100
19	About Phase-Covariant Quantum Cloning Machines 103
20	Cerf's Asymmetric Quantum Cloning Machine 103
21	Duan and Guo Probabilistic Quantum Cloning Machine 106
22	A Brief Summary 107

References 107
Index 109

Entanglement and Quantum Error Correction
Tohya Hiroshima, Masahito Hayashi 111
1 Introduction 111
2 Entanglement Distillation 111
    2.1 Background of Concentration............................. 112
    2.2 Exponents of Optimal Concentration 112
    2.3 Universal Entanglement Concentration 113
    2.4 Entanglement in Boson–Fock Space 114
    2.5 Computation of Distillable Entanglement of a Certain Class of Bipartite Mixed States 115
3 Quantum Error Correction 116
    3.1 Mathematical Formulation of Quantum Channel 117
    3.2 Background of Information Theory and Coding Theory 117
    3.3 Exponential Evaluation of Quantum Error Correcting Codes . 118
        3.3.1 Extensions 118
    3.4 Relation Between Teleportation and Entanglement Distillation 119
    3.5 Application to Quantum Key Distribution 120
4 Basic Characteristics of Bipartite Entanglement 121
    4.1 Concurrence Hierarchy 121
    4.2 Optimal Compression Rates and Entanglement of Purification 122
    4.3 Simultaneous Schmidt Decomposition and Maximally Correlated States 123
    4.4 Bell-Type Inequalities Via Combinatorial Approach 123
    4.5 Quantum Graph Coloring Game 124
5 SLOCC Convertibility 124
    5.1 Multipartite Entanglement............................... 124
    5.2 Bipartite Entanglement in Infinite-Dimensional Space 126
6 Protocols Assisted by Multipartite Entangled State 128
    6.1 Teleportation by W State 128
    6.2 Remote State Preparation of Entangled State 128
References 129

Index.. 132

On Additivity Questions
Keiji Matsumoto 133
1   Introduction 133
2   Additivity Questions: Definitions and Comments 134
    2.1 Holevo Capacity, Output Minimum Entropy, and Maximum
         Output p-Norm................................... 134
    2.2 Entanglement of Formation 137
3   Linking Additivity Conjectures......................... 138
    3.1 Channel States 138
    3.2 Strong Superadditivity and Additivity of Holevo Capacity ... 140
    3.3 Equivalence Theorem by Shor, and One More New
         Equivalent Additivity Question 141
    3.4 Group Symmetry 143
    3.5 Analysis of Examples.............................. 146
         3.5.1 Example 1.................................. 146
         3.5.2 Example 2.................................. 147
         3.5.3 Example 3.................................. 148
4   Additivity for Special Cases 149
    4.1 Additivity of WH Channel 149
5   Numerical Studies on Additivity Questions 153
    5.1 A Qubit Channel That Requires Four Input States .. 153
         5.1.1 Some Useful Facts 153
         5.1.2 Setup 154
         5.1.3 Heuristic Construction of a Four-State Channel 154
         5.1.4 Approximation Algorithm to Compute the Holevo
                Capacity 155
         5.1.5 Numerical Verification of Four-State Channel 156
         5.1.6 Numerical Check of Additivity 157
    5.2 Strong Superadditivity of EoF of Pure States...... 159
References 161
Index.. 164

Part III Quantum Security

Quantum Computational Cryptography
Akinori Kawachi, Takeshi Koshiba 167
1   Introduction 167
2   Quantum One-Wayness of Permutations.................... 168
    2.1 Notations and Basic Operators 170
    2.2 Worst-Case Characterization 170
    2.3 Average-Case Characterization 171
    2.4 Universal Tests 175

3   Quantum Public-Key Cryptosystem 176
    3.1   Cryptographic Properties of QSCD$_{ff}$ 177
    3.2   Trapdoor Property 178
    3.3   Reduction From the Worst Case to the Average Case 178
    3.4   Hardness of QSCD$_{ff}$ 179
    3.5   Construction 180
    3.6   Remarks 182
References 182
Index 184

Quantum Key Distribution: Security, Feasibility and Robustness

Xiang-Bin Wang 185

1   Introduction 185
2   Security Proof of BB84 QKD With Perfect Single-Photon Source .. 187
    2.1   Hashing and Error-Correction in Classical Communication ... 188
    2.2   The Main Idea of Entanglement Purification 190
    2.3   Error Test 192
    2.4   Entanglement Purification by Hashing 192
        2.4.1   Bit-Flip Error Correction 193
        2.4.2   Phase-Flip Error Correction 194
    2.5   Classicalization 195
3   Secure Key Distillation With a Known Fraction of Tagged Bits 196
    3.1   Final Key Distillation With a Fraction of Tagged Bits 196
    3.2   PNS Attack 198
4   The Decoy-State Method 200
    4.1   The Main Ideas and Results 201
    4.2   The Issue of Unconditional Security 206
    4.3   Robustness Analysis 206
    4.4   Final Key Rate and Further Studies 209
    4.5   Summary 210
5   QKD With Asymmetric Channel Noise 210
    5.1   Channel Error, Tested Error and Key-Bits Error 211
    5.2   QKD With One-Way Classical Communication 212
    5.3   Six-State Protocol With Two-Way Classical Communications . 214
6   Quantum Key Distribution With Encoded BB84 States 217
    6.1   A Protocol For Collective Channel Noise 219
        6.1.1   Protocol 1 and Security Proof 220
        6.1.2   Protocol 2 221
        6.1.3   Physical Realization 223
        6.1.4   Another Protocol For Robust QKD With Swinging
               Objects 223
        6.1.5   Summary and Discussion 224
    6.2   A Protocol With Independent Noise 225
        6.2.1   The Method and the Main Idea 225

        6.2.2 The QPFER Code 226
        6.2.3 The Protocol and Its Linear Optical Realization 227
        6.2.4 Security Proof 230
        6.2.5 Subtlety of the "Conditional Advantage" 230
7   Summary and Concluding Remarks 231
References 231
Index 233

Why Quantum Steganography Can Be Stronger Than Classical Steganography
Shin Natori 235
1   Introduction 235
2   Definitions 235
    2.1   General Model of Steganography System 235
    2.2   Classical Model of Steganography System 237
3   Related Works 237
4   Quantum Steganography 237
    4.1   Model of Quantum Steganography System 237
    4.2   Comparison Between Classical and Quantum Steganography . 238
5   Conclusions and Future Work 240
References 240
Index 240

Part IV Realization of Quantum Information System

Photonic Realization of Quantum Information Systems
Akihisa Tomita, Bao-Sen Shi 243
1   Introduction 243
2   Cryptography 244
    2.1   High-Sensitivity Photon Detector 245
          2.1.1   Requirement for Single-Photon Detectors 245
          2.1.2   Improved Single-Photon Detector for Fiber Transmission 246
    2.2   Single-Photon Transmission Over 150 km in a Unidirectional
          System With Integrated Interferometers 248
    2.3   Refinements Toward a Practical QKD System 251
          2.3.1   Temperature-Insensitive Interferometer 251
          2.3.2   High-Speed Operation 253
3   Quantum Computation 256
    3.1   Measured Quantum Fourier Transform 256
          3.1.1   Implementation and Experimental Results 256
          3.1.2   Effects of Imperfection 258
          3.1.3   Validity of Majority Voting 260
          3.1.4   Toward Quantum Computers 262
    3.2   Control-Unitary Gates 263

		3.2.1 Solid-State Bell State Measurement Devices by Two-Photon Absorption 264
4	\multicolumn{2}{l	}{Generation of Entangled Photon Pairs by SPDC .... 265}
	4.1	SPDC With Two-Crystal Geometry 266
	4.2	Interferometric Generation of Entangled Photon Pairs 269
	4.3	A New Material for SPDC: Periodically Poled KTP 270
5	\multicolumn{2}{l	}{Conclusion .... 272}

References 273

Index.. 275

Index 277

Part I

Quantum Computation

Quantum Identification of Boolean Oracles

Andris Ambainis[1], Kazuo Iwama[2], Akinori Kawachi[3], Rudy Raymond[2], and Shigeru Yamashita[4]

[1] Department of Combinatorics and Optimization, Univ. of Waterloo
    ambainis@math.uwaterloo.ca
[2] Graduate School of Informatics, Kyoto University
    {iwama,raymond}@kuis.kyoto-u.ac.jp
[3] Graduate School of Information Science and Engineering,
    Tokyo Institute of Technology
    kawachi@is.titech.ac.jp
[4] Graduate School of Information Science,
    Nara Institute of Science and Technology
    ger@is.naist.jp

Abstract. We introduce the Oracle Identification Problem (OIP), which includes many problems in oracle computation such as those of Grover search and Bernstein–Vazirani as its special cases. We give general upper and lower bounds on the number of oracle queries of OIP. Thus, our results provide general frameworks for analyzing the quantum query complexity of oracle computation. Our results are also related to exact learning in the computational learning theory.

1 Introduction

An *oracle* is given as a Boolean function of n variables, denoted by $f(x_0..x_{n-1})$, and so there are 2^{2^n} (or 2^N for $N = 2^n$) different oracles. An *oracle computation* is, given a specific oracle f which we do not know, to determine, through queries to the oracle, whether or not f satisfies a certain property. Note that f has N black-box 0/1 values, $f(0,\ldots,0)$ through $f(1,\ldots,1)$. ($f(0,\ldots,0)$ is also denoted as $f(0)$, $f(1,\ldots,1)$ as $f(N-1)$, and similarly for an intermediate $f(j)$.) So, in other words, we are asked whether or not these N bits satisfy the property. There are many such interesting properties: For example, it is called *OR* if the question is whether all the N bits are 0 and *PARITY* if the question is whether the N bits include an even number of 1's. The most general question (or task in this case) is how to obtain all the N bits. Our complexity measure is the so-called *query complexity*, i.e., the number of oracle calls, to get a right answer with bounded error. Note that the trivial upper bound is N since we can know all the N bits by asking $f(0)$ through $f(N-1)$. If we use a classical computer, this N is also a lower bound in most cases. If we use a quantum computer, however, several interesting speedups are obtained. For example, the previous three problems have (quantum) query complexities of $O(\sqrt{N})$, $\frac{N}{2}$ and $\frac{N}{2} + \sqrt{N}$, respectively [1, 2, 3, 4].

In this Chapter, we discuss the following problem, which we call the *oracle identification* problem: We are given a set S of M different oracles out of the

2^N ones for which we have the complete information (i.e., for each of the 2^N oracles, we know whether it is in S or not). Now we are asked to determine which oracle in S is currently in the black box. A typical example is the Grover search [1], where $S = \{f_0, \ldots, f_{N-1}\}$ and $f_i(j) = 1$ iff $i = j$. (Namely, exactly one bit among the N bits is 1 in each oracle in S. Finding its position is equivalent to identifying the oracle itself.) It is well known that its query complexity is $\Theta(\sqrt{N})$. Another example is the so-called Bernstein–Vazirani problem [5], where $S = \{f_0, \ldots, f_{N-1}\}$ and $f_i(j) = 1$ iff the inner product of i and j (mod 2) is 1. A little surprisingly, its query complexity is just 1.

Thus the oracle identification problem is a promise version of the oracle computation problem. For both oracle computation and oracle identification problems, the paper [6] developed a very general method for proving their lower bounds of the query complexity. Also, many nontrivial upper bounds are known, as mentioned above. However, all those upper bounds are for specific problems such as the Grover search; no general upper bounds for a wide class of problems are known so far.

Our Contribution.

In this Chapter, we give general upper and lower bounds for the oracle identification problem. More concretely, we prove: 1. The query complexity of the oracle identification for *any* oracle set S is $O(\sqrt{N \log M \log N} \log \log M)$ if $|S| = M > N$. 2. It is $O(\sqrt{N})$ for *any* S if $|S| = N$. 3. For a wide range of oracles ($M = N$), such as random oracles and balanced oracles, the query complexity is $\Theta\left(\sqrt{\frac{N}{K}}\right)$, where K is a parameter determined by S. The bound in 1 is better than the obvious bound N if $M < 2^{N/\log^3 N}$. Both algorithms for 1 and 2 are quite tricky, and the result 2 includes the upper bound for the Grover search as a special case. Result 1 is almost optimal, and results 2 and 3 are optimal; to prove their optimality we introduce a general lower bound theorem whose statement is simpler than that of [6].

Related Results.

Query complexity has consistently been one of the central topics in quantum computation; to cover everything is obviously impossible. For the upper bounds of query complexity, the most significant result, known as the Grover search, is due to [1], which also derived many applications and extensions [2, 7, 8, 9, 10]. In particular, some results showed efficient quantum algorithms by combining the Grover search with other (quantum and classical) techniques. For example, the quantum counting algorithm [11] gives an approximate counting method by combining the Grover search with the quantum Fourier transformation, and quantum algorithms for the claw-finding and the element distinctness problems [12] also exploit classical random sampling and sorting. Most recently, the paper [13] developed an optimal quantum algorithm with $O(N^{2/3})$ queries for the element distinctness problem, which makes use of quantum walk and matches to the lower bounds in [14]. The paper [15] used the element distinctness algorithm to design a quantum algorithm for finding triangles in a graph. The paper [16] also showed an efficient quantum search algorithm for spatial regions based on recursive Grover search, which is applicable to some geometrically structured problems such as search on a 2-D grid.

On the lower-bound side, there are two popular techniques to derive quantum lower bounds, i.e., the polynomial method and the quantum adversary method. The polynomial method was firstly introduced for quantum computation by *Beals* et al. [17], who borrowed the idea from the classical counterpart. For example, it was shown that for bounded error cases, evaluations of AND and OR functions need $\Theta(\sqrt{N})$ number of queries, while PARITY and MAJORITY functions need at least $N/2$ and $\Theta(N)$, respectively. Recently, *Shi* [14] and *Aaronson* [18] used the polynomial method to show the lower bounds for the collisions and element distinctness problems.

The classical adversary method, which is also called the hybrid argument, was used in [19, 20]. Their method can be used, for example, to show the lower bound of the Grover search. As mentioned above, the paper [6] introduced a quite general method, which is known as the quantum adversary argument, for obtaining lower bounds of various problems, e.g., the Grover search, AND of ORs and inverting a permutation. *Barnum* and *Saks* [21] recently established a lower bound of $\Omega(\sqrt{N})$ on the bounded-error quantum query complexity of read-once Boolean functions by extending the results in [6]. *Barnum* et al. [22] generalized the quantum adversary method from the aspect of semidefinite programming, and *Laplante* and *Magniez* [23] generalized the method from Kolmogorov complexity perspective. Furthermore, *Dürr* et al. [24] and *Aaronson* [25] showed the lower bounds for graph connectivity and local search problem, respectively, using the quantum adversary method. The paper [26] also gave a comparison between the quantum adversary method and the polynomial method.

2 Formalization

Our model is basically the same as standard ones (see, e.g., [6]). For a Boolean function $f(x_0, \ldots, x_{n-1})$ of n variables, an *oracle* maps $|x_0, \ldots, x_{n-1}\rangle|b\rangle$ to $(-1)^{b \cdot f(x_0, \ldots, x_{n-1})}|x_0, \ldots, x_{n-1}\rangle|b\rangle$. A *quantum computation* is a sequence of unitary transformations $U_0 \to O \to U_1 \to O \to \cdots \to O \to U_t$, where O is a single oracle call against our *black-box oracle* (sometimes called an *input oracle*), and U_j may be any unitary transformation without oracle calls. The above computation sequence involves t oracle calls, which is our measure of the complexity (*the query complexity*). Let $N = 2^n$, and hence there are 2^N different oracles.

Our problem is called the *oracle identification problem* (*OIP*). An OIP is given as an infinite sequence $S_1, S_2, S_4, \ldots, S_N, \ldots$. Each S_N ($N = 2^n, n = $

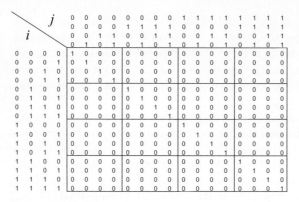

Fig. 1. The OIP matrix of Grover search: $f_i(j) = 1$ iff $i = j$

$0, 1, \ldots$) is a set of oracles (Boolean functions with n variables) whose size, $|S_N|$, is denoted by M ($\leq 2^N$). A (quantum) algorithm $\mathcal{A}$ which solves the OIP is a quantum computation as given above. $\mathcal{A}$ has to determine which oracle ($\in S_N$) is the current input oracle with bounded error. If $\mathcal{A}$ needs at most $g(N)$ oracle calls, we say that the query complexity of $\mathcal{A}$ is $g(N)$. It should be noted that $\mathcal{A}$ knows the set S_N completely; what is unknown for $\mathcal{A}$ is the current input oracle.

For example, the Grover search is an OIP whose S_N contains N (i.e., $M = N$) Boolean functions $f_1, \ldots, f_N$ such that

$$f_i(j) = 1 \text{ iff } i = j\,.$$

Note that $f(j)$ means $f(a_0, a_1, \ldots, a_{n-1})$ ($a_i = 0$ or 1) such that $a_0, \ldots, a_{n-1}$ is the binary representation of the number j. Note that S_N is given as an $M \times N$ Boolean matrix. More formally, the entry at row i ($0 \leq i \leq M-1$) and column j ($0 \leq j \leq N-1$) shows $f_i(j)$. Figure 1 shows such a matrix of the Grover search for $N = M = 16$. Each row corresponds to each oracle in S_N, and each column to its Boolean value. Figure 2 shows another famous example given by an $N \times N$ matrix, which is called the Bernstein–Vazirani problem [5]. It is well known that there is an algorithm whose query complexity is just 1 for this problem [5].

As described in the previous section, there are several similar, but subtly different settings. For example, the problem in [6, 27] is given as a matrix which includes all the rows (oracles), each of which contains $N/2$ 1's or $(1/2+\varepsilon)N$ 1's for $\varepsilon > 0$. We do not have to identify the current input oracle itself but have only to answer whether the current oracle has $N/2$ 1's or not. (The famous Deutsch–Jozsa problem [28] is its special case.) The l-target Grover search is given as a matrix consisting of all (or a part of) the rows containing l 1's. Again, we do not have to identify the current input oracle but have to answer with a column which has value 1 in the current input. Figure 3 shows an example, where each row contains $N/2 + 1$ ones. One can see that

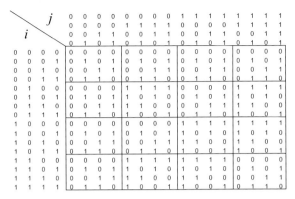

Fig. 2. The OIP matrix of the Bernstein–Vazirani problem: $f_i(j) = i \cdot j = \sum_x i_x \cdot j_x \mod 2$

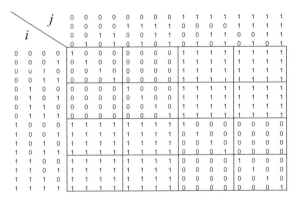

Fig. 3. An example of a harder case for OIP matrix

the multitarget Grover search is easy ($O(1)$ queries are enough since we have roughly one half 1's), but identifying the input oracle itself is much harder.

The paper [6] gave a very general lower bound for oracle computation. When applied to the OIP (the original statement is more general), it claims the following:

Proposition 1. *Let S_N be a given set of oracles, and X, Y be two disjoint subsets of S_N. Let $R \subset X \times Y$ be such that:*

1. *For every $f_a \in X$, there exist at least m different $f_b \in Y$ such that $(f_a, f_b) \in R$.*
2. *For every $f_b \in Y$, there exist at least m' different $f_a \in X$ such that $(f_a, f_b) \in R$.*

Let $l_{f_a,i}$ be the number of $f_b \in Y$ such that $(f_a, f_b) \in R$ and $f_a(i) \neq f_b(i)$, and $l_{f_b,i}$ be the number of $f_a \in X$ such that $(f_a, f_b) \in R$ and $f_a(i) \neq f_b(i)$. Let

$l_{\max}$ be the maximum of $l_{f_a,i} l_{f_b,i}$ over all $(f_a, f_b) \in R$ and $i \in \{0, \ldots, N-1\}$ such that $f_a(i) \neq f_b(i)$. Then, the query complexity for S_N is $\Omega\left(\sqrt{\frac{mm'}{l_{\max}}}\right)$.

In this Chapter, we always assume that $M \geq N$. If $M \leq N/2$, then we can select M columns out of the N ones while keeping the uniqueness property of each oracle. Then by changing the state space from n bits to at most $n-1$ bits, we have a new $M \times M$ matrix, i.e., a smaller OIP problem.

3 General Upper Bounds

As mentioned in the previous section, we have a general lower bound for the OIP. But we do not know any nontrivial general upper bounds. In this section, we give two general upper bounds for the case that $M > N$ and for the case that $M = N$. The former is almost tight as described after the theorem, and the latter includes the upper bound for the Grover search as a special case. An $M \times N$ OIP denotes an OIP whose S_N (or simply S by omitting the subscript) is given as an $M \times N$ matrix as described in the previous section. Before proving the theorems, we introduce a convenient technique called a *column flip*.

Column Flip.

Suppose that S is any $M \times N$ matrix (a set of M oracles). Then any quantum computation for S can be transformed into a quantum computation for an $M \times N$ matrix S' such that the number of 1's is less than or equal to the number of 0's in every column. We say that such a matrix is 1-*sensitive*. The reason is straightforward. If some column in S holds more 1's than 0's, then we "flip" all the values. Of course, we have to change the current oracle into the new ones but this can be easily done by adding an extra circuit to the output of the oracle.

Theorem 1. *If $M > N$, the query complexity of any $M \times N$ OIP is $O(\sqrt{N \log M \log N} \log \log M)$.*

Proof. To see the idea, we first prove an easier bound $O(\sqrt{N} \log M \log \log M)$. (Since M can be an exponential function in N, this bound is significantly worse than that of the theorem.) If necessary, we convert the given matrix S to be 1-sensitive by column flip. Then, just apply the Grover search against the input oracle. If we get a column j (the input oracle has 1 there), then we can eliminate all the rows having 0 in that column. The number of such removed rows is at least one half by the 1-sensitivity. Just repeat this (including the conversion to 1-sensitive matrices) until the number of rows becomes 1, which needs $O(\log M)$ rounds. Each Grover search needs $O(\sqrt{N})$ oracle calls. Since we perform many Grover searches, the $\log \log M$ term is added to take care of the success probability.

In this algorithm we counted $O(\sqrt{N})$ oracle calls for the Grover search, which is the target of our improvement. More precisely, our algorithm is the following quantum procedure. Let $S = \{f_0, ..., f_{M-1}\}$ be the given $M \times N$ matrix:

Step 1. Let $Z \subseteq S$ be a set of candidate oracles (or equivalently an $M \times N$ matrix, each row of which corresponds to each oracle). Set $Z = S$ initially.

Step 2. Repeat Steps 3–6 until $|Z| = 1$.

Step 3. Convert Z into a 1-sensitive matrix.

Step 4. Compute the largest integer K such that at least one half of the rows of Z contain K 1's or more. (This can be done simply by sorting the rows of Z with the number of 1's.)

Step 5. For the current (modified) oracle, perform the multitarget Grover search [2], where we set $\frac{9}{2}\sqrt{N/K}$ to the maximum number of oracle calls. Iterate this Grover search $\log \log M$ times (to increase the success probability).

Step 6. If we succeed in finding 1 by the Grover search in the previous step, i.e., a column j such that the current oracle actually has 1 in that column, then eliminate all the rows of Z having 0 in the column j. (Let Z be this reduced matrix.) Otherwise eliminate all the rows of Z having at least K 1's.

Now we estimate the number of oracle calls in this algorithm. Let M_r and K_r be the number of the rows of Z and the value of K in the rth repetition, respectively. Initially, $M_1 = M$. Note that the number of the rows of Z becomes $|Z|/2$ or less after step 6, i.e., $M_{r+1} \leq M_r/2$, even if the Grover search is successful or not in step 5 since the number of 1's in each column of the modified matrix is less than $|Z|/2$ and the number of the rows which have at least K 1's is $|Z|/2$ or more. Assuming that we need T repetitions to identify the current input oracle, the total number of the oracle calls is

$$\frac{9}{2}\left(\sqrt{\frac{N}{K_1}} + \cdots + \sqrt{\frac{N}{K_T}}\right)\log\log M.$$

We estimate the lower bounds of K_r. Note that there are no identical rows in Z and the number of possible rows that contain at most K_r 1's is $\sum_{i=0}^{K_r}\binom{N}{i}$ in the rth repetition. Thus, it must hold that $\frac{M_r}{2} \leq \sum_{i=0}^{K_r}\binom{N}{i}$. Since $\sum_{i=0}^{K_r}\binom{N}{i} \leq 2N^{K_r}$, $K_r = \Omega\left(\frac{\log M_r}{\log N}\right)$ if $M_r \geq N$, otherwise $K_r \geq 1$. Therefore the number of the oracle calls is at most

$$\frac{9}{2}\sqrt{N}\log\log M \sum_{i=1}^{T'}\sqrt{\frac{\log N}{\log M_i}} + \frac{9}{2}\sqrt{N}\log\log M \log N,$$

where the number of rows of Z becomes N or less after the T'-th repetition. For $\{M_1, \ldots, M_{T'}\}$, there exists a sequence of integers $\{k_1, \ldots, k_{T'}\}$ ($1 \le k_1 < \cdots < k_{T'} \le \log M$) such that

$$1 \le \frac{M}{2^{k_{T'}}} < M_{T'} \le \frac{M}{2^{k_{T'}-1}} \le \cdots \le \frac{M}{2^{k_2}} < M_2 \le \frac{M}{2^{k_1}} < M_1 = M,$$

since $M_r/2 \ge M_{r+1}$ for $r = 1, \ldots, T'$. Thus, we have

$$\sum_{i=1}^{T'} \frac{1}{\sqrt{\log M_i}} \le \sum_{i=1}^{T'} \frac{1}{\sqrt{\log(M/2^{k_i})}} \le \sum_{i=0}^{\log M - 1} \frac{1}{\sqrt{\log M - i}} \le 2\sqrt{\log M}.$$

Then, the total number of oracle calls is $O\left(\sqrt{N \log M} \log N \log \log M\right)$.

Next, we consider the success probability of our algorithm. By the analysis of the Grover search in [2], if the number of 1's of the current modified oracle is larger than K_r in the rth repetition, then we can find 1 in the current modified oracle with probability at least $1 - (3/4)^{\log \log M}$. This success probability worsens after T rounds of repetition but still keeps a constant as follows: $(1 - (3/4)^{\log \log M})^T \ge (1 - 1/\log M)^{\log M} = \Omega(1)$.

Theorem 2. *There is an OIP whose query complexity is $\Omega\left(\sqrt{\frac{N}{\log N}} \log M\right)$.*

Proof. This can be shown in the same way as Theorem 5.1 in [6] as follows. Let X be the set of all the oracles whose values are 1 at exactly K positions, and Y be the set of all the oracles that have 1's at exactly $K+1$ positions. We consider the union of X and Y for our oracle identification problem. Thus, $M = |X| + |Y| = \binom{N}{K} + \binom{N}{K+1}$, and therefore we have $\log M < K \log N$. Let also a relation R be the set of all (f, f') such that $f \in X$, $f' \in Y$ and they differ in exactly a single position. Then the parameters in Theorem 5.1 in [6] take values $m = \binom{N-K}{1} = N-K$, $m' = \binom{K+1}{1} = K+1$ and $l = l' = 1$. Thus the lower bound is $\Omega(\sqrt{(N-K)(K+1)})$. Since $\log M = O(K \log N)$, K can be as large as $\Omega(\frac{\log M}{\log N})$, which implies our lower bound.

Thus the bound in Theorem 1 is almost tight but not exactly. When $M = N$, however, we have another algorithm which is tight within a factor of a constant. Although we prove the theorem for $M = N$, it also holds for $M = poly(N)$, as shown in detail in [29].

Theorem 3. *The query complexity of any $N \times N$ OIP is $O(\sqrt{N})$.*

Proof. Let S be the given $N \times N$ matrix. Our algorithm is the following procedure:

1. Let $Z = S$. If there is a column in Z which has at least $\sqrt{N}$ 0's and at least $\sqrt{N}$ 1's, then perform a *classical* oracle call with this column. Eliminate all the inconsistent rows and update Z.

2. Modify Z to be 1-sensitive. Perform the multitarget Grover search [2] to obtain column j.
3. Find a column k which has 0 and 1 in some row while the column j obtained in step 2 has 1 in that row (there must be such a column because any two rows are different). Perform a *classical* oracle call with column k and remove inconsistent rows. Update Z. Repeat this step until $|Z| = 1$.

Since the correctness of the algorithm is obvious, we only prove the complexity. A single iteration of step 1 removes at least $\sqrt{N}$ rows, and hence we can perform at most $\sqrt{N}$ iterations (at most $\sqrt{N}$ oracle calls). Note that after this step each column of Z has at most $\sqrt{N}$ 0's or at most $\sqrt{N}$ 1's. Since we perform the column flip in step 2, we can assume that each column has at most $\sqrt{N}$ 1's. The Grover search in step 2 needs $O(\sqrt{N})$ oracle calls. Since column j has at most $\sqrt{N}$ 1's, the classical elimination in step 3 needs at most $\sqrt{N}$ oracle calls.

4 Relation With Learning Theory

Our technique to reduce rows of S (or oracle candidates) can be considered as the so-called *halving algorithm* as follows. Let us consider the algorithm in Theorem 1. To fix the black-box oracle, we construct a hypothesis $h : \{1, \ldots, N\} \to \{0, 1\}$ whose properties are satisfied by at most $1/2$ of the oracles in S (step 4) and can be verified with relatively few queries. Here, h is the hypothesis that the number of 1's in the black-box oracle is at least K. Now, step 5 is the verification of the above hypothesis such that only one half of the oracles in S satisfy the verification result.

The above idea is furtherly refined by *Atici* et al. in [30] in the context of quantum learning theory. They consider the hypothesis such as the value of the black-box oracle on some column set $T \in \{1, \ldots, N\}$ is "1". The column set T is constructed by column flipping and adding one of the columns with the largest number of 1's to T, repeatedly, until T covers at least a quarter of the rows of S. In this case, the size of T is at most $1/\gamma_S$, where γ_S is the combinatorial parameter denoting the smallest fraction of oracles in any subset of S that can be eliminated by knowing the value of the black box at a particular column. This directly results in a quantum algorithm for learning a concept class S with $O(\log |S| \log \log |S| / \sqrt{\gamma_S})$ queries, almost establishing a conjecture in [31] of $O(\log |S| / \sqrt{\gamma_S})$ queries. In contrast, our algorithm establishes another conjecture in [31], which states that any concept class S can be learned using $O(\sqrt{|S|})$ queries. Actually, for larger $|S|$ our result is much better than what was conjectured.

5 Tight Upper Bounds for Small M

In this section, we investigate the case that $M = N$ in more detail. Note that Theorem 3 is tight for the whole $N \times N$ OIP but not for its subfamilies. (For example, the Bernstein–Vazirani problem needs only $O(1)$ queries.) To seek optimal bounds for subfamilies, we introduce the following parameter: Let S be an OIP given as an $M \times N$ matrix. Then $\#(S)$ is the maximum number of 1's in a single column of the matrix. We first give a lower bound theorem in terms of this parameter, which is a simplified version of Proposition 1.

Theorem 4. *Let S be an $M \times N$ matrix and $K = \#(S)$. Then S needs $\Omega(\sqrt{M/K})$ queries.*

Proof. Without loss of generality, we can assume that S is 1-sensitive, i.e., $K \leq M/2$. We select X (Y, resp.) as the upper (lower, resp.) half of S (i.e., $|X| = |Y| = M/2$) and set $R = X \times Y$ (i.e., $(x,y) \in R$ for every $x \in X$ and $y \in Y$). Let δ_j be the number of 1's in the jth column of Y. Now it is not hard to see that we can set $m = m' = \frac{M}{2}$, $l_{x,j} l_{y,j} = \max\{\delta_j(\frac{M}{2} - K_j + \delta_j), (\frac{M}{2} - \delta_j)(K_j - \delta_j)\}$, where K_j is the number of 1's in column j. Since $K_j \leq K$, this value is bounded from above by $\frac{M}{2}K$. Hence, Proposition 1 implies $\Omega\left(\sqrt{\frac{mm'}{l_{\max}}}\right) \geq \Omega\left(\sqrt{\frac{(\frac{M}{2})^2}{\frac{M}{2}K}}\right) = \Omega\left(\sqrt{\frac{M}{K}}\right)$.

Although this lower bound looks much simpler than Proposition 1, it is equally powerful for many cases. For example, we can obtain $\Omega(\sqrt{N})$ lower bound for the OIP given in Fig. 3, which we denote by X. Note in general that if we need t queries for a matrix S, then we also need at least t queries for any $S' \supseteq S$. Therefore it is enough to obtain a lower bound for the matrix X' which consists of the $N/2$ upper-half rows of X, and all the 1's of the right half can be changed to 0's by the column flip. Since $\#(X') = 1$, Theorem 4 gives us a lower bound of $\Omega(\sqrt{N})$.

Now we give tight upper bounds for three subfamilies of $N \times N$ matrices. The first one is not a worst-case bound but an average-case bound: Let $AV(K)$ be an $N \times N$ matrix where each entry is 1 with the probability K/N.

Theorem 5. *The query complexity for $AV(K)$ is $\Theta(\sqrt{N/K})$ with high probability if $K = N^\alpha$ for $0 < \alpha < 1$.*

Proof. Suppose that X is an $AV(K)$. By using a standard Chernoff-bound argument, we can show that the following three statements hold for X with high probability (proofs are omitted): 1. Let c_i be the number of 1's in column i. Then for any i, $1/2K \leq c_i \leq 2K$. 2. Let r_j be the number of 1's in row j. Then for any j, $1/2K \leq r_j \leq 2K$. 3. Suppose that D is a set of any d columns in X (d is a function in α which is constant since α is a constant). Then the number of rows which have 1's in all the columns in D is at most $2 \log N$. Our lower bound is immediate from 1 by Theorem 4. For the

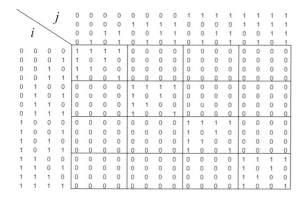

Fig. 4. An example of hybrid matrix $H(k)$ with $n = 4$ and $k = 2$

upper bound, our algorithm is quite simple: just perform the Grover search independently d times. Each single round needs $O(\sqrt{N/K})$ oracle calls by 2. After that the number of candidates is decreased to $2 \log N$ by 3. Then we simply perform the classical elimination, just as step 3 of the algorithm in the proof of Theorem 3, which needs at most $2 \log N$ oracle calls. Since d is a constant, the overall complexity is $O(\sqrt{N/K}) + \log N = \sqrt{N/K}$ if $K = N^\alpha$.

The second subfamily is called a *balanced matrix*. Let $B(K)$ be a family of $N \times N$ matrices in which every row and every column has exactly K 1's. (Again the theorem holds if the number of 1's is $\Theta(K)$.)

Theorem 6. *The query complexity for $B(K)$ is $\Theta(\sqrt{N/K})$ if $K \leq N^{1/3}$.*

Proof. The lower-bound part is obvious by Theorem 4. The upper-bound part is to use a single Grover search + K classical elimination. Thus the complexity is $O(\sqrt{N/K} + K)$, which is $O(\sqrt{N/K})$ if $K \leq N^{1/3}$.

The third subfamily is somewhat artificial. Let $H(k)$, called a *hybrid matrix* because it is a combination of Grover and Bernstein–Vazirani matrices, be a matrix defined as follows: Let $a = (a_1, a_2, \ldots, a_{n-k}, a_{n-k+1}, \ldots, a_n)$ and $x = (x_1, x_2, \ldots, x_{n-k}, x_{n-k+1}, \ldots, x_n)$. Then $f_a(x) = 1$ iff 1. $(a_1, \ldots, a_{n-k}) = (x_1, \ldots, x_{n-k})$ and 2. $(a_{n-k+1}, \ldots, a_n) \cdot (x_{n-k+1}, \ldots, x_n) = 0 \pmod{2}$. Figure 4 shows the case that $k = 2$ and $n = 4$.

Theorem 7. *The query complexity for $H(k)$ is $\Theta(\sqrt{N/K})$, where $K = 2^k$.*

Proof. We combine the Grover search [1, 2] with the Bernstein–Vazirani algorithm [5] to identify the oracle f_a by determining the hidden value a of f_a. We first can determine the first $n - k$ bits of a. Fixing the last k bits to $|0\rangle$, we apply the Grover search using oracle f_a for the first $n - k$ bits to determine $a_1, \ldots, a_{n-k}$. It should be noted that $f_a(a_1, \ldots, a_{n-k}, 0, \ldots, 0) = 1$ and $f_a(x_1, \ldots, x_{n-k}, 0, \ldots, 0) = 0$ for any $x_1, \ldots, x_k \neq a_1, \ldots, a_k$. Next, we apply the Bernstein–Vazirani algorithm to determine the remaining k bits of a. This algorithm requires $O(\sqrt{N/K})$ queries for the Grover search and $O(1)$ queries for the Bernstein–Vazirani algorithm to determine a. Therefore we can identify the oracle f_a using $O(\sqrt{N/K})$ queries.

6 Classical Lower and Upper Bounds

The lower bound for the general $M \times N$ OIP is obviously N if $M > N$. When $M = N$, we can obtain bounds being smaller than N for some cases.

Theorem 8. *The deterministic query complexity for $N \times N$ OIP S with $\#(S) = K$ is at least $\lfloor \frac{N}{K} \rfloor + \lfloor \log K \rfloor - 2$.*

Proof. Let f_a be the current input oracle. The following proof is due to the standard adversary argument. Let A be any deterministic algorithm using the oracle f_a. Suppose that we determine $a \in \{0,1\}^n$ to identify the oracle f_a. Then the execution of A is described as follows: 1. In the first round, A calls the oracle with the predetermined value x_0, and the oracle answers with $d_0 = f_a(x_0)$. 2. In the second round, A calls the oracle with value x_1, which is determined by d_0, and the oracle answers with $d_1 = f_a(x_1)$. 3. In the $(i+1)$th round, A calls the oracle with x_i, which is determined by $d_0, d_1, \ldots, d_{i-1}$, and the oracle answers with $d_i = f_a(x_i)$. 4. In the mth round A outputs a, which is determined by $d_0, d_1, \ldots, d_{m-1}$, and stops. Thus, the execution of A is completely determined by the sequence $(d_0, d_1, \ldots, d_{m-1})$ which is denoted by $A(a)$. (Obviously, if we fix a specific a, then $A(a)$ is uniquely determined).

Let $m_0 = \lfloor N/K \rfloor + \lfloor \log K \rfloor - 3$ and suppose that A halts in the m_0-th round. We compute the sequence $(c_0, c_1, \ldots, c_{m_0})$, $c_i \in \{0,1\}$, and another sequence $(L_0, L_1, \ldots, L_{m_0})$, $L_i \subseteq \{a | a \in \{0,1\}^n\}$, as follows (note that $c_0, \ldots, c_{m_0}$ are similar to $d_0, \ldots, d_{m-1}$ above and are chosen by the adversary): 1. $L_0 = \{0,1\}^n$. 2. Suppose that we have already computed $L_0, \ldots, L_i$, and $c_0, \ldots, c_{i-1}$. Let x_i be the value with which A calls the oracle in the $(i + 1)$th round. (Recall that x_i is determined by $c_0, \ldots, c_{i-1}$.) Let $L^0 = \{s | f_s(x_i) = 0\}$ and $L^1 = \{s | f_s(x_i) = 1\}$. Then if $|L_i \cap L^0| \geq |L_i \cap L^1|$ we set $c_i = 0$ and $L_{i+1} = L_i \cap L^0$. Otherwise, i.e., if $|L_i \cap L^0| < |L_i \cap L^1|$, then we set $c_i = 1$ and $L_{i+1} = L_i \cap L^1$. Now we can make the following two claims:

Claim 1. $|L_{m_0}| \geq 2$. (Reason: Note that $|L_0| = N$ and the size of L_i decreases as i increases. By the construction of L_i, one can see that until $|L_i|$ becomes $2K$, its size decreases additively by at most K in a single round, and after that it decreases multiplicatively at most one half. The claim then follows by a simple calculation.)

Claim 2. If $a \in L_{m_0}$, then $(c_0, \ldots, c_{m_0}) = A(a)$. (Reason: Obvious, since $a \in L_0 \cap L_1 \cap \cdots \cap L_{m_0}$.)

Now it follows that there are two different a_1 and a_2 in L_{m_0} such that $A(a_1) = A(a_2)$ by claims 1 and 2. Therefore A outputs the same answer for two different a_1 and a_2, a contradiction.

For the classical upper bounds, we only give the bound for the hybrid matrix. Similarly for $AV(K)$ and $B(K)$.

Theorem 9. *The deterministic query complexity for $H(k)$ is $O(\frac{N}{K} + \log K)$.*

Proof. Let f_a be the current input oracle. The algorithm consists of an exhaustive and a binary search to identify the oracle f_a by determining the hidden value a of f_a. First, we determine the first $n - k$ bits of a by fixing the last k bits to all 0's and using exhaustive search. Second, we determine the last k bits of a by using binary search. This algorithm needs $2^{n-k}(= \frac{N}{K})$ queries in the exhaustive search, and $O(k)(= O(\log K))$ queries in the binary search. Therefore, the total complexity of this algorithm is $O(2^{n-k} + k) = O(\frac{N}{K} + \log K)$.

7 Concluding Remarks

A natural question is the possibility of improving Theorem 1, for which our target is an algorithm whose number of queries matches that in Theorem 2 for small M, and $O(\sqrt{N \log M})$ for sufficiently large M. Recently, it turned out that by combining the idea of column flip with a more refined hypothesis construction, we can construct such an algorithm as shown in [29].

It is also important to note that our problem OIP is equivalent to *exact learning*, which is a well-studied model of computational learning, by comments from Servedio. *Gortler* and *Servedio* [32] have shown interesting results on the quantum exact learning that are independent of our main result on the quantum upper bound of any OIP. More precisely, the paper [32] defined a natural quantum version of two learning models and proved the equivalence up to polynomial factors between classical and quantum query complexity for the models. Interpreting the result on the exact learning into the context of our OIP, if there exists a quantum algorithm that solves an OIP S with Q queries then there exists a deterministic algorithm that solves S with $O(Q^3 \log N)$ queries.

Recently, two papers, by *Høyer* et al. [33] and *Buhrman* et al. [34], raised the question of how to cope with *imperfect oracles* for the quantum case using the following model: The oracle returns, for the query to bit a_i, a quantum pure state from which we can measure the correct value of a_i with a constant probability. This noise model naturally fits the motivation that a similar mechanism should apply when we use bounded-error quantum subroutines. In [33] *Høyer* et al. gave a quantum algorithm that robustly computes the Grover's problem with $O(\sqrt{N})$ queries, which is only a constant factor worse than the noiseless case. *Buhrman* et al. [34] also gave a robust quantum algorithm to output all the N bits by using $O(N)$ queries. This obviously implies that $O(N)$ queries are enough to compute the PARITY of the N bits, which contrasts with the classical $\Omega(N \log N)$ lower bound given in [35]. These raise the question if the algorithm for OIP can be made robust as shown for some special cases by the above two papers. Our most recent results in [29] answer the question in the positive flavour.

Acknowledgements

The authors would like to thank Rocco Servedio for his comments on the relationships between our problem and computational learning.

References

[1] L. K. Grover: A fast quantum mechanical algorithm for database search, in *Proceedings of the 28th ACM Symposium on Theory of Computing* (1996) pp. 212–219
[2] M. Boyer, G. Brassard, P. Høyer, A. Tapp: Tight bounds on quantum searching, Fortschritte der Physik **46**, 493–505 (1998)
[3] E. Farhi, J. Goldstone, S. Gutmann, M. Sipser: A limit on the speed of quantum computation in determining parity, Phys. Rev. Lett. **81**, 5442–5444 (1998)
[4] W. van Dam: Quantum oracle interrogation: Getting all information for almost half the price, in *Proceedings of the 39th IEEE Symposium on the Foundation of Computer Science* (1998) pp. 362–367
[5] E. Bernstein, U. Vazirani: Quantum complexity theory, SIAM J. Comput. **26**, 1411–1473 (1997)
[6] A. Ambainis: Quantum lower bounds by quantum arguments, J. Comput. Syst. Sci. **64**, 750–767 (2002)
[7] D. Biron, O. Biham, E. Biham, M. Grassl, D. A. Lidar: Generalized grover search algorithm for arbitrary initial amplitude distribution, in *Proceedings of the 1st NASA International Conference on Quantum Computing and Quantum Communication*, vol. 1509, LNCS (Springer, Berlin, Heidelberg 1998) pp. 140–147
[8] D. P. Chi, J. Kim: Quantum database searching by a single query, in *Proceedings of the 1st NASA International Conference on Quantum Computing and Quantum Communication*, vol. 1509, LNCS (Springer, Berlin, Heidelberg 1998) pp. 148–151
[9] L. K. Grover: A framework for fast quantum mechanical algorithms, in *Proceedings of the 30th ACM Symposium on Theory of Computing* (1998) pp. 53–62
[10] L. K. Grover: Rapid sampling through quantum computing, in *Proceedings of the 32th ACM Symposium on Theory of Computing* (2000) pp. 618–626
[11] G. Brassard, P. Høyer, M. Mosca, A. Tapp: Quantum amplitude amplification and estimation, in *AMS Contemporary Mathematics Series Millennium Volume entitled "Quantum Computation & Information"*, vol. 305 (2002) pp. 53–74

[12] H. Buhrman, C. Dürr, M. Heiligman, P. Høyer, F. Magniez, M. Santha, R. de Wolf: Quantum algorithms for element distinctness, in *Proceedings of the 16th IEEE Annual Conference on Computational Complexity* (2001) pp. 131–137
[13] A. Ambainis: Quantum walks and a new quantum algorithm for element distinctness, in *ERATO Conference on Quantum Information Science 2003* (2003)
[14] Y. Shi: Quantum lower bounds for the collision and the element distinctness problems, in *Proceedings of the 43rd IEEE Symposium on the Foundation of Computer Science* (2002) pp. 513–519
[15] F. Magniez, M. Santha, M. Szegedy: An $o(n^{1.3})$ quantum algorithm for the triangle problem, in *Proceedings of Symposium on Discrete Algorithms* (2005) pp. 1109–1117, quant-ph/0310134
[16] S. Aaronson, A. Ambainis: Quantum search of spatial regions, in *Proceedings of the 44th Symposium on Foundations of Computer Science* (2003) pp. 200–209
[17] R. Beals, H. Buhrman, R. Cleve, M. Mosca, R. de Wolf: Quantum lower bounds by polynomials, in *Proceedings of 39th IEEE Symposium on Foundation of Computer Science* (1998) pp. 352–361
[18] S. Aaronson: Quantum lower bound for the collision problem, in *Proceedings of the 34th Symposium on Theory of Computing* (2002) pp. 635–642
[19] C. Bennett, E. Bernstein, G. Brassard, U. Vazirani: Strengths and weaknesses of quantum computing, SIAM J. Comput. **26**, 1510–1523 (1997)
[20] U. Vazirani: On the power of quantum computation, Philos. Transact. Roy. Soc. London pp. 1759–1768 (1998)
[21] H. Barnum, M. Saks: A lower bound on the quantum complexity of read-once functions, J. Comput. Syst. Sci. **69**, 244–258 (2004)
[22] H. Barnum, M. Saks, M. Szegedy: Quantum query complexity and semi-definite programming, in *Proceedings of the 18th IEEE Conference on Computational Complexity* (2003) pp. 179–193
[23] L. Laplante, F. Magniez: Lower bounds for randomized and quantum query complexity using Kolmogorov arguments, in *Proceedings of the 19th IEEE Conference on Computational Complexity* (2004) quant-ph/0311189
[24] C. Dürr, M. Mhalla, Y. Lei: Quantum query complexity of graph connectivity, in *ERATO Conference on Quantum Information Science 2003* (2003) quant-ph/0303169
[25] S. Aaronson: Lower bounds for local search by quantum arguments, in *Proceedings of the 36th Symposium on Theory of Computing* (2004) pp. 465–474
[26] A. Ambainis: Polynomial degree vs. quantum query complexity, in *Proceedings of the 44th IEEE Symposium on Foundations of Computer Science* (2003) pp. 230–239
[27] A. Nayak, F. Wu: The quantum query complexity of approximating the median and related statistics, in *Proceedings of the 31th ACM Symposium on Theory of Computing* (1999) pp. 384–393
[28] D. Deutsch, R. Jozsa: Rapid solutions of problems by quantum computation, in *Proceedings of the Royal Society*, Series A **439** (London 1992) pp. 553–558
[29] A. Ambainis, K. Iwama, A. Kawachi, R. Raymond, S. Yamashita: Robust quantum algorithms for oracle identification (2005) quant-ph/0411204
[30] A. Atici, R. A. Servedio: Improved bounds on quantum learning algorithms, Quantum Information Processing **4**, 335–386 (2005)

[31] M. Hunziker, D. A. Meyer, J. Park, J. Pommersheim, M. Rothstein: The geometry of quantum learning (2003)
URL quant-ph/0309059
[32] S. Gortler, R. Servedio: Quantum versus classical learnability, in *Proceedings of the 16th Annual Conference on Computational Complexity* (2001) pp. 138–148
[33] P. Høyer, M. Mosca, R. de Wolf: Quantum search on bounded-error inputs, in *Proceedings of the 30rd International Colloquium on Automata, Languages and Programming*, vol. 2719, LNCS (2003) pp. 291–299
[34] H. Buhrman, I. Newman, H. Röhrig, R. de Wolf: Robust quantum algorithms and polynomials, in *Proceedings of the 22nd Symposium on Theoretical Aspects of Computer Science*, vol. 3404, LNCS (2005) pp. 593–604
[35] U. Feige, P. Raghavan, D. Peleg, E. Upfal: Computing with noisy information, SIAM J. Comput. **23**, 1001–1018 (1994)

Index

1-sensitive, 8

balanced matrix, 13
Bernstein–Vazirani problem, 4

column flip, 8

exact learning, 15

Grover search, 4

halving algorithm, 11
hybrid argument, 5
hybrid matrix, 13

imperfect oracle, 15
input oracle, 5

multitarget Grover search, 9

oracle, 3
oracle computation, 3
oracle identification, 3
oracle identification problem, 5

polynomial method, 5

quantum adversary argument, 5
quantum adversary method, 5
quantum computation, 5
quantum learning theory, 11
query complexity, 3, 5

robust quantum algorithm, 16

Query Complexity of Quantum Biased Oracles

Kazuo Iwama[1], Rudy Raymond[2], and Shigeru Yamashita[3]

[1] Graduate School of Informatics, Kyoto University, Kyoto 606-8501, Japan
   iwama@kuis.kyoto-u.ac.jp
[2] Graduate School of Informatics, Kyoto University, Kyoto 606-8501, Japan
   raymond@kuis.kyoto-u.ac.jp
[3] Graduate School of Information Science, Nara Institute of Science and Technology, Nara 630-0192, Japan
   ger@is.naist.jp

Abstract. In this Chapter, our focus is the efficiency of computation with quantum oracles whose answers are correct only with probability $1/2 + \epsilon$. In real-world applications, quantum oracles might be realized from quantumization of probabilistic algorithms which are object to error in the success probability. Thus, designing efficient quantum algorithms for biased oracles is important. The first result to discuss such biased oracles was by Adcock and Cleve, who relate the efficiency of computation with biased oracles with the difficulty of inverting one-way functions. They showed a quantum algorithm for solving the so-called Goldreich–Levin problem, a result which has a special implication in the cryptographic setting. In this Chapter, we prove the optimality of their algorithm and show a general method for designing robust algorithms querying biased oracles for solving various problems. The method is optimal in the sense that the additional number of queries to biased oracles matches the lower bounds, which are also part of our results.

1 Introduction

The classical oracle computation is the following scenario: We want to compute a designated Boolean function $f(x_1, x_2, \ldots, x_N)$, but the value of a_i ($= 0$ or 1) of each x_i can be obtained only by making a query to a black box called an *oracle*. Then, we often consider the smallest number of necessary oracle calls, which we call the *query complexity*, to obtain the value of $f(a_1, a_2, \ldots, a_N)$ with a high (say, constant) probability. Suppose we want to compute the Boolean OR of input variables, i.e., $f = x_1 \vee x_2 \vee \ldots \vee x_N$. In the case of classical computation, we need $\Theta(N)$ queries to compute the function.

By contrast, we need far fewer queries in the case of quantum computation. For instance, we need only $O(\sqrt{N})$ queries to find an index i such that $a_i = 1$ (Grover search [1]). This is one of the major examples of quantum superiority. Therefore, quantum query complexity has been intensively studied as a central issue of quantum computation. Indeed, there have been a number of applications and extensions of Grover search, e.g., [2, 3, 4, 5, 6]. Also, many results on efficient quantum algorithms are shown by sophisticated ways of using Grover search. *Brassard* et al. [7] showed a quantum counting algorithm that gives an approximate counting method by combining Grover search with quantum Fourier transformation. Quantum algorithms for the claw-finding and the element distinctness problems given by *Buhrman* et al. [8] also exploited classical random and sorting methods with Grover search. (*Ambainis* [9] developed an optimal quantum algorithm with $O(N^{2/3})$ queries for element distinctness problem, which makes use of quantum walk and matches to the lower bounds shown by *Shi* [10].) *Ambainis* and *Aaronson* [11] constructed quantum search algorithms for spatial regions by combining Grover search with the divide-and-conquer method. *Magniez, Santha* and *Szegedy* [12] showed efficient quantum algorithms to find a triangle in a given graph by using combinatorial techniques with Grover search. *Dürr, Heiligman, Høyer* and *Mhalla* [13] also investigated quantum query complexity of several graph-theoretic problems. In particular, they exploited Grover search on some data structures of graphs for their upper bounds. *Ambainis* et al. [14] studied the query complexity of the most general problem, which they call the oracle identification problem (OIP). An OIP is given a set S of M Boolean oracles out of 2^N ones, to determine which oracle in S is the current black-box oracle. We can exploit the information that candidates of the current oracle are restricted to S. They provide almost an optimal algorithm whose query complexity is $O(\sqrt{N \log M \log N} \log \log M)$.

For oracle computation, there are several situations where we can get only a *noisy* Boolean value for each variable. Suppose again that we want to compute the Boolean OR of input variables, i.e., $f = x_1 \vee x_2 \vee \ldots \vee x_N$, by asking an input oracle. However, this time the oracles is *noisy* in the sense that it returns us the correct a_i ($= 0$ or 1) with probability $\frac{1}{2} + \epsilon$. In this Chapter, we call this oracle an ϵ-biased oracle. For the above particular example, one simple algorithm is to call the biased oracle for each x_i many times and to guess the value of a_i by majority. It is not hard to see that we need $\Omega(\frac{1}{\epsilon^2})$ oracle calls to know the correct value of each a_i with constant probability. Thus, the query complexity obviously depends on the value of ϵ. Note that many studies assume that ϵ is a constant, which disappears in the query complexity under the big-O notation [15, 16]. Note also that we can get all the values of N bits with high probability by querying each a_i $O(\log N)$ times instead of once. Thus, we can make any algorithm robust, i.e., resilient against biased oracles, at the cost of an $O(\log N)$-factor overhead. In some cases, this factor of $O(\log N)$ is actually needed: *Feige* et al. [17] proved that any classical *robust* algorithm to compute the parity of the N bits needs $\Omega(N \log N)$ queries. On the other hand, the same paper also gives a nontrivial classical algorithm which computes OR of the N bits with $O(N)$ queries.

Recently, two papers, by *Høyer* et al. [18] and *Buhrman* et al. [19], raised the question of how to cope with biased oracles in the quantum case. For the quantum setting, both papers are based on the following model: The oracle returns, for the query to bit a_i, a quantum pure state from which we can measure the correct value of a_i with a constant probability. This noise model naturally fits the motivation that a similar mechanism should apply when we use bounded-error quantum subroutines.

Based on the above biased oracle model, *Høyer* et al. gave a quantum algorithm that robustly executes Grover search with $O(\sqrt{N})$ queries, which is only a constant factor worse than the perfect oracle case [18]. *Buhrman* et al. [19] also adopted the same model and gave a robust quantum algorithm to output all the N bits by using $O(N)$ queries. This obviously implies that $O(N)$ queries are enough to compute the parity of the N bits, which contrasts with the classical $\Omega(N \log N)$ lower bound mentioned earlier. Thus, robust quantum computation does not need a serious overhead, at least for several important problems. They consider the bias factor as constant, and therefore they do not show any analysis how the bias factor affects the query complexity.

In this Chapter, we discuss upper and lower bounds on the quantum query complexity of oracles with an *explicit* bias factor ϵ. *Adcock* et al. [20] were the first to define oracles with an explicit bias factor, although their oracles are restricted to the so-called inner product oracles. On the other hand, we consider wider types of quantum oracles (but with some conditions) and show that for any quantum algorithm solving some problem with high probability and using T queries to perfect oracles, there exists a quantum algorithm solving the same problem with high probability using $O(T/\epsilon)$ queries to the corresponding ϵ-biased oracles. As one of its applications, it can be shown that there exists a quantum algorithm for computing the parity of N-bit biased inputs with only $O(N/\epsilon)$ queries while any classical algorithms need at least $\Omega(N \log N/\epsilon^2)$ queries, as shown by *Feige* et al. [17]. With another special condition, we show that the query complexity does not change even if the oracle is biased, while the same thing does not occur in the classical corresponding situation.

Having the general upper bounds, the next interesting question is how optimal those bounds are. To answer this, we generalize *Ambainis'* quantum adversary argument [21] and obtain lower bounds with an explicit bias factor. Our result is that if a problem can be shown to require $\Omega(T)$ queries to perfect oracles by quantum adversary argument, then our generalization implies that the same problem needs $\Omega(T/\epsilon)$ queries to biased oracles. This also implies that our general upper bounds are optimal in terms of bias factor. Furthermore, since $\Omega(1)$ is an obvious lower bound for all oracular problems, our generalization implies that $\Omega(1/\epsilon)$ is an obvious lower bound for all biased oracular problems. For the so-called quantum Goldreich–Levin (GL) problem, this immediately gives us the matching lower bound as a corollary.

2 Goldreich–Levin Problem and Biased Oracles

The Goldreich–Levin Theorem is a cryptographic reduction which enables a cryptographically hard predicate to be based on the computational difficulty of a one-way function [22]. It can be abstracted as the following problem, which we henceforth refer to as the *GL problem*: Let $a \in \{0,1\}^n$ and ε satisfy $0 < \varepsilon \leq 1$. Let information about a be available only from IP (inner product) and EQ (equivalence) oracle queries. The IP oracle has the property that, for a uniformly distributed random $x \in \{0,1\}^n$, $\Pr[\text{IP}(x) = a \cdot x] \geq \frac{1}{2}(1+\varepsilon)$. The EQ oracle, on input $x \in \{0,1\}^n$, returns a bit specifying whether or not $x = a$. The task is to determine a.

For an algorithm solving the GL problem, its efficiency corresponds to the overhead in the underlying cryptographic reduction. The more efficient an algorithm for the GL problem is, the tighter the correspondence is between the cryptographic primitives to which it is applied. Determining the most efficient algorithm for the GL problem is therefore a matter of interest in complexity theory-based cryptography in both classical and quantum frameworks (see, e.g., [20] for further discussion).

When there are no errors (i.e., $\varepsilon = 1$), it is straightforward to show that n queries are necessary and sufficient for any classical algorithm; however, with a quantum algorithm, one query suffices [23,24]. For smaller ε, *Levin* [25] shows how to solve the problem classically with $O(n/\varepsilon^2)$ IP and EQ queries; however, the approach requires superpolynomial (in n/ε) auxiliary operations. *Goldreich* and *Levin* [22] show how to solve this problem classically with a number of queries *and auxiliary operations* that is polynomial in n/ε, and this can be refined into an efficient algorithm that makes $O(n/\varepsilon^2)$ IP queries followed by $O(1/\varepsilon^2)$ EQ queries [26, 27].

Adcock and *Cleve* [20] show that the classical IP query complexity for solving the GL problem with bounded-error probability is $\Omega(n/\varepsilon^2)$ whenever the number of EQ queries is at most $\sqrt{2^n}$ (for a reasonable range of values of ε). It can also be shown that $\Omega(1/\varepsilon^2)$ EQ queries are necessary classically.

For quantum algorithms, *Adcock* and *Cleve* [20] show that $O(1/\varepsilon)$ IP queries, $O(1/\varepsilon)$ EQ queries, and $O(n/\varepsilon)$ auxiliary one- and two-qubit gates are sufficient to solve the GL problem; however, they do not address the question whether these costs are necessary. We address this question by showing the following:

Theorem 1. *Any quantum algorithm solving the GL problem with constant success probability requires $\Omega(1/\varepsilon)$ EQ queries, whenever $\varepsilon \geq (\frac{1}{2})^{n/2}$.*

It is not possible to lower-bound the number of IP queries independently of the number of EQ queries, because $O(\sqrt{2^n})$ EQ queries would eliminate the need for any IP queries [1]. The next theorem implies that whenever the number of EQ queries is $o(\sqrt{2^n})$ the number of IP queries must be $\Omega(1/\varepsilon)$.

Theorem 2. *Any quantum algorithm solving the GL problem with constant success probability requires either $\Omega(\sqrt{2^n})$ EQ queries or $\Omega(1/\varepsilon)$ IP queries, whenever $\varepsilon \geq (\frac{1}{2})^{n/2}$.*

For the quantum case, a query that, on input $x \in \{0,1\}^n$, returns one bit can be regarded as a unitary operation U, where the output bit is understood to be the last qubit of $U\,|x\rangle\,|0\rangle$. The stochastic property of IP queries is in terms of the measured result of the output qubit (see [20] for further discussion about formalizing quantum IP queries).

Our proof technique for the former theorem is by combining a lower bound arising in the list decoding of Hadamard codes (which we show explicitly), in conjunction with known lower bounds for quantum searching [28]. The latter theorem is proved by considering a special class of amplitude amplification problems that easily reduce to the GL problem and can be lower-bounded by a standard hybrid argument.

Proof of Theorem 1.

For any even k such that $0 < k \le n$, define $f_k\colon \{0,1\}^n \to \{0,1\}$ as

$$f_k(x_1, x_2, \ldots, x_n) = x_1 x_2 \oplus x_3 x_4 \oplus \cdots \oplus x_{k-1} x_k.$$

Let $\varepsilon \ge (\frac{1}{2})^{n/2}$ be given, and set k to the unique even number such that $(\frac{1}{2})^{k/2+1} < \varepsilon \le (\frac{1}{2})^{k/2}$. Now fix the IP oracle to $\mathrm{IP}(x) = f_k(x)$. Note that fixing the IP oracle makes all IP queries in the algorithm redundant. We will show that this particular setting of the IP oracle has the interesting property that there are $\Omega(1/\varepsilon^2)$ different $a \in \{0,1\}^n$ that are consistent with it in the sense that $\Pr_x[f_k(x) = a \cdot x] \ge \frac{1}{2}(1+\varepsilon)$. Since there are $\Omega(1/\varepsilon^2)$ candidates for the actual solution – which must be found using EQ queries – the well-known lower bound for searching [28] implies that the number of EQ queries necessary (for constant success probability) is $\Omega(\sqrt{1/\varepsilon^2}) = \Omega(1/\varepsilon)$.

We now provide the technical details of the proof, starting with the following simple lemma:

Lemma 1. *Let k be even and $x_1, \ldots, x_k$ be independent uniformly distributed random bits. Then*

$$\Pr[x_1 x_2 \oplus \cdots \oplus x_{k-1} x_k = 0] = \tfrac{1}{2}(1 + (\tfrac{1}{2})^{k/2}).$$

Proof. Define $Y = (-1)^{x_1 x_2 \oplus \cdots \oplus x_{k-1} x_k}$. Then

$$\mathrm{E}[Y] = \mathrm{E}[(-1)^{x_1 x_2}] \cdots \mathrm{E}[(-1)^{x_{k-1} x_k}] = (\tfrac{1}{2})^{k/2},$$

from which it follows that $\Pr[x_1 x_2 \oplus \cdots \oplus x_{k-1} x_k = 0] = \frac{1}{2}(1 + \mathrm{E}[Y]) = \frac{1}{2}(1 + (\frac{1}{2})^{\frac{k}{2}})$.

The following proposition provides a characterization of several $a \in \{0,1\}^n$ that are consistent with the IP oracle.

Proposition 1. *For all $a \in \{0,1\}^n$ such that $f_k(a) = 0$ and $a_{k+1} = a_{k+2} = \cdots = a_n = 0$, if $x \in \{0,1\}^n$ is randomly chosen then $\Pr[f_k(x) = a \cdot x] \ge \frac{1}{2}(1+\varepsilon)$.*

Proof.

$$\Pr[f_k(x) = a \cdot x]$$
$$= \Pr[(x_1 x_2 \oplus \cdots \oplus x_{k-1} x_k) \oplus (a_1 x_1 \oplus \cdots \oplus a_k x_k) = 0]$$
$$= \Pr[(x_1 x_2 \oplus a_1 x_1 \oplus a_2 x_2) \oplus \cdots \oplus (x_{k-1} x_k \oplus a_{k-1} x_{k-1} \oplus a_k x_k) = 0]$$
$$= \Pr[(x_1 \oplus a_2)(x_2 \oplus a_1) \oplus \cdots \oplus (x_{k-1} \oplus a_k)(x_k \oplus a_{k-1}) \oplus f_k(a) = 0]$$
$$= \Pr[x_1 x_2 \oplus \cdots \oplus x_{k-1} x_k = 0]$$
$$= \tfrac{1}{2}(1 + (\tfrac{1}{2})^{k/2}) \quad \text{(by Lemma 1)}$$
$$\geq \tfrac{1}{2}(1 + \varepsilon).$$

The following proposition, in conjunction with Proposition 1, lower-bounds the number of $a \in \{0,1\}^n$ that are consistent with the IP oracle.

Proposition 2. *The number of $a \in \{0,1\}^n$ such that $f_k(a) = 0$ and $a_{k+1} = a_{k+2} = \cdots = a_n = 0$ is at least $\tfrac{1}{8}(1/\varepsilon^2)$.*

Proof. Lemma 1 implies that the number of $a \in \{0,1\}^k$ such that $f_k(a) = 0$ is $\tfrac{1}{2}(1 + (\tfrac{1}{2})^{k/2}) 2^k = 2^{k-1} + 2^{k/2 - 1} > \tfrac{1}{8} 2^{k+2} > \tfrac{1}{8}(1/\varepsilon^2)$.

Proof of Theorem 2.

Let $\varepsilon > (\tfrac{1}{2})^{n/2}$ be given. For each $a \in \{0,1\}^n$ such that $a \neq 0$, define two oracles. The first is the aforementioned EQ oracle (that, on input $x \in \{0,1\}^n$, returns a bit specifying whether or not $x = a$). To define the second type of oracle, first define the unitary operation A acting on n qubits such that, for all $y \in \{0,1\}^n$,

$$A|y\rangle = \sqrt{1 - \varepsilon^2}\,|y\rangle + i\varepsilon\,|a \oplus y\rangle . \tag{1}$$

Note that $|\langle a|\,A\,|0\rangle| = \varepsilon$. The second type of query is a *controlled-A* operation, denoted as cont-A, where cont-$A\,|y\rangle\,|b\rangle = (A^b\,|y\rangle)\,|b\rangle$, for all $y \in \{0,1\}^n$ and $b \in \{0,1\}$.

Consider the following *amplitude amplification* problem. There is an unknown $a \in \{0,1\}^n$ such that $a \neq 0$. Information about a is available by EQ, cont-A, and cont-$A^\dagger$ queries. The goal is to determine a. The well-known amplitude amplification algorithm [7] solves this problem using $O(1/\varepsilon)$ EQ, cont-A, and cont-$A^\dagger$ queries. We first show that this is optimal in the following sense:

Lemma 2. *The amplitude amplification problem requires either $\Omega(\sqrt{2^n})$ EQ queries or $\Omega(1/\varepsilon)$ cont-A or cont-$A^\dagger$ queries, whenever $\varepsilon \geq (\tfrac{1}{2})^{n/2}$.*

Proof. This is straightforward to prove by modifying the quantum lower bound for searching that uses the hybrid method [28]. That lower bound proof shows that there is a state $|\phi\rangle$ such that, if only t EQ queries are available, then, averaging over all values of a, the final state of the algorithm has

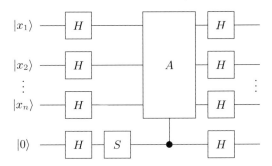

Fig. 1. Method to simulate an IP query using a cont-A query. The last qubit, when measured, is biased toward $a \cdot x$

distance only $t(2/\sqrt{2^n-1})$ from $|\phi\rangle$ (note that, since $a \neq 0$, the size of the search space is $2^n - 1$).

The present scenario is different in that cont-A and cont-$A^\dagger$ queries can be interleaved into the computation. This is addressed by showing that each cont-A and cont-$A^\dagger$ query can have a limited effect on a quantum state. The precise result is that, for *any* quantum state $|\psi\rangle$, $\| |\psi\rangle - \text{cont-}A\,|\psi\rangle \| \leq \sqrt{2}\varepsilon$. This inequality can be proven by noting that the eigenvalues of cont-A are all either 1 or $\sqrt{1-\varepsilon^2} \pm i\varepsilon$. Thus, each eigenvalue is distance at most $\sqrt{2}\varepsilon$ away from 1. It follows that, if there are s cont-A and cont-$A^\dagger$ queries and t EQ queries, then, averaging over all values of a, the final state of the algorithm has distance only $s(\sqrt{2}\varepsilon) + t(2/\sqrt{2^n-1})$ from $|\phi\rangle$, from which the result follows.

Next, we observe that a cont-A query can be used to simulate an IP query. The simulation is given by the circuit in Fig. 1, denoted as C, where H denotes the Hadamard gate and S is defined as $S|b\rangle = (-i)^b |b\rangle$, for $b \in \{0,1\}$.

Lemma 3. *If the last output qubit in the above circuit is measured then the probability that the outcome is $a \cdot x$ is $\frac{1}{2}(1+\varepsilon)$.*

Proof. It is sufficient to show that

$$\langle x, a \cdot x | C | x, 0 \rangle = \frac{1+\varepsilon - i(-1)^{a \cdot x}\sqrt{1-\varepsilon^2}}{2}, \qquad (2)$$

for all $x \in \{0,1\}^n$, since this implies that $|\langle x, a \cdot x | C | x, 0 \rangle|^2 = \frac{1}{2}(1+\varepsilon)$.

One way of establishing (2) is as follows. If circuit C is executed up to the stage of the cont-A gate on state $|x, 0\rangle$, the resulting state is

$$\frac{1}{\sqrt{2}}\left(\frac{1}{\sqrt{2^n}} \sum_{y \in \{0,1\}^n} (-1)^{x \cdot y} |y\rangle\right) |0\rangle$$
$$+ \frac{1}{\sqrt{2}}\left(\frac{1}{\sqrt{2^n}} \sum_{y \in \{0,1\}^n} (-1)^{x \cdot y} \left(-i\sqrt{1-\varepsilon^2} + (-1)^{a \cdot x} \varepsilon\right) |y\rangle\right) |1\rangle \quad (3)$$

Also, if the last stage of circuit C is executed on state $|x, a \cdot x\rangle$, the resulting state is

$$\frac{1}{\sqrt{2}}\left(\frac{1}{\sqrt{2^n}} \sum_{y \in \{0,1\}^n} (-1)^{x \cdot y} |y\rangle\right) |0\rangle$$
$$+ \frac{1}{\sqrt{2}}\left(\frac{1}{\sqrt{2^n}} \sum_{y \in \{0,1\}^n} (-1)^{x \cdot y} (-1)^{a \cdot x} |y\rangle\right) |1\rangle \quad (4)$$

Equation (2) is obtained as the inner product between the states in (3) and (4).

Since Lemma 3 implies that a violation of Theorem 2 leads to a violation of Lemma 2, this completes the proof.

2.1 The Model of Quantum Biased Oracles

In the computer science community, we usually consider that the quantum state generated by a quantum biased oracle is a *pure state*. In other words, a biased oracle is considered to be a unitary transformation. Recently a great deal of research [18, 19, 20, 29, 30] has been based on this model. If we consider a quantum subroutine as an oracle, the oracle can be considered as this model; therefore, the motivation of this model is also natural. *Adcock* and *Cleve* first discussed quantum biased oracles of this model [20], and their definition can be written as follows:

Definition 1. *A quantum ϵ-biased oracle is a unitary transform (denoted by O_ϵ hereafter) on $n + m + 1$ qubits which satisfies the following two properties:*

1. *If the last qubit of $O_\epsilon |x\rangle |0^m\rangle |0\rangle$ is measured, yielding the value $w \in \{0, 1\}$, then $Pr[w = f(x)] \geq \frac{1}{2} + \epsilon$ for any $x \in \{0, 1\}^n$.*
2. *For any $x \in \{0, 1\}^n$ and $y \in \{0, 1\}^{m+1}$, the state of the first n qubits of $O_\epsilon |x\rangle |y\rangle$ is $|x\rangle$. For simplicity, we just assume $N = 2^n$ in the rest of this Chapter. (Otherwise we consider an oracle whose input size is $N' = 2^n (2N > N' > N)$ by adding some dummy inputs. It is obvious that this does not change the query complexity in the big-O notation.)*

The second property is for technical convenience, and any unitary operation without this property can be converted to one that has this property by first producing a copy of the classical basis state $|x\rangle$. Note that we use the bias (ϵ in the definition) of the success probability from $1/2$ to denote the parameter for a biased oracle.

Since O_ϵ is a unitary transform, $O_\epsilon |x\rangle |0^m\rangle |0\rangle$ must be written as

$$|x\rangle \left(\alpha_x |v_x\rangle |f(x)\rangle + \beta_x |w_x\rangle \left|\overline{f(x)}\right\rangle \right).$$

Generally we should consider that v_x, w_x, and α_x are different according to x.

3 Upper Bounds of the Query Complexity of Biased Oracles With Special Conditions

In this section, we consider two special cases of quantum biased oracles where we relax the conditions in Definition 1. Note that our relaxations seem to be fair from the view point of classical computation, i.e., it seems almost impossible to utilize the relaxation in the classical case. However, as we will see, the query complexity changes dramatically in the quantum world.

3.1 Basic Tools for Quantum Computation

Before describing our algorithms, we show some basic tools for quantum computation.

Theorem 3 (*Brassard* et al. [7]). *Let $\mathcal{O}$ be any quantum algorithm that uses no measurements, and let $\chi\colon Z \to \{0,1\}$ be any Boolean function. There exists a quantum algorithm that, given the initial success probability $p > 0$ of $\mathcal{O}$, finds a good solution with certainty using a number of applications of $\mathcal{O}$ and $\mathcal{O}^{-1}$ which is in $\Theta(\frac{1}{\sqrt{p}})$ in the worst case.*

When we have no knowledge of the success probability of $\mathcal{O}$, we can have a good estimation, as shown in [7]. The idea is to estimate the phase θ_p for $p = \sin^2(\theta_p)$. Once we have the estimate value of θ_p, we can apply Q to find a good solution with high probability. Note that the following theorem appeared in [7] as a theorem to estimate the amplitude of success probability. We rewrite it in terms of phase estimation for simplifying our discussion in this Chapter.

Theorem 4 (*Brassard* et al.). *For any integer $k > 0$, there exists a quantum algorithm $Est\_Phase(\mathcal{O}, \chi, M)$ which outputs $\tilde{\theta}_p$ ($0 \le \tilde{\theta}_p \le \frac{\pi}{2}$) such that*

$$|\theta_p - \tilde{\theta}_p| \le \frac{k\pi}{M}$$

with probability at least $\frac{8}{\pi^2}$ when $k = 1$, and with probability greater than $1 - \frac{1}{2(k-1)}$ for $k \geq 2$. It uses exactly M evaluations of χ. If $\theta_p = 0$ then $\tilde{\theta}_p = 0$ with certainty, and if $\theta_p = \frac{\pi}{2}$ and M is even, then $\tilde{\theta}_p = \frac{\pi}{2}$ with certainty.

3.2 Quantum Biased Oracles With the Same Bias Factor

We consider the following quantum biased oracles where the bias factor is the same for all inputs:

Definition 2. *A quantum oracle of a Boolean function f with bias ϵ is a unitary transformation O_f^ϵ or its inverse $O_f^\epsilon \dagger$ such that:*

1. *For all $x \in \{0,1\}^n$, the measurement on the last qubit of $O_f^\epsilon |x, 0^m\rangle$ results in $w \in \{0,1\}$ such that $Pr[w = f(x)] = \frac{1}{2} + \epsilon$.*
2. *For any $x \in \{0,1\}^n$ and $y \in \{0,1\}^m$, the first n qubits of $O_f^\epsilon |x, y\rangle$ are $|x\rangle$.*

Formally, we can write O_f^ϵ as follows:

$$O_f^\epsilon |x\rangle |0^{m-1}\rangle |0\rangle = |x\rangle (\alpha_x |v_x\rangle |f(x)\rangle + \beta_x |w_x\rangle |\overline{f(x)}\rangle), \tag{5}$$

where $|\alpha_x|^2 = 1/2 + \epsilon$. Mostly, quantum algorithms using O_f^ϵ should produce the output a which is hidden in the Boolean function f. To emphasize this, we also denote f with input x as $f_a(x)$ or $f(a,x)$. For example, in the GL problem, $f(x) = a \cdot x$ for some fixed a, and the algorithm solving it must output this a with high probability. For this reason, we will interchangeably use $f(x)$, $f_a(x)$, and $f(a,x)$ in the hereafter.

Now, suppose that we have a quantum algorithm $\mathcal{A}$ solving some problem with high probability using a perfect oracle O_f. In this section, we construct a quantum algorithm $\mathcal{A}'$ solving the same problem when a quantum oracle with bias ϵ, O_f^ϵ, is given instead. The following lemma and Fig. 1 show how to simulate O_f with O_f^ϵ by converting the biased oracles in Definition 2 to the form of oracles returning identifiable good and bad states.

Lemma 4. *There exists a quantum oracle $\tilde{O}_f^\epsilon$ which uses one O_f^ϵ and one $O_f^\epsilon \dagger$ and acts on $n+m+1$ qubits such that for all $x \in \{0,1\}^n$,*

$$\tilde{O}_f^\epsilon |x, 0^m, 0\rangle = (-1)^{f(x)} 2\epsilon |x, 0^m, 0\rangle + |x, \psi_x\rangle,$$

where $|x, \psi_x\rangle$ is orthogonal to $|x, 0^m, 0\rangle$ and its norm is $\sqrt{1 - 4\epsilon^2}$.

Proof. We show the construction of $\tilde{O}_f^\epsilon$ in Fig. 2. X denotes the NOT gate which flips the state $|0\rangle$ to $|1\rangle$, and vice versa. The circuit in the middle

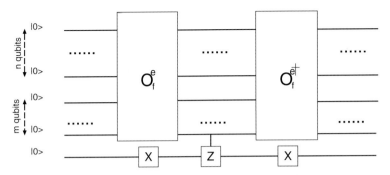

Fig. 2. This figure shows how to construct flipping $\tilde{O}_f^\epsilon$ oracles from the standard noisy oracles O_f^ϵ

of O_f^ϵ and $O_f^\epsilon\dagger$ is the controlled Z which transforms the state $|x\rangle|y\rangle$ to $(-1)^{x \cdot y}|x\rangle|y\rangle$ for $x, y \in \{0, 1\}$. It can be shown easily that

$$\langle x, 0^m, 0| \tilde{O}_f^\epsilon |x, 0^m, 0\rangle = (-1)^{f(x)} 2\epsilon.$$

This implies that

$$\tilde{O}_f^\epsilon |x, 0^m, 0\rangle = (-1)^{f(x)} 2\epsilon |x, 0^m, 0\rangle + |x, \psi_x\rangle,$$

where $|x, \psi_x\rangle$ is perpendicular to $|x, 0^m, 0\rangle$ and its norm is $\sqrt{1 - 4\epsilon^2}$.

Note that a perfect quantum oracle O_f returns the state $(-1)^{f(x)}|x, 0^m, 0\rangle$ on input $|x, 0^m, 0\rangle$. Thus, it is natural to consider the amplitude amplification to simulate O_f by O_f^ϵ. With regard to the previous lemma, we can define the good states as the states in which the last $m + 1$ qubits are 0. It is easy to build a circuit which recognizes such good states. Thus, we can construct $\mathcal{A}'$ from $\mathcal{A}$ using the same number of qubits, as that of $\mathcal{A}$ plus the additional $m + 1$ qubits. Any unitary transformations beside the oracle query can be used as they are in $\mathcal{A}$, and a query of O_f in $\mathcal{A}$ is simulated by querying $\tilde{O}_f^\epsilon$ combined with the amplitude amplification method. The following theorem is straightforward from Theorem 3 by replacing $\mathcal{O}$ with O_f^ϵ and p with $(2\epsilon)^2$:

Theorem 5. *Let $\mathcal{A}$ be any quantum algorithm solving some problem with probability p and using an oracle O_f for T number of queries. Then, given ϵ there exists a quantum algorithm $\mathcal{A}'$ solving the same problem also with probability p and using an oracle $\tilde{O}_f^\epsilon$ for $O(T/\epsilon)$ number of queries.*

Next, we want to consider the case when the value ϵ is not given. The following lemma states that if we have an estimated value of θ_ϵ such that $\epsilon^2 = \sin^2 \theta_\epsilon$ is the initial success probability of $\mathcal{O}$, we can use it to amplify the success probability of $\mathcal{O}$ close to 1:

Lemma 5. *Let $\mathcal{O}$ be any quantum algorithm that uses no measurements and $\chi: Z \to \{0,1\}$ be any Boolean function and T be any integer at least 2. If a $\tilde{\theta}_\epsilon$ is given such that*

$$|\theta_\epsilon - \tilde{\theta}_\epsilon| \leq \frac{\theta_\epsilon}{(\pi+1)T},$$

where $\epsilon^2 = \sin^2 \theta_\epsilon$ is the initial success probability of $\mathcal{O}$, then there exists a quantum algorithm that finds a good solution with probability at least $\left(1 - \frac{1}{T^2}\right)$ using a number of applications of $\mathcal{O}$ and $\mathcal{O}^{-1}$ which is in $O(\frac{1}{\epsilon})$.

Proof. Consider the following cases:

- If $\exists \theta_\epsilon^* \in \mathbb{R}$ such that $|\tilde{\theta}_\epsilon - \theta_\epsilon^*| \leq \frac{\theta_\epsilon}{(\pi+1)T}$ and $m^* = \frac{1}{2}\left(\frac{\pi}{2\theta_\epsilon^*} - 1\right)$ is an integer. In this case, we can apply the amplification operator Q for m^* times to amplify the success probability of $\mathcal{O}$. Note that since $|\theta_\epsilon - \theta_\epsilon^*| \leq \frac{2\theta_\epsilon}{(\pi+1)T}$, after applying Q for m^* times the success probability is

$$\sin^2((2m^*+1)\theta_\epsilon) = \sin^2\left(\frac{\pi}{2} \cdot \frac{\theta_\epsilon}{\theta_\epsilon^*}\right). \tag{6}$$

It can be shown that

$$1 - \frac{\frac{2}{(\pi+1)T}}{1 + \frac{2}{(\pi+1)T}} \leq \frac{\theta_\epsilon}{\theta_\epsilon^*} \leq 1 + \frac{\frac{2}{(\pi+1)T}}{1 - \frac{2}{(\pi+1)T}}.$$

Therefore, the success probability in (6) is

$$\sin^2\left(\frac{\pi}{2}\frac{\theta_\epsilon}{\theta_\epsilon^*}\right) \geq \sin^2\left(\frac{\pi}{2}\left(1 + \frac{\frac{2}{(\pi+1)T}}{1 - \frac{2}{(\pi+1)T}}\right)\right)$$

$$= \cos^2\left(\frac{\pi}{2}\frac{\frac{2}{(\pi+1)T}}{1 - \frac{2}{(\pi+1)T}}\right) \quad \text{since } \sin(\pi/2 + x) = \cos x$$

$$\geq 1 - \left(\frac{\pi}{2}\frac{\frac{2}{(\pi+1)T}}{1 - \frac{2}{(\pi+1)T}}\right)^2 \quad \text{since } \cos^2 x \geq 1 - x^2$$

$$\geq 1 - \frac{1}{T^2}.$$

- Otherwise: We apply the derandomization idea in [7]. Since $m^* = \frac{1}{2}\left(\frac{\pi}{2\tilde{\theta}_\epsilon} - 1\right)$ is not an integer, choose $\overline{m^*} = \lceil m^* \rceil$ and set $\theta_\epsilon^* = \frac{\pi}{4\overline{m^*}+2}$. Let $p^* = \sin^2(\theta_\epsilon^*)$ and $\tilde{p} = \sin^2(\tilde{\theta}_\epsilon)$. We add a new register r of 1 qubit initialized to 0 and apply a unitary transformation on it to obtain the state $\sqrt{\frac{p^*}{\tilde{p}}}|0\rangle + \sqrt{1 - \frac{p^*}{\tilde{p}}}|1\rangle$. It is easy to construct such a unitary transformation if the value of p^* and $\tilde{p}$ are known. A good solution is now defined as the one in which $\mathcal{O}$ produces a good solution and r is 0. This means that we have a new quantum algorithm $\mathcal{O}^*$ with success probability $\frac{\epsilon^2 p^*}{p}$. Let $\sin^2 \alpha = \frac{\epsilon^2 p^*}{p}$. We can assume that $\alpha = \frac{\theta_\epsilon^* \theta_\epsilon}{\theta_\epsilon}$ for sufficiently small θ_ϵ ($\theta_\epsilon \ll 1$). Thus, after $\overline{m^*}$ repetitions of Q on $\mathcal{O}^*$, we obtain a good solution with success probability,

$$\sin^2(\overline{2m^*}+1)\alpha = \sin^2\left(\frac{\pi}{2}\frac{\theta_\epsilon}{\tilde{\theta}_\epsilon}\right).$$

Again, since

$$1 - \frac{\frac{1}{(\pi+1)T}}{1+\frac{1}{(\pi+1)T}} \leq \frac{\theta_\epsilon}{\tilde{\theta}_\epsilon} \leq 1 + \frac{\frac{1}{(\pi+1)T}}{1-\frac{1}{(\pi+1)T}},$$

by analysis similar to that of the previous case it can be shown that

$$\sin^2(\overline{2m^*}+1)\alpha \geq 1 - \frac{1}{T^2}.$$

From the above lemma we know that if only we can estimate θ_ϵ within some relative error then we can amplify the success probability of $\mathcal{O}$ close to 1. Moreover, we have the algorithm $Est\_Phase(\mathcal{O}, \chi, M)$, which estimates θ_ϵ within $O(1/M)$. It turns out that we can utilize it to estimate θ_ϵ within some relative error. The main idea is again shown in [7]. Here, we present it to complete our discussion.

Algorithm 1 ($Est\_Phase\_Rel(\mathcal{O}, \chi, T)$).

1. Set $l = 0$.
2. Increase l by 1.
3. Set $\tilde{\theta}_\epsilon = Est\_Phase(\mathcal{O}, \chi, 2^l)$.
4. If $\tilde{\theta}_\epsilon = 0$ and $2^l \leq \frac{\sqrt{N}}{10}$ then go to step 2.
5. Output $\tilde{\theta}_\epsilon = Est\_Phase(\mathcal{O}, \chi, \lceil 5\pi(\pi+1)T2^l \rceil)$.

Lemma 6. *$Est\_Phase\_Rel(\mathcal{O}, \chi, T)$ finds a $\tilde{\theta}_\epsilon$ with probability at least $\frac{2}{3}$ such that*

$$|\theta_\epsilon - \tilde{\theta}_\epsilon| \leq \frac{\theta_\epsilon}{(\pi+1)T},$$

where $\sin^2 \theta_\epsilon = \epsilon^2$ is the success probability of $\mathcal{O}$. Moreover, it uses $\mathcal{O}$ and its inverse for $O(\frac{T}{\theta_\epsilon})$ times.

Proof. Let $m = \lfloor \log \frac{1}{5\theta_\epsilon} \rfloor$. By Theorem 11 of [7], the probability that step 3 outputs $\tilde{\theta}_\epsilon = 0$ for $l = 1, 2, \ldots, m$ is at least $\cos^2(\frac{2}{5})$.

Then, since step 3 has outputted $\tilde{\theta}_\epsilon = 0$ at least m times, we have $2^m \leq \frac{1}{5\theta_\epsilon} \leq \frac{\sqrt{N}}{5}$. Therefore, at step 5 we call $Est\_Phase\_Rel(\mathcal{O}, \chi, M)$ with

$M \geq \frac{\pi(\pi+1)T}{\theta_\epsilon}$. By Theorem 12 of [7], the probability of Est_Phase outputting $\tilde{\theta}_\epsilon$ such that $|\theta_\epsilon - \tilde{\theta}_\epsilon| \leq \frac{\theta_\epsilon}{(\pi+1)T}$ is at least $\frac{8}{\pi^2}$. The overall probability is therefore at least $\cos^2(2/5) \times \frac{8}{\pi^2} > \frac{2}{3}$. Since we increase the repetition of Est_Phase exponentially and by Theorem 4, the query complexity is $O(\frac{T}{\theta_\epsilon})$.

Now, we are ready for the following main theorem.

Theorem 6. *Let $\mathcal{A}$ be any quantum algorithm solving some problem with probability p and using an oracle O_f for $T(\geq 2)$ number of queries. Then, there exists a quantum algorithm $\mathcal{A}'$ solving the same problem with probability at least $\frac{p}{6}$ and using an oracle O_f^ϵ for $O(\frac{T}{\epsilon})$ number of queries.*

Proof Sketch.

We construct $\mathcal{A}'$ by using the same idea as that in Theorem 5. By Lemma 6, i.e., $Est\_Phase\_Rel(O_f^\epsilon, \chi, T)$, we can obtain an appropriate estimation of θ_ϵ, i.e., $\tilde{\theta}_\epsilon$ with probability at least $\frac{2}{3}$, and moreover, we only use $O(\frac{T}{\epsilon})$ number of queries. Then, we simulate one query in $\mathcal{A}$ by using $\tilde{\theta}_\epsilon$. Note that we need to use an additional $\lceil \log T \rceil$-bit register to prevent destructive interferences. We initialize it to zero and increase its value by 1, when we finish simulating one query in $\mathcal{A}$, on condition that the content of the other registers is good.

By Lemma 5, i.e., replacing $\mathcal{O}$ with O_f^ϵ, we know that each simulation of one query in $\mathcal{A}$ results in the success probability at least $1 - \frac{1}{T^2}$. Since for all $t, n \in \mathbb{R}$ such that $n \geq 1$ and $|t| \leq n$, $\left(1 + \frac{t}{n}\right)^n \geq e^t \left(1 - \frac{t^2}{n}\right)$, at the end the success probability is

$$\left(1 - \frac{1}{T^2}\right)^T p \geq \left(1 - \frac{1}{T}\right)^2 p.$$

The overall probability is therefore $> \frac{2}{3} \cdot \frac{p}{4} \geq \frac{p}{6}$ for $T \geq 2$, while the overall complexity is $O(\frac{T}{\epsilon})$. □

Remark.

Careful readers might wonder if the above results are trivial applications of the amplitude amplification. They are not quite so, since the amplitude amplification only guarantees the amplification up to a constant probability, say, $2/3$. Since in $\mathcal{A}$ O_f^ϵ might be used for more than a constant time, this is not enough for our purpose of simulating the perfect oracles. Lemmas 2 and 3 show that it is possible to amplify the amplitude very close to 1 using only $O(T/\epsilon)$ queries to O_f^ϵ.

Next, we show gaps between quantum and classical algorithms using biased oracles. Since the parity of N-bit inputs can be solved in $O(N)$ queries by a quantum algorithm if oracles are perfect, the following proposition is straightforward. Note that the classical lower bound of computing N-bit parity with $\log N$ overhead was proved in [17]. The paper [17] showed that for functions like N-bit OR, the log-overhead can be avoided.

Proposition 3. *For any $0 < \epsilon \leq 1/6$, it holds that any classical algorithm that computes the N-bit parity function with high probability requires $\Omega(N \log N/\epsilon^2)$ queries to oracles with bias ϵ, while there exists a quantum algorithm that needs only $O(N/\epsilon)$ queries to the corresponding quantum oracles.*

3.3 Quantum Biased Oracles With Resettable Condition

In addition to the relaxation mentioned in the previous section, if the quantum biased oracle does not have a work space, it is essentially the same as the perfect oracle. That is, although a work space does not matter in the classical case, we cannot ignore the work space in Definition 1 for the model of the quantum biased oracles.

To discuss the above matter, we introduce the following special quantum ϵ-biased oracle.

Definition 3. *The following quantum ϵ-biased oracle is called a resettable biased oracle:*

$$O_\epsilon \ket{x}\ket{0^m}\ket{0} = \ket{x}\ket{0^m}\left(\alpha\ket{f(x)} + \beta\ket{\overline{f(x)}}\right),$$

where $\alpha = \sqrt{\frac{1}{2} + \epsilon}$ and $\beta = \sqrt{\frac{1}{2} - \epsilon}$.

The above oracle is essentially the same as the following one. (It is easy to verify that $\tilde{O}_\epsilon$ can be constructed by O_ϵ and two Hadamard gates.)

$$\tilde{O}_\epsilon \ket{x}\ket{0} = \ket{x}\left((-1)^{f(x)}\alpha\ket{0} + \beta\ket{1}\right), \tag{7}$$

$$\tilde{O}_\epsilon \ket{x}\ket{1} = \ket{x}\left(\alpha\ket{1} - (-1)^{f(x)}\beta\ket{0}\right), \tag{8}$$

where $\alpha = \sqrt{\frac{1}{2} + \epsilon}$ and $\beta = \sqrt{\frac{1}{2} - \epsilon}$.

Let V be any perfect quantum oracle which maps

$$\ket{x, b, z} \to (-1)^{b \cdot f(x)} \ket{x, b, z},$$

where $x \in \{0,1\}^n$ and z be any qubit strings. Note that V is the standard definition for perfect oracles which often appears in the literature [1, 21, 31].

Theorem 7. *If there exists a quantum algorithm A solving some problem with probability $1 - \delta$ by querying V T times, then instead of querying V, A can solve the same problem with probability $1-\delta$ by querying O_ϵ $O(T)$ times, where O_ϵ is a resettable biased oracle for V.*

Proof. For simplicity, we omit the description of z since it is left unchanged by the oracle transformation. Suppose that we have a quantum state $\ket{\psi} = \sum_x \gamma_x \ket{x}\ket{0}$ at some moment of the algorithm, where $\sum_x |\gamma_x|^2 = 1$. Then it follows that applying oracle O_ϵ to this $|\psi\rangle$ results in $O_\epsilon \sum_x \gamma_x |x\rangle |0\rangle = \sum_x (-1)^{f(x)} \alpha \gamma_x |x\rangle |0\rangle + \sum_x \beta \gamma_x |x\rangle |1\rangle$.

Now here comes our key technique, namely, to use a measurement: If the measurement on the last qubit results in the state $|0\rangle$, we know that the quantum state after this measurement is exactly the same as the quantum state after calling V. Otherwise, if the state $|1\rangle$ is measured, we simply need to flip the last qubit to 0 and repeat querying O_ϵ since the previous state $|\psi\rangle$ is completely preserved. Note that the expected number of iterations is constant. Thus, A can query O_ϵ instead of V and the query complexity is roughly the same.

The two relaxations discussed in this section alter the query complexities, unlike the classical case.

4 Lower Bounds of the Query Complexity of Biased Oracles

We have showed how to use quantum amplitude amplification and estimation to obtain quantum algorithms which are not only resilient to imperfect oracles, i.e., do not have log-overhead additional query complexity but also have query complexity which is proportional to $1/\epsilon$. In this section, we consider the optimality of algorithms of the previous section by deriving general lower bounds on the number of queries to ϵ-biased oracles.

We will modify the quantum adversary method to deal with general biased oracles. First, we present the main result of quantum adversary argument [21] with regard to perfect oracles.

Theorem 8 (Ambainis). *Let $f(x_1, \ldots, x_N)$ be a function of N variables with values from some finite set and X, Y be two sets of inputs such that $f(x_1, \ldots, x_N) \neq f(y_1, \ldots, y_N)$ if $x = x_1, \ldots, x_N \in X$ and $y = y_1, \ldots, y_N \in Y$. Let $R \subset X \times Y$ be such that*

1. *For every $x \in X$, there exist at least m different $y \in Y$ such that $(x, y) \in R$.*
2. *For every $y \in Y$, there exist at least m' different $x \in X$ such that $(x, y) \in R$.*

Let $l_{x,i}$ be the number of $y \in Y$ such that $(x, y) \in R$ and $x_i \neq y_i$, and $l_{x,i}$ be that of $x \in X$ such that $(x, y) \in R$ and $x_i \neq y_i$. Let $l_{\max}$ be the maximum of $l_{x,i} l_{y,i}$ over all $(x, y) \in R$ and $i \in \{1, \ldots, N\}$ such that $x_i \neq y_i$. Then, any quantum algorithm computing f uses $\Omega\left(\sqrt{\frac{mm'}{l_{\max}}}\right)$ queries.

With regard to biased oracles whose unitary transformation follows (5), we can show the corresponding lower bound in Theorem 9. To prove it using the quantum adversary argument, we need some technical details as shown in the following.

First, for simplicity we consider the case when $m = 1$, i.e., there are no working qubits (or equivalently, the ancilla qubits v_x and w_x are always cleared to 0^{m-1} after calling biased queries.) Next, we need to address one of the obstacles in using the quantum adversary argument: It also requires the definition of the oracles' unitary transformation on input $|x\rangle |1\rangle$ which is not uniquely defined from (5). Note that for the perfect oracles, the definition of unitary transformation on input $|x\rangle |0\rangle$ implies a unique definition, up to the phase factor, on the corresponding input $|x\rangle |1\rangle$ due to the unitarity constraint. Thus, to use the quantum adversary argument we should consider adding behavior of biased oracles on input $|x\rangle |1\rangle$ *without* implying additional power to the oracles. Formally, we have to consider the following biased oracles:

$$O_f^\epsilon |x, 0\rangle = |x\rangle \left(\alpha_x |f(x)\rangle + \beta_x \left|\overline{f(x)}\right\rangle\right), \tag{9}$$

$$O_f^\epsilon |x, 1\rangle = e^{i\theta_x} |x\rangle \left(-\alpha_x \left|\overline{f(x)}\right\rangle + \beta_x |f(x)\rangle\right), \tag{10}$$

with the phase $e^{i\theta_x}$ defined appropriately.

The next lemma states that it is safe to assume $e^{i\theta_x} = (-1)^{f(x)}$ in the sense that we are not using more powerful oracles. (Note that if we assume $e^{i\theta_x} = 1$ for all x, then the oracles become stronger such that the best lower bound one can get is Theorem 8, i.e., no explicit bias factor in the lower bound. Indeed, one can design algorithms achieving this lower bound.)

Lemma 7. *Let O_f^ϵ be a unitary transformation that is only defined on input $|x, 0\rangle$ such that*

$$O_f^\epsilon |x, 0\rangle = |x\rangle \left(\alpha_x |f(x)\rangle + \beta_x \left|\overline{f(x)}\right\rangle\right).$$

Let also $\tilde{O}_f^\epsilon$ be a unitary transformation that is defined on input $|x, 0\rangle$ and $|x, 1\rangle$ such that

$$\tilde{O}_f^\epsilon |x, 0\rangle = |x\rangle \left(\alpha_x |f(x)\rangle + \beta_x \left|\overline{f(x)}\right\rangle\right),$$

$$\tilde{O}_f^\epsilon |x, 1\rangle = (-1)^{f(x)} |x\rangle \left(-\alpha_x \left|\overline{f(x)}\right\rangle + \beta_x |f(x)\rangle\right).$$

Then, $\tilde{O}_f^\epsilon$ is not more powerful than O_f^ϵ in the sense that if the quantum query complexity with $\tilde{O}_f^\epsilon$ is $O(T)$, then the quantum query complexity with O_f^ϵ is also $O(T)$.

Proof. Details are omitted. The idea is to simulate $\tilde{O}_f^\epsilon$ with O_f^ϵ using its second input, which is used by the oracle to write the answer to queries, as the control to the swap and Z gates. $\tilde{O}_f^\epsilon$ can be realized by a quantum circuit which consists of two O_f^ϵ, one $O_f^\epsilon\dagger$, and a constant number of elementary quantum gates.

We now have all the tools to prove the following general lower bounds for biased oracles.

Theorem 9. *Let O_{f_a} be a perfect oracle whose unitary transformation is $O_{f_a}|x,b\rangle = |x, b \oplus f(a,x)\rangle$ and $O_{f_a}^\epsilon$ be the corresponding biased oracle. If the quantum query complexity of perfect oracles is $\Omega\left(\sqrt{\frac{mm'}{l_{\max}}}\right)$, then the quantum query complexity of the corresponding biased oracles is $\Omega\left(\frac{1}{\epsilon}\sqrt{\frac{mm'}{l_{\max}}}\right)$.*

Proof. Similar to the proof of Theorem 8, we consider the input set $S = X \cup Y$ and the initial superposition

$$|0\rangle \otimes \frac{1}{\sqrt{2|X|}} \sum_{a \in X} |a\rangle + \frac{1}{\sqrt{2|Y|}} \sum_{b \in Y} |b\rangle,$$

where $|0\rangle$ denotes the initial state of algorithm's registers and $|a\rangle$ (and $|b\rangle$ on the second set) denote the oracle's input. Thus, the quantum system lies in the Hilbert spaces $\mathcal{H}_A \otimes \mathcal{H}_O$, $\mathcal{H}_A$ for denoting that of the quantum algorithm and $\mathcal{H}_O$ for that of the quantum oracle.

Let $|\psi_k\rangle = \sum_{x,a} \alpha_{x,a}^k |x,a\rangle$ be the quantum state of the algorithm after querying the oracle for k times, and $\rho_k = Tr_{\mathcal{H}_A}(|\psi_k\rangle\langle\psi_k|)$ be the density matrix obtained by tracing out the algorithm's part. Let us define the Ambainis measure at time k as $S_k = |\sum_{a,b:(a,b)\in R}(\rho_k)_{ab}|$, where $(\rho)_{ab}$ denotes the element at row a and column b. Thus, S_0 and S_T are the Ambainis measures at the beginning and at the end of the computation, respectively. The theorem follows from showing:

1. $S_0 - S_T \geq (1 - 2\sqrt{\delta(1-\delta)})\sqrt{mm'}$, where $1 - \delta$ is the success probability of the algorithm. This inequality follows similarly from that of Theorem 8 and therefore is omitted.
2. $S_{k-1} - S_k \leq 8\epsilon\sqrt{l_{\max}}$. This inequality follows from tedious calculation which principally follows the proof of Theorem 8 as the following.

Note that, unlike the upper bound case, we do not need to impose $|\alpha_x|^2 = 1/2 + \epsilon$ for all x in Lemma 7. However, we will assume so since it greatly simplifies the proof.

Assume that before querying the biased oracle for the $(k-1)$-th time, the quantum state is $|\psi_{k-1}\rangle$ such that

$$|\psi_{k-1}\rangle = \sum_{x,z \in \{0,1\}} \sqrt{p_{x,z}} |x,z\rangle \otimes \sum_a \gamma_{x,z,a} |a\rangle, \tag{11}$$

where $\sum_{x,z \in \{0,1\}} p_{x,z} = 1$ and $\sum_a |\gamma_{x,z,a}|^2 = 1$. It is easy to verify $\rho_{k-1} = Tr_{\mathcal{H}_A}(|\psi_{k-1}\rangle\langle\psi_{k-1}|)$ such that

$$(\rho_{k-1})_{ab} = \sum_x p_{x,0} \gamma_{x,0,a} \gamma_{x,0,b}^* + p_{x,1} \gamma_{x,1,a} \gamma_{x,1,b}^*. \tag{12}$$

From (11), the quantum state $|\psi_k\rangle$ right after calling the biased oracle O_f^ϵ is

$$\begin{aligned}|\psi_k\rangle &= O_f^\epsilon |\psi_{k-1}\rangle \\ &= \sum_{x,a} (\sqrt{p_{x,0}}\gamma_{x,0,a} O_f^\epsilon |x,0,a\rangle + \sqrt{p_{x,1}}\gamma_{x,1,a} O_f^\epsilon |x,1,a\rangle), \\ &= \sum_{x,a} \Gamma_{x,f(a,x)} |x, f(a,x), a\rangle + \Gamma_{x,\overline{f(a,x)}} \left|x, \overline{f(a,x)}, a\right\rangle,\end{aligned}$$

where

$$\Gamma_{x,f(a,x)} = (\sqrt{p_{x,0}}\gamma_{x,0,a}\alpha_x - \sqrt{p_{x,1}}(-1)^{f(a,x)}\gamma_{x,1,a}\beta_x),$$
$$\Gamma_{x,\overline{f(a,x)}} = (\sqrt{p_{x,0}}\gamma_{x,0,a}\beta_x + \sqrt{p_{x,1}}(-1)^{f(a,x)}\gamma_{x,1,a}\alpha_x).$$

It can also be verified that $\rho_k = Tr_{\mathcal{H}_A}(|\psi_k\rangle\langle\psi_k|)$ satisfies

$$\begin{aligned}(\rho_k)_{ab} &= \sum_{x:f(a,x)=f(b,x)} (\Gamma_{x,f(a,x)}\Gamma^*_{x,f(b,x)} + \Gamma_{x,\overline{f(a,x)}}\Gamma^*_{x,\overline{f(b,x)}}) \\ &+ \sum_{x:f(a,x)\neq f(b,x)} (\Gamma_{x,\overline{f(a,x)}}\Gamma^*_{x,f(b,x)} + \Gamma_{x,f(a,x)}\Gamma^*_{x,\overline{f(b,x)}}). \quad (13)\end{aligned}$$

Note that the first group of summation on the right-hand side evaluates to

$$\sum_{x:f(a,x)=f(b,x)} \left(p_{x,0}\gamma_{x,0,a}\gamma^*_{x,0,b} + p_{x,1}\gamma_{x,1,a}\gamma^*_{x,1,b}\right), \quad (14)$$

while the second one evaluates to

$$\begin{aligned}\sum_{x:f(a,x)\neq f(b,x)} & 2\alpha_x\beta_x \left(p_{x,0}\gamma_{x,0,a}\gamma^*_{x,0,b} + p_{x,1}\gamma_{x,1,a}\gamma^*_{x,1,b}\right) \\ &+ \sqrt{p_{x,0}p_{x,1}}(-1)^{f(a,x)}(\alpha_x^2 - \beta_x^2)\left(\gamma_{x,1,a}\gamma^*_{x,0,b} - \gamma_{x,0,a}\gamma^*_{x,1,b}\right). \quad (15)\end{aligned}$$

Since $\alpha_x = \sqrt{\frac{1}{2} + \epsilon}$ for all x and from (12), (14), and (15) we can see that

$$\begin{aligned}&(\rho_{k-1})_{ab} - (\rho_k)_{ab} \\ &= \sum_{x:\ f(a,x)\neq f(b,x)} (1-\sqrt{1-4\epsilon^2}) \cdot (p_{x,0}\gamma_{x,0,a}\gamma^*_{x,0,b} + p_{x,1}\gamma_{x,1,a}\gamma^*_{x,1,b}) \\ &\qquad - \sqrt{p_{x,0}p_{x,1}}(-1)^{f(a,x)}2\epsilon(\gamma_{x,1,a}\gamma^*_{x,0,b} - \gamma_{x,0,a}\gamma^*_{x,1,b}). \quad (16)\end{aligned}$$

To bound $S_{k-1} - S_k \leq |\sum_{a,b:(a,b)\in R}(\rho_{k-1})_{ab} - (\rho_k)_{ab}|$, let us bound the absolute value of the sum of the first term over all a and b on the right-hand side of (16), that is,

$$\sum_{a,b:\,(a,b)\in R}\sum_{x:\,f(a,x)\neq f(b,x)} (1-\sqrt{1-4\epsilon^2})p_{x,0}|\gamma_{x,0,a}|\,|\gamma_{x,0,b}|$$

$$= \sum_x (1-\sqrt{1-4\epsilon^2})p_{x,0} \sum_{a,b:\,(a,b)\in R \wedge f(a,x)\neq f(b,x)} |\gamma_{x,0,a}|\,|\gamma_{x,0,b}|$$

$$\leq \sum_x (1-\sqrt{1-4\epsilon^2})p_{x,0}\sqrt{l_{a,x}l_{a,y}}$$

$$\leq \sum_x 4\epsilon^2 p_{x,0}\sqrt{l_{a,x}l_{a,y}},$$

where the first inequality in the above equation is due to the Cauchy–Schwartz inequality in bounding $\sum_{a:\,f(a,x)=0}|\gamma_{x,0,a}|\sum_{b:\,f(b,x)=1}|\gamma_{x,0,b}|$. Similarly, we can bound the absolute value of the sum of the second term on the right-hand side of (16), that is,

$$\sum_{a,b:\,(a,b)\in R}\sum_{x:\,f(a,x)\neq f(b,x)}(1-\sqrt{1-4\epsilon^2})p_{x,1}|\gamma_{x,1,a}|\,|\gamma_{x,1,b}|$$

$$\leq \sum_x 4\epsilon^2 p_{x,1}\sqrt{l_{a,x}l_{a,y}}.$$

To bound the third and the fourth terms, note that since $\sqrt{p_{x,0}p_{x,1}} \leq \frac{p_{x,0}+p_{x,1}}{2}$, and by following the bound of the absolute sum of the first term, it can be shown that

$$\sum_{a,b:\,(a,b)\in R}\sum_{x:\,f(a,x)\neq f(b,x)}\sqrt{p_{x,0}p_{x,1}}\,2\epsilon|\gamma_{x,1,a}|\,|\gamma_{x,0,b}|$$

$$\leq \sum_x 2(p_{x,0}+p_{x,1})\epsilon\sqrt{l_{a,x}l_{a,y}},$$

$$\sum_{a,b:\,(a,b)\in R}\sum_{x:\,f(a,x)\neq f(b,x)}\sqrt{p_{x,0}p_{x,1}}\,2\epsilon|\gamma_{x,0,a}|\,|\gamma_{x,1,b}|$$

$$\leq \sum_x 2(p_{x,0}+p_{x,1})\epsilon\sqrt{l_{a,x}l_{a,y}}. \quad (17)$$

Therefore, we have

$$S_{k-1} - S_k \leq 4\sqrt{l_{\max}}\sum_x(\epsilon^2+\epsilon)(p_{x,0}+p_{x,1})$$

$$\leq 8\epsilon\sqrt{l_{\max}}\sum_x(p_{x,0}+p_{x,1}) \leq 8\epsilon\sqrt{l_{\max}}.$$

This proves the theorem.

From Theorem 9, we can show the query complexity for quantum search with biased EQ oracles whose upper bound is considered by Høyer et al. [18]; the lower bound $\Omega(\sqrt{N}/\epsilon)$ follows directly from the lower bound of the Grover search by Theorem 8.

Moreover, Theorem 9 also gives the tight lower bounds for the GL problem, which simplifies the proof of Theorem 2 as follows:

Proof. In the following, $N = 2^n$ for n is the length of the hidden string a of the GL problem. Consider oracles IP_a as described in Definition 1. Similar to [21], let ρ_k be the density matrix of $\mathcal{H}_I$ after k queries. Let the Ambainis measure at time k be $S_k = \sum_{a,b:a\neq b} |(\rho_k)_{a,b}|$, X and Y be S (the set of all inputs), and $(x, y) \in R$ iff $x \neq y$. Then, it can be showed that the following (in)equalities hold:

1. $S_0 = N - 1$.
2. $S_T \leq 2\sqrt{\delta(1-\delta)}(N-1)$ since $m = m' = (N-1)$, while δ is the error probability of the algorithm and T is the number of query.
3. $S_{k-1} - S_k \leq 2\sqrt{N-1}$, if EQ_a is called at time k.
4. $S_{k-1} - S_k \leq 4\epsilon N$, if IP_a is called at time k.

Therefore if an algorithm queries IP_a for T_{IP} times and EQ_a for T_{EQ} times, then $S_0 - S_T \leq 4T_{IP}\epsilon N + 2T_{EQ}\sqrt{N-1}$. Hence, in order to solve the GL problem, either $T_{IP} = \Omega(\frac{1}{\epsilon})$ or $T_{EQ} = \Omega(\sqrt{N})$, which proves the theorem.

The first and second (in)equalities are identical to the proof in [21]. The third one holds since with regard to EQ_a, $l_{x,i}l_{y,i} \leq (N-1)$. The fourth one can be obtained in two ways. Either since with regard to IP_a, $l_{x,i} \leq N/2$ and $l_{y,i} \leq N/2$, or directly from the proof of Theorem 9 in bounding $S_{k-1} - S_k$, substituting $f(a, x)$ and $l_{\max}$ with $a \cdot x$ and $N^2/4$, respectively.

5 Concluding Remarks

In this Chapter we have discussed the upper and lower bounds of quantum query complexity of oracles with an explicit bias factor ϵ. With regard to upper bounds, we have shown nontrivial query complexities for the bias oracles if the oracles have some conditions. With regard to lower bounds, we have generalized the quantum adversary method to bound the number of queries with an explicit bias factor term. Besides proving that our upper bounds are mostly optimal, we can also show the trade-off on the number of queries between two types of oracles used in solving the GL problem.

References

[1] O. Goldreich: A fast quantum mechanical algorithm for database search, in *Proc. 28th Ann. ACM Symp. on Theory of Computing (STOC 1996)* (1996) pp. 212–219

[2] D. Biron, O. Biham, E. Biham, M. Grassl, D. A. Lidar: Generalized grover search algorithm for arbitrary initial amplitude distribution, in *Proceedings of the 1st NASA International Conference on Quantum Computing and Quantum Communication*, vol. 1509, LNCS (Springer, Berlin, Heidelberg 1998) pp. 140–147

[3] M. Boyer, G. Brassard, P. Høyer, A. Tapp: Tight bounds on quantum searching, Fortschritte der Physik **46**, 493–505 (1998)

[4] D. P. Chi, J. Kim: Quantum database searching by a single query, in *Proceedings of the 1st NASA International Conference on Quantum Computing and Quantum Communication*, vol. 1509, LNCS (Springer, Berlin, Heidelberg 1998) pp. 148–151

[5] L. K. Grover: A framework for fast quantum mechanical algorithms, in *Proceedings of the 30th ACM Symposium on Theory of Computing* (1998) pp. 53–62

[6] L. K. Grover: Rapid sampling through quantum computing, in *Proceedings of the 32th ACM Symposium on Theory of Computing* (2000) pp. 618–626

[7] G. Brassard, P. Høyer, M. Mosca, A. Tapp: Quantum amplitude amplification and estimation, in *Quantum Computation and Quantum Information: A Millennium Volume*, AMS Contemporary Mathematics Series (2002) p. 305

[8] H. Buhrman, C. Dürr, M. Heiligman, P. Høyer, F. Magniez, M. Santha, R. de Wolf: Quantum algorithms for element distinctness, in *Proceedings of the 16th IEEE Annual Conference on Computational Complexity (CCC'01)* (2001) pp. 131–137

[9] A. Ambainis: Quantum walk algorithm for element distinctness, in *Proceedings of the 45th Symposium on Foundations of Computer Science* (2004) pp. 22–31

[10] Y. Shi: Quantum lower bounds for the collision and the element distinctness problems, in *FOCS'02 Proceedings* (2002) pp. 513–519

[11] S. Aaronson, A. Ambainis: Quantum search of spatial regions, in *Proceedings of the 44th Symposium on Foundations of Computer Science* (2003) pp. 200–209

[12] F. Magniez, M. Santha, M. Szegedy: Quantum algorithms for the triangle problem, in *Proceedings of the 16th ACM-SIAM Symposium on Discrete Algorithms* (2005)

[13] C. Dürr, M. Heiligman, P. Høyer, M. Mhalla: Quantum query complexity of some graph problems, in *Proceedings of the 31st International Colloquium on Automata, Languages and Programming*, vol. 3142, LNCS (2004) pp. 481–493

[14] A. Ambainis, K. Iwama, A. Kawachi, H. Masuda, R. H. Putra, S. Yamashita: Quantum identification of Boolean oracles, in *Proceedings of the 21st Annual Symposium on Theoretical Aspects of Computer Science*, vol. 2996, LNCS (2004) pp. 105–116

[15] A. Blum, A. Kalai, H. Wasserman: Noise-tolerant learning, the parity problem and the statistical query model, in *Proc. 32nd Ann. ACM Symp. on Theory of Computing (STOC)* (2000) pp. 435–440

[16] M. Szegedy, X. Chen: Computing Boolean functions from multiple faulty copies of input bits, in S. Rajsbaum (Ed.): *Proc. 5th Latin American Symp. Theoretical Informatics (LATIN)* (2002) pp. 539–553

[17] U. Feige, P. Raghavan, D. Peleg, E. Upfal: Computing with noisy information, SIAM J. Comput. **23**, 1001–1018 (1994)
[18] P. Høyer, M. Mosca, R. de Wolf: Quantum search on bounded-error inputs, in *ICALP '03 Proceedings*, vol. 2719, LNCS (Springer, Berlin, Heidelberg 2003) pp. 291–299
[19] H. Buhrman, I. Newman, H. Röhrig, R. de Wolf: Robust quantum algorithms and polynomials, in *Proc. of the 22nd Symp. on Theoretical Aspects of Computer Science*, vol. 3404, LNCS (2005) pp. 593–604
[20] M. Adcock, R. Cleve: A quantum Goldreich-Levin theorem with cryptographic applications, in *STACS'02 Proceedings*, vol. 2285, LNCS (Springer, Berlin, Heidelberg 2002) pp. 323–334
[21] A. Ambainis: Quantum lower bounds by quantum arguments, J. Comput. Syst. Sci. **64**, 750–767 (2002)
[22] O. Goldreich, L. Levin: Hard-core predicates for any one-way function, in *Proc. 21st Ann. ACM Symp. on Theory of Computing (STOC 1989)* (1998) pp. 25–32
[23] E. Bernstein, U. Vazirani: Quantum complexity theory, SIAM J. Comput. **26**, 1411–1473 (1997)
[24] B. Terhal, J. Smolin: Single quantum querying of a database, Phys. Rev. A **58**, 1822–1826 (1998)
[25] L. A. Levin. One-way functions and pseudorandom generators, Combinatorica **7**, 357–363 (1987)
[26] O. Goldreich: *Modern Cryptography, Probabilistic Proofs and Pseudorandomness* (Springer, Berlin, Heidelberg 1999)
[27] L. A. Levin: Randomness and non-determinism, J. Symbolic Logic **58**, 1102–1103 (1993)
[28] C. Bennett, E. Bernstein, G. Brassard, U. Vazirani: Strengths and weaknesses of quantum computing, SIAM J. Comput. **26**, 1510–1523 (1997)
[29] K. Iwama, A. Kawachi, R. Raymond, S. Yamashita: Robust quantum algorithms for oracle identification (2004) quant-ph/0411204
[30] K. Iwama, R. Raymond, S. Yamashita: Quantum query complexity of biased oracles, in *Booklet of Workshop on ERATO Quantum Information Science 2003* (2003) pp. 33–34
[31] H. Barnum, M. Saks: A lower bound on the quantum query complexity of read-once functions (2002) quant-ph/0201007

Index

ϵ-biased oracle, 20

amplitude amplification, 23

equivalence (EQ), 22

Goldreich–Levin (GL) problem, 21

Grover search, 19

inner product (IP), 22

oracle identification problem, 20

quantum query complexity, 19

quantum adversary argument, 21

robust, 20

Part II

Quantum Information

Quantum Statistical Inference

Masahito Hayashi

ERATO Quantum Computation and Information Project, Japan Science and Technology Agency 201 Daini Hongo White Bldg. 5-28-3, Hongo, Bunkyo-ku, Tokyo 113-0033, Japan
masahito@qci.jst.go.jp

Abstract. We studied state estimation for several quantum statistical models and for estimation of unitary evolution. We also researched the hypothesis testing and state discrimination for entangled states from theoretical and experimental viewpoints. Moreover, we discussed the measurement theory. These results are reviewed in this Chapter.

1 Introduction

In order to obtain information from the quantum system of interest, we need to perform a quantum measurement and extract the desired information from the obtained data. Needless to say, we must optimize the above two processes for obtaining the maximal amount of information of the quantum system. Such a research area is called quantum statistical inference. Our project obtained the following results in this research area.

- quantum state estimation
    - state estimation in pure state family
    - state estimation in quantum Gaussian States family
    - state estimation in nonregular family
    - estimation of eigenvalue of density matrix in qubit system
- estimation of SU(2) action with entanglement
- hypothesis testing and discrimination
    - hypothesis testing of entangled state
    - distinguishability and indistinguishability by local operations and classical communications (LOCC)
    - application of quantum hypothesis testing
- application to experimental setting
- quantum measurement with negligible state demolition

When a huge number (over 1 000 000) of data are available,[1] we can easily estimate the quantum state in the system. In this case, the statistical inference theory is not required. However, when the number of obtained data is not so large, and is less or equal 100 $\sim$ 1000, we need the help of statistical inference theory for precise decisions. For example, in the remote state preparation, it is difficult to obtain many data, so the theory presented here is required.

When we estimate the unknown state, we usually have several knowledges about the unknown state a priori. In this case, we often assume that this unknown state belongs to a subset of set of all states, which is called a state family. When every state in a family is commutative with each other, the optimal measurement is the measurement concerning the common basis of this family. Then, the estimation can result in the estimation of the probability distribution corresponding to eigenvalues. Hence, the difficulty of state estimation appears in the noncommutative case.

The estimation for the probability distribution has been well established. In particular, when the number of obtained data is sufficiently large (but not so huge), its asymptotic theory can be applied. In this case, the maximum likelihood estimator is almost optimal, and its minimum error can be characterized by the Fisher information matrix, which is defined for a probability distribution family containing the unknown distribution.

In the noncommutative case, our task can be divided into two parts: One is the appropriate choice of the quantum measurement, the other is that of the function estimating the estimated parameter(s) from obtained data. While the latter belongs to the problem of classical statistics, the former is the central issue in quantum estimation. Even if the state family has several parameters, we can independently choose an estimating function for every parameter, but we have to choose a common quantum measurement. When the optimal quantum measurement depends on the parameter, the choice of the quantum measurement is crucial. In quantum theory, any measurement is described by positive operator-valued measure (POVM). We discussed the problem in two cases: One is the case where the state family consists of pure states (Sects. 2.1 and 2.2), and the other is the quantum Gaussian state family, which consists of Gaussian mixture of coherent states (Sect. 2.3).

Further, when the state family has only one parameter, the estimation error can be asymptotically described by the symmetric logarithmic derivative (SLD) Fisher information, which was defined by *Helstrom* [1,2]. However, this discussion cannot be applied to the case when the family has singularities. In order to resolve this problem, we established a general estimation theory

---

[1] Precisely the number where we do not need statistical inference theory depends on the complexity of system. In the typical quantum systems (qubit system and two times the tensor product of it), at least, if we have over 1 000 000 data, statistical inference theory is not required. In this case, the averages of the obtained data give almost true parameters.

that can be applied to more general cases, i.e., it can be applied to a family with singularities (Sect. 2.4).

We also treated the two-dimensional space, in which states can be characterized by three parameters. These parameters can be divided into two parts: One corresponds to the eigenvalue of density matrix, the other to the unitary angle of this density matrix. We focused on the estimation of the eigenvalue of density matrix and compared collective measurements with separable measurements regarding the estimation error of the optimal case. That is, we compared estimation errors of two cases: One is the case when we can perform quantum measurement with interference between samples, and the other is the case where we cannot use such a quantum measurement (Sect. 2.5).

Moreover, we treated the problem estimating unknown unitary evolution. When an unknown unitary evolution is given, we can estimate it by choosing the input state. We discussed this problem in two-dimensional case (Sect. 3).

We also treated the case where the unknown state is assumed to belong to a discrete set. In this case, the decision problem of the unknown state is called discrimination or hypothesis testing. We treated the hypothesis testing and the possibility of the discrimination in the case of entangled states. In these settings, it is natural to limit our measurement to a class of local operation and classical communication (LOCC). This is because it is very difficult to realize measurement using quantum correlation between two parties when these parties are far from one another (Sect. 4). We also discussed the application of quantum hypothesis testing to several topics in quantum information theory (Sect. 4.3).

We also applied the statistical inference theory to an experimental setup with polarization states of biphotons. One application is estimating the unknown state by using Akaike's information criterion. The other application is testing the maximally entangled state (Sect. 5).

Further, we propose quantum measurement with negligible state demolition. By smearing our measurement, the degree of state demolition can be decreased. We also applied this method to universal quantum data compression (Sect. 6.2).

2 Quantum State Estimation

2.1 State Estimation in Pure State Family

First, we consider the quantum state estimation of the pure state family [3]. That is, we treat the estimation problem when the unknown state is assumed to belong to the state family of pure states. As a result of this restriction, the analysis of estimation error becomes quite simple.

In this research, we focus on the locally unbiased condition for our estimator and minimize the estimation error under this condition. More precisely, the weighted sum of mean square errors of respective parameters is minimized under this conditions. This minimum value is called the Cramér–Rao bound. In fact, this condition is essentially equivalent to assuming that the true state is known to belong to the local neighborhood of a fixed state. Since the local neighborhood can be approximated by tangent space, this minimization problem can be defined at the tangent space at each point. However, even though this assumption is not valid, the minimum error under this condition is equal to the minimum error in the asymptotic framework.

Our main contribution to this problem is simplification of this minimizing problem in the pure states case. For this purpose, we focused on the linear transform $\boldsymbol{D}$, which is defined in relation with complex structure. We showed that the Cramér–Rao bound can be written as a quite simple minimization problem characterized by the linear transform $\boldsymbol{D}$. This produces several good byproducts. As the first byproduct, we found that any collective measurement cannot improve the Cramér–Rao bound in the pure states model. In fact, in the pure states full model, it is known that the multiple trial of optimal measurement of simple copy state has an equivalent performance with the optimal collective measurement in the first-order asymptotics [3]. We extended this result to a more general pure states model.

As the second byproduct, we found that the Cramér–Rao bound is closely related to the eigenvalue of the linear transform $\boldsymbol{D}$. In particular, when these values are larger, the Cramér–Rao bound is larger. In other words, when these value are close to zero, all parameters can independently be measured by the respective optimal measurement. However, when these values are large, it is impossible to simultaneously realize the optimal measurement for each parameter. Hence, we can regard these values as the degree of noncommutativity.

For example, we consider the two-parameter case. In the maximally noncommutative case, it is proven that an optimal measurement for one parameter does not bring about any information of the other parameter. That is, an optimal measurement for two parameters is randomly performing an optimal measurement for the respective parameter.

Also, if a unitary-invariant distance (e.g., one minus fidelity) is chosen as a risk function, the asymptotic error of state estimation is shown to be an increasing function of noncommutativity. In two-parameter state family, this fact is proven for all the distance functions.

2.2 State Estimation for Covariant Pure States Family

Following the above result, we studied state estimation for several covariant pure states families with two parameters. For example, the all pure states family in two-level system has a group symmetry for action SU(2), and is parameterized by the sphere. The boson coherent states family has that for the Weyl–Heisenberg group, and is parameterized by the complex plane. Further, when only the squeezing parameter is unknown, the squeezed states family has that for SU(1,1) (SL(2,R)), and is parameterized by the unit disk (the upper half plane).

We derived the optimal covariant estimator for the finite number of copies of the unknown state in these three covariant models [4, 5]. The optimal estimator is obtained only assuming the covariance of the risk function. That is, the optimal estimator depends only on the covariance, not on the form of the risk function. However, there is no optimal covariant estimator for the squeezed states family when the number of copies is one or two [6, 7]. We need at least three copies for performing the optimal covariant estimator.

Further, we focus on the average of $1 - $ the fidelity, and check that the first-order asymptotic behavior coincides with the above Cramér–Rao approach. We also focus on the second-order asymptotic behavior, and compare it and geometrical curvature in the above three models [5].

The optimal covariant (universal) cloning machine is known when the initial state belongs to the boson coherent states family or the n-copy of pure spin states family. We derived the optimal cloning of squeezed states family [6, 7]. In this case, there is no optimal cloning when the initial number of copies is one or two. Further, we showed that the state family with the SU(1,1) covariance is given as the family of the output states of optimal cloning of squeezed states family if and only if the family is quasi-classical [6, 7].

2.3 State Estimation in Gaussian States Family

Any single-mode photon system is described by the Boson–Fock space, i.e., a Hilbert space spanned by $|0\rangle, |1\rangle, \ldots, |n\rangle, \ldots$, where the state $|n\rangle$ refers to the n-photon state, and is called the number state with n photons. In quantum optics, it is not so hard to realize the coherent state $|\alpha\rangle \equiv e^{-|\alpha|^2/2} \sum_n \sqrt{\frac{\alpha^k}{n}} |n\rangle$, while it is very hard to realize the number state with nonzero photons. If the number state is influenced by the thermal noise, it eventually becomes a Gaussian state, which is described by the Gaussian mixture of coherent states, i.e., its density matrix is written by $\rho_{\zeta,N} \equiv \frac{1}{\pi N} \int e^{-|\alpha-\zeta|^2/N} |\alpha\rangle\langle\alpha| \, d\alpha$. In this system, we have two typical quantum measurements. One is number detection corresponding to the resolution by number states; the other is the heterodyne detection corresponding to the resolution by coherent states. In the present work, we compared estimation errors in this family in two settings:

1. One is the case where we can use quantum interference between several particles.
2. The other is the case where we cannot use it.

When ζ equals 0, the density is written by mixture of number states, i.e., $\rho_{0,N} = \sum_{n=0}^{\infty} \frac{N^n}{(N+1)^{n+1}}$. Since all densities in the family $\{\rho_{0,N}\}$ are commutative with each other, the number detection is optimal. On the other hand, the heterodyne detection is optimal for estimating the parameter ζ in the family $\{\rho_{\zeta,N}\}$.

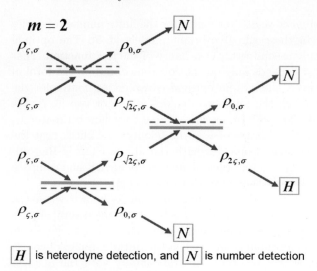

Fig. 1. Tournament method

Although we cannot simultaneously realize the heterodyne detection and the number detection, we can construct a scheme to realize almost simultaneously both of them by using beam splitters in the following steps [8]:

1. We perform unitary evolution as $\rho_{\zeta,N}^{\otimes n} \to \rho_{\sqrt{n}\zeta,N} \otimes \rho_{0,N}^{\otimes n-1}$.
2. We perform the heterodyne detection on the first quantum system whose state is $\rho_{\sqrt{n}\zeta,N}$ and obtain the data ζ'.
3. We perform the number detection on the rest of quantum systems and obtain the data $k_1, \ldots, k_{n-1}$.
4. We infer that the unknown parameters ζ and N are $\frac{\zeta'}{\sqrt{n}}$ and $\frac{1}{n-1}\sum_{i=1}^{n-1} k_i$.

By using this method, we can simultaneously estimate two parameters with errors that are almost as small as the optimal case. In particular, we can realize the unitary evolution $\rho_{\zeta,N}^{\otimes n} \to \rho_{\sqrt{n}\zeta,N} \otimes \rho_{0,N}^{\otimes n-1}$ in the $n = 2^m$ case as described in Fig. 1.

However, it is very hard to realize the perfect number detection. In particular, it is difficult to discriminate one-photon state from two-photon states. Therefore, we can only realize approximate number detection which discriminates the zero-photon state (the vacuum state) from other number states. Moreover, the quantum efficiency t does not equal 100%, and it is usually 30% to 90%, where the quantum efficiency t denotes the percentage of detected photons. By numerical analysis, we checked the advantage of this method with such an approximate number detection.

However, there is noise in the unitary evolution by the beam splitter. We studied the effect of this noise, and improved the estimator by taking it into account [9].

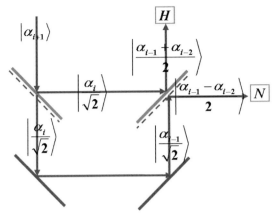

H is heterodyne detection, and N is number detection

Fig. 2. Markov type correlation

The realization scheme given in Fig. 1 has another problem. This scheme needs the preparation of separate quantum information sources. In order to resolve this problem, we propose a Markov type correlation scheme (Fig. 2), where we need to prepare only a single quantum information source. Using this scheme, we can realize a quantum measurement with quantum correlation.

2.4 State Estimation in Non-Regular Family

In the regular classical/quantum estimation theory, it is assumed that families of probability distributions or quantum states are smooth. In such a case, it is possible to define Fisher information or its quantum analogue based on SLD [or right logarithmic derivative (RLD)].[2] These quantities give tight bounds for the estimation error. On the other hand, in the nonregular cases, the families are not necessarily smooth [10]. For example, the probability distribution family forms the devil's staircase. In classical estimation, Hammersley, Chapman and Robbins (HCR) [11, 12] gave a generalized Fisher information based on the difference of two probability distributions.

For nonregular quantum estimation, we generalized HCR's argument as follows:

---

[2] In quantum estimation theory, there are several quantum analogues of Fisher information One is based on symmetric logarithmic derivative (SLD) L_θ^S: $\frac{d\rho_\theta}{d\theta} = \frac{1}{2}(\rho_\theta L_\theta^S + L_\theta^S \rho_\theta)$. Another is on right logarithmic derivative (RLD) L_θ^R: $\frac{d\rho_\theta}{d\theta} = \rho_\theta L_\theta^R$. Indeed, in the one-parameter regular case, the SLD-type Fisher information gives the tight bound, but the RLD-type Fisher information does not give the tight bound.

1. We introduced SLD-type and RLD-type bounds for the nonregular case.
2. We also introduced the asymptotic type bound for the large sample case.
3. When the family of quantum states is discrete, we showed that the RLD-type bound exponentially decreases asymptotically.

The phenomenon of point 3 is related to the fact that the estimation error for a discrete model exponentially decreases.

We gave the following examples of nonregular models:

1. a noncommutative discrete uniform distribution model
2. the concurrence of entanglement in 2×2 system
3. a model that forms the devil's staircase

2.5 Estimation of Eigenvalue of Density Matrix in Qubit System

When a quantum system is in the two-dimensional space, the state can be characterized by three parameters as

$$\rho_{r,\theta} := \frac{1}{2}\begin{pmatrix} 1 + r\cos\theta_1 & r\sin\theta_1 e^{i\theta_2} \\ r\sin\theta_1 e^{-i\theta_2} & 1 - r\cos\theta_1 \end{pmatrix}, \quad 0 \le \theta_1 \le \frac{\pi}{2}, \quad 0 \le \theta_2 \le 2\pi. \quad (1)$$

We treated the estimation of the parameter r corresponding to the eigenvalue of the density matrix [13].

This problem is very easy when we know the parameter θ because it results in the classical case. However, it becomes difficult when the parameter θ is unknown. In the present work, in order to discuss the efficiency of quantum correlation in the quantum measurement, we considered two settings. In the first setting, we assume that we can only perform the same quantum measurement on each quantum system. In the second setting, we divide n quantum systems into $n/2$ pairs of quantum systems, and we can only perform the same quantum measurement on each pair of quantum systems, i.e., we can only use a quantum correlation between two quantum systems. *Hendrich* et al. [14] realized a quantum measurement that belongs to the second setting, and they pointed out that we can estimate the parameter r by their measurement. We showed that in the first setting, the optimal measurement is independent of the unknown parameter r. We proved that the performance of their measurement is better than the optimal one in the first setting when the parameter r is larger than 0.7. We also proposed several measurements in the second setting and compared their performance.

3 Estimation of SU(2) Action With Entanglement

In a quantum system, when the state evolution is free from noise, it is described by a unitary matrix. If the unitary matrix is not known and needs to be known, we estimate it by examining output states of several input states.

Fujiwara [15] treated this problem in the two-dimensional system, showing that there is a great advantage by entangling input state with the reference system like the dense coding. He showed that the error of his method is much smaller than the conventional one. In this setting, we estimate the unknown unitary matrix from n data after we perform the same experiment n times. Therefore, we assume that we can arrange the unknown unitary evolution on n two-level systems. In his method, any input state entangled on n two-level systems was not discussed. In the present work, we discuss whether we can gain an advantage by using an input state entangled on n two-level systems. Our answer of this question is *yes*, i.e., we have a great advantage by using such an entangled state.

Further, estimating the unknown unitary action is closely related to query complexity (See Part I: Quantum Computation). In the query complexity, we estimate the unknown query, which is essentially equal to the unknown unitary action. Further, in the query complexity, we can choose input state adaptively. Hence, query complexity has a wider choice than estimation of unitary action discussed in this section.

3.1 One-Parameter Case

A similar problem has been discussed by *Bužek, Derka* and *Massar* [16]. Their problem was estimating the eigenvalue $e^{i\theta}$ of the unknown unitary matrix in a two-level system with the knowledge of the eigenvectors in the same setting. They focused on the mean value of $\sin^2(\hat{\theta}-\theta)$, where $\hat{\theta}$ is the estimated error. They proved that the error goes to 0 in proportion to $1/n^2$ when the optimal input state and optimal measurement is chosen. Since the estimation error usually goes to 0 in proportion to the inverse of the number of samples, their results are very surprising and indicate the importance of the entangled input state.

3.2 Three-Parameter Case

In our work [17], we discussed whether such a phenomena happens in the estimation of SU(2) unitary action. When the error between the true SU(2) action U and the estimated one $\hat{U}$ is given by $1 - |\operatorname{Tr}\frac{U^{-1}\hat{U}}{2}|^2$, we obtained a surprising correspondence between our problem and that of *Bužek, Derka* and *Massar* [16]. By using this correspondence, we can trivially show that our error goes to 0 in proportion to $1/n^2$, and its coefficient is π^2. We also clarified that we can neglect the advantage of entangling the input system with the reference system in this method. We can also regard a part of the composite system of n input systems as the tensor product of the system of interest and the reference system. In other words, there is a "self-entanglement" effect in this method. On completion of this research, the author found that the same results were obtained by two other groups [18, 19, 20].

4 Hypothesis Testing and Discrimination

4.1 Hypothesis Testing of Entangled State

Entangled states are important resources for quantum information processing [21]. When we experimentally realize the quantum information processing, we must artificially prepare entangled states that are close to the maximally entangled state.

Many experimentalists have invented devices to produce entangled states, however, the quality of these generated states must be verified by a systematic method. Hence, it is desired to establish a systematic method for verifying the quality of generated entangled state experimentally. Many researchers proposed entanglement witness for this purpose (*Barbieri* et al. [22]). However, their arguments are not satisfactory and sometimes are ad hoc from the viewpoint of statistical hypothesis testing.

Usually the quality of industrial products is verified by the method of statistical hypothesis testing based on the probabilistic treatment of random sampling. Since the measurement outcome is obtained probabilistically in the quantum system, it is worthwhile to verify the quality of entangled states based on the method of statistical hypothesis testing. In the presented work, we investigated this method as a systematic method for verifying the quality of generated entangled states. In particular, we focus on the following two hypotheses:

– Null hypothesis: The state is not close to the maximally entangled state versus.
– Alternative hypothesis: The state is not close to the maximally entangled state.

Our purpose is to find a good method to test the hypotheses. For practical convenience, it is required that:

(R1) The POVM should be implemented by local operations and classical communications (LOCCs) between two parties (Alice and Bob).

We discussed this problem in the following cases:

1. One independent sample is given: While *Virmani* et al. [23] obtained the optimal solution in this case by using invariance and the positve partial transpose (PPT) condition, we simplified their proof, and proposed a realizable method for the optimal test.
2. Two independent samples are given: We derived optimal solutions in two settings by a group invariance method. In the first setting, we require only **(R1)**, but in the second setting, we require not only **(R1)** but also **(R2)**:

(R2) The POVM should be implemented by LOCC between the first sample and the second one.

As a result, we found that the optimal test in the second setting improves the optimal test in the first setting.

4.2 Distinguishability and Indistinguishability by LOCC

It is a fundamental and interesting question to consider the distinguishability of entangled states shared by distant parties if only LOCC is allowed. Not only entangled states, but also the local discrimination of any quantum states shared by distant parties has been attracting considerable attention recently. It is clear that orthogonal quantum states can be distinguished, while nonorthogonal states can only be distinguished probabilistically if there are no restrictions for measurements. If the quantum states are shared by two distant parties, say Alice and Bob, and only LOCC is allowed, the possibility of distinguishing these quantum states may decrease since considerable restrictions are imposed for the measurements. Interestingly, *Walgate* et al. [24] showed that any two orthogonal pure states shared by Alice and Bob can be distinguished by LOCC. On the other hand, there is a set of orthogonal bipartite pure product states that cannot be distinguished with certainty by LOCC. Recently, *Horodecki* et al. [25] showed a phenomenon of "more nonlocality with less entanglement." It differentiates nonlocality from entanglement. A number of other interesting and often counterintuitive results have been obtained. Our results can also be added to this list of counterintuitive results. We obtained two main results in the following:

1. Indistinguishability 1 [26]: A set of linearly independent quantum states $\{(U_{m,n} \otimes I)\rho^{AB}(U_{m,n}^\dagger \otimes I)\}_{m,n=0}^{d-1}$ cannot be discriminated deterministically or probabilistically by LOCC, where $U_{m,n}$ are generalized Pauli matrices.
2. Indistinguishability 2 [27]: The number of locally distinguishable maximally entangled states is equal to or less than the dimension of the local space.
3. Indistinguishability 3 [28]: We also investigate the upper bound of the maximal number of locally distinguishable entangled states not only in the bipartite case but also in the multipartite case.
4. Distinguishability [26]: On the other hand, any l maximally entangled states from this set are locally distinguishable if $l(l-1) \leq 2d$. The explicit projecting measurements are obtained to locally discriminate these states.

4.3 Application of Quantum Hypothesis Testing

In quantum information theory, several information processing protocols were proposed and their asymptotic performances were discussed. These topics contain quantum channel coding, quantum compression and entanglement concentration. Among the classical information theory community, it is known that the asymptotic performances of several types of information

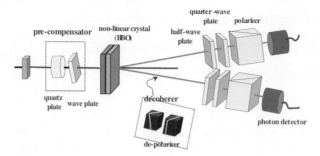

Fig. 3. Experimental setup for producing various polarization states of biphotons and measuring them

processing are closely related to hypothesis testing. By using the information spectrum method, *Han* [29] pointed out its relation more clearly, i.e., he obtained the general relations between the asymptotic behavior of the probability distribution function of the likelihood and the asymptotic performances of respective information processing. *Nagaoka* and *Hayashi* [30] gave the quantum analogues of the information spectrum method and the likelihood. In the present work [31,32], we unifiedly treated the bound of respective information processing, i.e., quantum channel coding, quantum compression and entanglement concentration through the insight of this quantum analogue of likelihood.

5 Experimental Application of Quantum Statistical Inference

In this work, we obtained theoretical and experimental analyses of errors in quantum state estimation and hypothesis testing of entangled states, putting a special emphasis on their asymptotic behavior. In particular, we focused on the state of two qubits (two 2-level quantum systems). The two-qubit system in four-dimensional Hilbert space is the simplest one where a peculiar characteristic of quantum mechanics, *entanglement*, emerges. Since entanglement plays a critical role in the mysterious phenomena in the quantum world, it is interesting to ask whether entanglement affects the accuracy of the estimation. Various kinds of two qubits (including entangled states) are practically realizable as polarization states of biphotons produced via spatially nondegenerate, spontaneous parametric down conversion (SPDC) with type-I phase matching. Thus, in our experiments, we followed the above methods for producing the ensembles of the biphoton polarization states with the following setup for measuring them, and for estimating or testing their density matrices.

5.1 State Estimation in the Two-Qubit System

The procedure to estimate the state of two qubits has been well established by *James, Kwiat, Munro* and *White* [33]. Hence, the main purpose of our work [34] is to quantitatively show the limit of accuracy of quantum-state estimation of their method.

For this purpose, we demonstrated that the accuracy depends on the state to be estimated and also on the measurement strategy. In order to do that, we introduced a strategy of quantum-state estimation utilizing Akaike's information criterion (AIC) [35] for eliminating numerical problems in the estimation procedures, especially in estimating (nearly) pure quantum states. While the number of parameters used for characterizing density matrices of quantum states is fixed in the conventional estimation strategies [33, 36, 37], the number is varied in the strategy for eliminating redundant parameters.

Further, we pointed out in the biphoton case, the measurement outcome obeys the Poisson distribution. That is, our estimation strategy is based on the stochastic behavior of the number of detected photons of each measurement in the fixed time.

Consequently, we can quantitatively compare experimentally evaluated errors in the estimation with their asymptotic lower bound derived from the Cramér–Rao inequality without bothering about the delicate numerical problem accompanying the redundant parameters. It was shown that the errors of the experimental results nearly achieve their lower bounds for all quantum states we examined. Moreover, because of the reduction of the parameters, the AIC-based new estimation strategy slightly decreases the lower bounds.

Our results reveal that when measurements are performed locally (i.e., separately) on each qubit, the existence of entanglement may degrade the accuracy of estimation. Thus, while the measurements in our experiments are local ones, we numerically examined the performance of an alternative measurement strategy, which includes inseparable measurements on two qubits.

5.2 Testing of Entangled State in the SPDC System

Further, we investigated the testing method of entangled states in the biphoton system with the same two hypotheses as Sect. 4.1 [38]. Many experimentalists used the visibility for experimentally checking the quality of the generated entangled states. However, its optimality is not proved, and it is expected that this method can be improved. In this experimental setting, the measurement outcome obeys the Poisson distribution. Hence, it is necessary to establish an estimating method based on the stochastic behavior of the number of detected photons of each measurement in fixed time. Hence, a treatment different from that in Sect. 4.1 is needed.

Consequently, we obtain a testing method improving the visibility. Further, we experimentally demonstrate this method with the SPDC system.

However, in order to further improve it, we have to optimize the time allocation of each measurement. Unfortunately, the optimal time allocation depends on the true state. Of course, if we have no knowledge concerning the true state, the equal allocation is optimal, and this method is the same as above method. For further improvement, we proposed the two-step method, in which we first estimate the best allocation in the first step, and perform the obtained allocation in the second method. We also experimentally demonstrated it. Further, we compare this two-step method with the equal allocation method both from the theoretical viewpoint and in the experimental data.

6 Analysis on Quantum Measurement

6.1 Quantum Measurement With Negligible State Demolition

In our papers [39, 40], we considered the optimal measurement for the decision of the coding length (the estimation of the entropy rate) in the sense of the large deviation (also optimal in the sense of mean square error, in many cases). Since such an optimal measurement demolishes the state, an unsharp measurement, generated by smearing out the optimal measurement, was considered. Such an unsharp measurement is also optimal in the sense of the large deviation, while it is no longer optimal in the sense of the mean square error. Therefore, this measurement can be used for estimating state with a negligible state demolition. In addition, by constructing a quantum variable-length code from a quantum universal fixed-length code, we can clarify the trade-off between the compression rate and the nondemolition.

6.2 Quantum Universal Compression

In the classical data compression theory, there are two types of universal codes. One is a fixed-length universal code, and the other is a variable-length source code. The former depends on the compression rate, but the latter is independent of it. Therefore, when we compress our data in a classical computer, we usually do not use a fixed-length code, but instead use a variable-length source code like gzip. In the quantum case, according to *Schumacher's* result [41], when our quantum data obey independent identical distribution (i.i.d.) of a probability p of quantum states, we can compress our data up to the entropy rate of the average density operator defined as the mixture with probability p. However, his protocol is not applicable to the case where we do not know the average density operator because the construction of the protocol depends on it. Using the representation theory of unitary groups, *Jozsa* et al. [42] constructed a quantum universal fixed-length code, and it is efficient in the i.i.d. case when the entropy rate of the source is less than the rate of the code. Otherwise, this protocol demolishes the state unrecoverably.

In order to avoid state demolition, we need a quantum universal variable-length code that does not depend on the rate. Of course, in such a code, the coding rate must not be determined a priori, and it must be decided from the input state. While this decision does not change the source in the classical case, it does cause the destruction in the quantum case because this decision requires a quantum measurement. Therefore, we treated the trade-off between the compression rate and the degree of the nondemolition. While this type of code was thought to be impossible by some researchers [43], it was constructed by the strategy given in Sect. 6.1 [39, 40].

References

[1] C. W. Helstrom: Minimum mean-square error estimation in quantum statistics, Phys. Lett. **25A**, 101–102 (1967)
[2] C. W. Helstrom: *Quantum Detection and Estimation Theory* (Academic, New York 1976)
[3] K. Matsumoto: A new approach to the Cramer-Rao-type bound of the pure-state model, J. Phys. A: Math Gen. **35**, 3111–3123 (2002)
[4] M. Hayashi: Asymptotic estimation theory for a finite dimensional pure state model, J. Phys. A: Math. Gen. **31**, 4633–4655 (1998)
[5] M. Hayashi: On the second order asymptotics for pure states family, IEICE Transactions on Fundamentals A **J88-A**, 903–916 (2005) in Japanese
[6] M. Hayashi: Estimation of squeezed state, in *9th International Conference on Squeezed States and Uncertainty Relations, (ICSSUR 2005)* (Besancon, France 2005)
[7] M. Hayashi: Lectures on quantum estimation and information, Workshop on Quantum Measurments and Operations for Criptography and Information Processing (2005)
URL http://www.qubit.it/upcoming/gseminars/Hayashi/_SQUEEZED.PDF
[8] M. Hayashi: *Quantum Statistical Inferene and Quantum Correlation*, vol. 80-4 (Bussei Kenkyu 2003) pp. 368–391, in Japanese
[9] Y. Tsuda, K. Matsumoto, M. Hayashi: Disturbance of operation in quantum estimation for noised coherent light, in *ERATO Conference on Quantum Information Science 2001 (EQIS01)* (2001) p. 61
[10] Y. Tsuda, K. Matsumoto: Quantum estimation for non-differentiable models, J. Phys. A: Math. Gen. **38**, 1593–1613 (2005)
[11] D. G. Chapman, H. Robbins: Minimum variance estimation without regularity assumptions, Phys. Rev. Lett. **22**, 581–586 (1951)
[12] J. M. Hammersley: On estimating restricted parameters, J. Roy. Statist. Soc. Ser. **12**, 192–240 (1950)
[13] M. Hayashi: Quantum effect ob eigen-value estimation in quantum two-leve system, J. IPSJ **46(10)**, 2447–2467 (2005) in Japanese
[14] M. Hendrych, M. Dusek, R. Filip, J. Fiurasek: Simple optical measurement of the overlap and fidelity of quantum states: An experiment, Phys. Lett. A **310**, 95 (2003) quant-ph/0208091

[15] A. Fujiwara: Estimation of SU(2) operation and dense coding: An information geometric approach, Phys. Rev. A **65**, 012316 (2002)
[16] V. Bužek, R. Derka, S. Massar: Optimal quantum clocks, Phys. Rev. Lett. **82**, 2207 (1999)
[17] M. Hayashi: Estimation of SU(2) action by using entanglement, Phys. Lett. A **354**, 183–189 (2006) quant-ph/0407053
[18] E. Bagan, M. Baig, R. Muñoz-Tapia: Entanglement-assisted alignment of reference frames using a dense covariant coding, Phys. Rev. A **69**, 050303(R) (2004)
[19] E. Bagan, M. Baig, R. Muñoz-Tapia: Quantum reverse-engineering and reference-frame alignment without nonlocal correlations, Phys. Rev. A **70**, 030301(R) (2004)
[20] G. Chiribella, G. M. D'Ariano, P. Perinotti, M. F. Sacchi: Efficient use of quantum resources for the transmission of a reference frame, Phys. Rev. Lett. **93**, 180503 (2004)
[21] Y. Tsuda, K. Matsumoto, M. Hayashi: Hypothesis testing for a maximally entangled state, J. Phys. A: Math. Gen. quant-ph/0504203
[22] M. Barbieri, F. D. Martini, G. Di Nepi, P. Mataloni, G. M. D'Ariano, C. Macchiavello: Experimental detection of entanglement with polarized photons, Phys. Rev. Lett. **91**, 227901 (2003)
[23] S. Virmani, M. B. Plenio: Construction of extremal local positive-operator-valued measures under symmetry, Phys. Rev. A **67**, 062308 (2003)
[24] J. Walgate, A. J. Short, L. Hardy, V. Vedral: Local distinguishability of multipartite orthogonal quantum states, Phys. Rev. Lett. **85**, 4972 (2000)
[25] M. Horodecki, A. Sen De, U. Sen, K. Horodecki: Local indistinguishability: More nonlocality with less entanglement, Phys. Rev. Lett. **90**, 047902 (2003)
[26] H. Fan: Distinguishability and indistinguishability by local operations and classical communication, Phys. Rev. Lett. **92**, 177905 (2004)
[27] M. Owari, M. Hayashi: Local copying and local discrimination as a study for non-locality of a set, Phys. Rev. A submitted; quant-ph/0509062
[28] M. Hayashi, D. Markham, M. Murao, M. Owari, S. Virmani: Bounds on multipartite entangled orthogonal state discrimination using local operations and classical communication, Phys. Rev. Lett. **96**, 040501 (2006)
[29] T. S. Han: *Information-Spectrum Methods in Information Theory* (Springer 2003) the original Japanese edition was published from Baifukan-Press, Tokyo, in 1998
[30] H. Nagaoka, M. Hayashi: An information-spectrum approach to classical and quantum hypothesis testing, IEEE Trans. Info. Theor. (2002) submitted; LANL e-print quant-ph/0206185, (2002)
[31] M. Hayashi: Hypothesis testing approach to quantum information theory, Quantum Information Theory Satellite Workshop (QIT-EQIS03) (2003)
[32] M. Hayashi: *Quantum Information: An Introduction* (Springer 2006) the original Japanese edition was published from Saiensu-sha, Tokyo, in 2004
[33] D. F. V. James, P. G. Kwiat, W. J. Munro, A. G. White: Measurement of qubits, Phys. Rev. A **64**, 052312 (2001)
[34] K. Usami, Y. Nambu, Y. Tsuda, K. Matsumoto, K. Nakamura: Accuracy of quantum-state estimation utilizing Akaike's information criterion, Phys. Rev. A **68**, 022314 (2003)

[35] H. Akaike: Stochastic theory of minimal realization, IEEE Trans. Automat. Contr. **19**, 667 (1974)
[36] K. Banaszek, G. M. D'Ariano, M. G. A. Paris, M. F. Sacchi: Maximum-likelihood estimation of the density matrix, Phys. Rev. A **61**, 010304(R) (2000)
[37] J. Řeháček, Z. Hradil, M. Ježek: Iterative algorithm for reconstruction of entangled states, Phys. Rev. A **63**, 040303(R) (2001)
[38] M. Hayashi, B. Shi, A. Tomita, K. Mastumoto, Y. Tsuda, Y. Jinag: Hypothesis testing for an entangled state produced by spontaneous parametric down conversion, Phys. Rev. A submitted; quant-pW0603254
[39] M. Hayashi, K. Matsumoto: Quantum universal veriable-length source cording, Phys. Rev. A **66**, 022311 (2002)
[40] M. Hayashi, K. Matsumoto: Simple construction of quantum universal variable-length source coding, Quantum Informatin and Computation **2**, 519–529 (2002)
[41] B. Schumacher: Quantum coding, Phys. Rev. A **51**, 2738–2747 (1995)
[42] R. Jozsa, M. Horodecki, P. Horodecki, R. Horodecki: Universal quantum information compression, Phys. Rev. Lett. **81**, 1714 (1998) quant-ph/9805017 (1998)
[43] K. Bostroem, T. Felbinger: Lossless quantum data compression and variable-length coding, Phys. Rev. A **65**, 032313 (2002)

Index

Akaike's information criterion (AIC), 57

beam splitter (BS), 50
Boson–Fock space, 49

channel coding, 55
cloning machine, 49
compression, 55
concurrence, 52
covariant, 48

entanglement, 52, 57
entanglement concentration, 55

Gaussian state, 49

LOCC (local operation and classical communication), 47, 54, 55

POVM (positive operator-valued measure), 46, 54

spontaneous parametric down conversion (SPDC), 56
squeezed state, 48
state entangled, 53

Quantum Cloning Machines

Heng Fan

Beijing National Laboratory for Condensed Matter Physics, Institute of Physics
Chinese Academy of Sciences, Beijing 100080, China
hfan@aphy.iphy.ac.cn

Abstract. This Chapter is a review of various quantum cloning machines. The author focuses on several well-studied quantum cloning machines: universal quantum cloning machines, phase-covariant quantum cloning machines, asymmetric quantum cloning machines, and probabilistic quantum cloning machines.

1 Introduction

Quantum information theory [1, 2] has been attracting a great deal of interests. The no-cloning theorem describes one of the most fundamental nonclassical properties of quantum systems. It states that an unknown quantum state or an arbitrary state cannot be cloned exactly [3, 4], but only approximately or probabilistically. And the no-cloning theorem for pure states is extended to other cases [5, 6, 7]. However, the no-cloning theorem does not forbid imperfect cloning. We are interested in knowing how we can clone quantum states as well as possible. A great deal of effort has been put into developing optimal cloning processes. Approximate cloning is interesting not only from the viewpoint of the foundation of quantum mechanics, it is also applicable to other interesting quantum processes, e.g., quantum network coding [8], etc. Hence, it can be expected that the bound of approximate cloning contributes an important part of the foundation of quantum information.

Here we will review several quantum cloning machines. Compared with recent review papers about quantum cloning machines [9, 10], this paper will not cover the cloning of continuous variable states. But we will present self-contained and more detailed results on cloning of discrete quantum systems, at the expense of completeness in our references.

2 Bužek and Hillery Universal Quantum Cloning Machine

Bužek and *Hillery* [11] proposed a 1 to 2 universal quantum cloning machine (UQCM) which produces two identical copies from one qubit (two-level system), and the quality of each copy is independent of the input qubit. This UQCM was later proved to be optimal if the measure of quality is the fidelity between the input and the output [12]. The transformations of this UQCM are written as:

$$U|0\rangle_1|0\rangle_2|0\rangle_a = \sqrt{\frac{2}{3}}|0\rangle_1|0\rangle_2|0\rangle_a + \sqrt{\frac{1}{6}}(|0\rangle_1|1\rangle_2 + |1\rangle_1|0\rangle_2)|1\rangle_a, \quad (1)$$

$$U|1\rangle_1|0\rangle_2|0\rangle_a = \sqrt{\frac{2}{3}}|1\rangle_1|1\rangle_2|1\rangle_a + \sqrt{\frac{1}{6}}(|0\rangle_1|1\rangle_2 + |1\rangle_1|0\rangle_2)|0\rangle_a, \quad (2)$$

where the first qubit $|0\rangle_1, |1\rangle_1$ is the input state. After the copy process it will be changed as implied in the no-cloning theorem. The second qubit is for the copy; it is first set to $|0\rangle_2$. The third qubit is the ancillary state that is part of the UQCM and will be traced out to obtain the output state. U is a unitary transformation which is demanded by quantum mechanics. We can find that the output state for the first qubit and the second qubit are symmetric, so the two copies are the same.

For an arbitrary pure input state $|\psi\rangle = x_0|0\rangle + x_1|1\rangle$, where $|x_0|^2 + |x_1|^2 = 1$, we can apply the UQCM transformations for states $|0\rangle$ and $|1\rangle$, respectively. Here linear quantum mechanics is implied. After tracing out the ancillary state, we obtain the output state as follows:

$$\rho_{\text{out}} = \frac{2}{3}|\psi\rangle\langle\psi| \otimes |\psi\rangle\langle\psi| + \frac{1}{6}(|\psi\rangle|\psi^\perp\rangle + |\psi^\perp\rangle|\psi\rangle)(\langle\psi|\langle\psi^\perp| + \langle\psi^\perp|\langle\psi|), \quad (3)$$

where $|\psi^\perp\rangle = x_1^*|0\rangle - x_0^*|1\rangle$, which is orthogonal with $|\psi\rangle$. Each single copy in the output state is written as:

$$\rho_{\text{out, 1}} = \rho_{\text{out, 2}} = \frac{2}{3}|\psi\rangle\langle\psi| + \frac{1}{6}I, \quad (4)$$

where $I = |0\rangle\langle 0| + |1\rangle\langle 1|$ is the identity.

There is no unique criterion to quantify the quality of the copies. We general use fidelity to measure it. Other quantities that can measure distance between two quantum states can also be used. Here two different fidelities are reasonable measures for the copies. One is the fidelity between single qubit for input state and the output state, $F = \langle\psi|\rho_{\text{out, 1}}|\psi\rangle$. Another one is the fidelity between the two-qubit output state and the ideal copies, $F_{\text{global}} = ^{\otimes 2}\langle\psi|\rho_{\text{out}}|\psi\rangle^{\otimes 2}$. For the Bužek and Hillery UQCM, the two fidelities can be calculated as:

$$F = \frac{5}{6}, \quad F_{\text{global}} = \frac{2}{3}. \quad (5)$$

Those two fidelities are independent of the input state, and in this sense this kind of quantum cloning machines are called *universal* cloning machines. We may notice that the single qubit output state has two terms: the original input state and the identity I. We know that $\frac{1}{2}I$ is the complete mixed state that acts as the noise for the output state. So the factor 2/3 in (4) can also quantify the quality of the copies, which is called the *shrinking factor*. The Bužek and Hillery UQCM is optimal in the sense that the fidelity between single qubit of input and output reaches the upper bound, see [12].

3 N to M UQCM (Gisin and Massar)

We may notice that if we change the ancillary state in the Bužek and Hillery state as $|0\rangle \to |1\rangle, |1\rangle \to |0\rangle$ (we will exchange these notations hereafter), the transformation (1,2) can be written as a concise form,

$$|\psi\rangle \to \sqrt{\frac{2}{3}}|\psi\rangle_1|\psi\rangle_2|\psi^\perp\rangle_a + \sqrt{\frac{1}{6}}(|\psi\rangle_1|\psi^\perp\rangle_2 + |\psi^\perp\rangle_1|\psi\rangle_2)|\psi\rangle_a. \tag{6}$$

This cloning machine is for the 1 to 2 case, i.e., two qubits output state with one qubit input. *Gisin* and *Massar* [13] considered a general case in which M identical copies are generated from N ($M \geq N$) identical qubits. Their cloning transformation is a direct generalization of that in (6),

$$|N\psi\rangle \to \sum_{j=0}^{M-N} \alpha_j |(M-j)\psi, j\psi^\perp\rangle |R_j(\psi)\rangle_a, \tag{7}$$

where

$$\alpha_j = \sqrt{\frac{N+1}{M+1}}\sqrt{\frac{(M-N)!(M-j)!}{(M-N-j)!M!}}. \tag{8}$$

The fidelity between single qubit for input and output is

$$F_{N,M} = \frac{M(N+1)+N}{M(N+2)}. \tag{9}$$

The state $|(M-j)\psi, j\psi^\perp\rangle$ is a normalized symmetric state with $M-j$ states of $|\psi\rangle$ and j states of $|\psi^\perp\rangle$. For example, if we denote $|\uparrow\rangle \equiv |0\rangle$ and $|\downarrow\rangle \equiv |1\rangle$, we have $|2\uparrow,\downarrow\rangle = (|\uparrow\uparrow\downarrow\rangle + |\uparrow\downarrow\uparrow\rangle + |\downarrow\uparrow\uparrow\rangle)/\sqrt{3}$. The optimality of this fidelity is proved for cases $N = 1, 2, \ldots, 7$. Later, the connection between optimal quantum cloning and the optimal state estimation was introduced in [14], and a tight upper bound for the fidelity of N to M UQCM was obtained. *Bruß* et al. found that the fidelity (9) achieves the upper bound; thus they proved that it is optimal for general N. So the Gisin and Massar UQCM is optimal.

4 Universal Quantum Cloning Machine for General d-Dimensional System, Werner Cloning Machine

Bužek and Hillery, Gisin and Massar cloning machines are for the qubit case. Bužek and Hillery also proposed a 1 to 2 UQCM for the d-dimensional quantum state [15], which is generally called *qudit*. A general N to M UQCM for general qudits was proposed by *Werner* [16]. Suppose the input state is N pure state $|\psi\rangle$, Werner's cloning transformation is presented as:

$$\rho_{\text{out}} = \frac{d[N]}{d[M]} S_M \left((|\psi\rangle\langle\psi|)^{\otimes N} \otimes I^{\otimes(M-N)} \right) S_M, \tag{10}$$

where $d[N] = \binom{d+n-1}{n}$. S_M is a symmetrization operator. The process of the Werner cloning machine is that we add $M - N$ identities, then make symmetrization on these M-qudit states. We can obtain the output state after normalization. The tensor product of $M - N$ identities before symmetrization can be understood if we know nothing of the input state. And thus identity is a reasonable choice since $|\psi\rangle$ can be arbitrary.

The quality of the Werner cloning machine is quantified by two fidelities as for the qubit case: the fidelity between single qudit input and output states and the fidelity between M-qudit output state ρ_{out} with the ideal copies $|\psi\rangle^{\otimes M}$. Both fidelities are proved to be optimal for this cloning machine. The results are obtained by Werner [16] and Keyl and Werner [17]. These two optimal fidelities are

$$F_{N,M} = \frac{N(M+d) + M - N}{M(N+d)}, \tag{11}$$

$$F_{\text{global}} = \frac{M!(N+d-1)!}{N!(M+d-1)!}. \tag{12}$$

When $d = 2$, the result (11) recovers the result (9) for the Gisin and Massar cloning machine. And for $N = 1, M = 2, d = 2$, we rederive the result for the Bužek and Hillery UQCM.

We next present a simple example to show how to use Werner UQCM. We consider the case $N = 1, M = 2, d = 2$; the symmetrization operator can be written as:

$$S_2 = |\uparrow\uparrow\rangle\langle\uparrow\uparrow| + |\downarrow\downarrow\rangle\langle\downarrow\downarrow| + |\uparrow,\downarrow\rangle\langle\uparrow,\downarrow|. \tag{13}$$

We remark that $\{|\uparrow\uparrow\rangle, |\downarrow\downarrow\rangle, |\uparrow,\downarrow\rangle\}$ is a complete basis of symmetric subspace for the two-qubit state. Suppose the input state is $|\psi\rangle = \alpha|\uparrow\rangle + \beta|\downarrow\rangle$; our aim is to use S_2 to symmetrize $|\psi\rangle\langle\psi| \otimes I$. As already mentioned, the identity is $I = |\uparrow\rangle\langle\uparrow| + |\downarrow\rangle\langle\downarrow|$. After normalization, we can find

$$\rho_{\text{out}} = \frac{2}{3}|\psi\rangle\langle\psi||\psi\rangle\langle\psi| + \frac{1}{6}(|\psi\rangle|\psi^\perp\rangle + |\psi^\perp\rangle|\psi\rangle)(\langle\psi|\langle\psi^\perp| + \langle\psi^\perp|\langle\psi|).$$

Really we recover the output state obtained by the Bužek and Hillery UQCM.

5 A UQCM for *d*-Dimensional Quantum State Proposed by Fan et al.

Werner proposed a concise form for the general UQCM. *Fan, Matsumoto* and *Wadati* [18] later proposed a different version of UQCM that follows the method of Bužek, Hillery, Gisin and Massar, i.e., the cloning transformations with ancillary states are presented.

Let's introduce some notations. A d-level quantum system is spanned by the orthonormal basis $|i\rangle$ with $i = 1, \cdots, d$; vector $\bm{n}$ denotes $n_1, \cdots, n_d$; $|\bm{n}\rangle = |n_1, \cdots n_d\rangle$ is a completely symmetric and normalized state with n_i systems in $|i\rangle$; this state is invariant under permutations of all N d-level qubits. And an arbitrary pure state takes the form $|\psi\rangle = \sum_{i=1}^{d} x_i |i\rangle$ with $\sum_{i=1}^{d} |x_i|^2 = 1$. N identical pure states $|\psi\rangle^{\otimes N}$ can be expanded in terms of the basis of symmetric subspace

$$|\Psi\rangle^{\otimes N} = \sum_{\bm{n}=0}^{N} \sqrt{\frac{N!}{n_1! \cdots n_d!}} x_1^{n_1} \cdots x_d^{n_d} |\bm{n}\rangle . \tag{14}$$

Thus to copy N identical pure states to M copies, we only need to propose the cloning transformations for the basis of symmetric subspace.

Fan et al. proposed the N to M quantum cloning transformation for a d-level quantum system as follows [18]:

$$U_{NM} |\bm{n}\rangle \otimes R = \sum_{j=0}^{M-N} \alpha_{\bm{n}\bm{j}} |\bm{n}+\bm{j}\rangle \otimes R_{\bm{j}} , \tag{15}$$

where $\bm{n} + \bm{j} = \bm{m}$, i.e., $\sum_{k=1}^{d} j_k = M - N$, $R_{\bm{j}}$ denotes the orthogonal normalized internal states of QCM, and

$$\alpha_{\bm{n}\bm{j}} = \sqrt{\frac{(M-N)!(N+d-1)!}{(M+d-1)!}} \sqrt{\prod_{k=1}^{d} \frac{(n_k + j_k)!}{n_k! j_k!}} . \tag{16}$$

R denotes $M - N$ blank copies and the initial state of the cloning machine, $R_{\bm{n}\bm{m}}$ are internal states of the cloning machine, where $\sum_{\bm{n}=0}^{N}$ means sum over all variables under the condition $\sum_{i=1}^{d} n_i = N$. We also have $\sum_{i=1}^{d} m_i = M$. Because all kinds of symmetric states $|\bm{n}\rangle$ can be allowed as input states in this quantum cloning transformation, this quantum cloning machine actually not only can copy identical pure states but also arbitrary quantum states restricted to symmetric subspace. We can find two fidelities of the quantum cloning transformation (15,16) are the same as Werner UQCM (11,12). Thus this UQCM is also optimal; it can be understood as a realization of the Werner UQCM. For other results about UQCM, please see [19] by *Zanardi* and some related results in [19, 20].

6 Further Results About the UQCM

Bruß, Ekert and *Macchiavello* [14] have presented the optimal shrinking factor, which is related to the fidelity of single d-level quantum states between input and output. With shrinking factor to define the quality of the copies, it can work not only for pure states quantum cloning but also for cloning of all mixed and/or entangled states in symmetric subspace. We next review works of cloning quantum states in symmetric subspace. The cloning processing can admit arbitrary states in symmetric subspace as input and still the copy processing is optimal in the sense it achieves the optimal shrinking factor. The cloning machine thus can be concatenated together, i.e., the second cloning machine uses the output of the first cloning machine as input and produces more copies, and still the final copies are optimal. Suppose the available quantum cloning machine can just produce limited additional copies. With a cloning machine for arbitrary states in symmetric subspace, we can concatenate several cloning machines together and thus produce more copies with best quality. We next review some results about a generalization of UQCM proposed in [21].

6.1 UQCM for 2-Level System

We first restrict our discussions to the two-level $(|\uparrow\rangle, |\downarrow\rangle)$ quantum system. The input is an arbitrary density operator of M qubits in symmetric subspace,

$$\rho^{\text{in}}(M) = \sum_{j,j'=0}^{M} x_{jj'} |(M-j)\uparrow, j\downarrow\rangle\langle j'\downarrow, (M-j')\uparrow|. \quad (17)$$

Here $|(M-j)\Psi, j\Psi^\perp\rangle$ is the symmetric and normalized state with $M-j$ qubits in the state Ψ and j in the orthonormal state $\Psi^\perp$ which is invariant under all permutations. We take $|\Psi\rangle = |\uparrow\rangle, |\Psi^\perp\rangle = |\downarrow\rangle$, $x_{jj'}$ is an arbitrary matrix, and we let $\sum_{j=0}^{M} x_{jj} = 1$, which is the trace condition for density operators. We remark that M identical pure input states $|M\Psi\rangle \equiv |\Psi\rangle^{\otimes M}$ is a special case of (17). The reduced density operators of (17) at each qubit are the same and take the form

$$\rho^{\text{in}}_{\text{red.}}(M) = |\uparrow\rangle\langle\uparrow| \sum_{j=0}^{M} x_{jj} \frac{M-j}{M} + |\uparrow\rangle\langle\downarrow| \sum_{j=0}^{M-1} x_{jj+1} \frac{\sqrt{(M-j)(j+1)}}{M}$$
$$+ |\downarrow\rangle\langle\downarrow| \sum_{j=0}^{M} x_{jj} \frac{j}{M} + |\downarrow\rangle\langle\uparrow| \sum_{j=0}^{M-1} x_{j+1j} \frac{\sqrt{(M-j)(j+1)}}{M}. \quad (18)$$

Our goal here is to find the optimal cloning transformation with input (17) and output $\rho^{\text{out}}(M, L)$ in L qubits, so that the fidelity between $\rho^{\text{in}}_{\text{red.}}(M)$ in (18) and the output reduced density operator at each qubit $\rho^{\text{out}}_{\text{red.}}(M, L)$ can achieve the upper bound. We call the cloning transformation with (17) as input a generalized UQCM (g-UQCM) to distinguish it from UQCM which takes identical pure states as input [11, 13, 18]. The relation between input and output reduced density operators can be written in a scaling form

$$\rho_{\text{red.}}^{\text{out}} = \eta(M,L)\rho_{\text{red.}}^{\text{in}} + \frac{1}{2}(1-\eta(M,L)), \tag{19}$$

where $\eta(M,L)$ is the shrinking factor of the Bloch vector characterizing the operation of universal quantum cloning transformation. The optimal g-UQCM refer to maximal $\eta(M,L)$. By identifying the optimal fidelity of M to ∞ cloning with optimal fidelity of quantum state estimation for M identical unknown pure states [22, 23], Bruß, Ekert and Macchiavello [14] obtained the tight upper bound of the shrinking factor, $\eta(M,L) = \frac{M(L+2)}{L(M+2)}$.

With M identical pure qubits $|M\Psi\rangle$ as input, Bužek and Hillery ($1 \to 2$) and Gisin and Massar ($M \to L$) UQCM which achieve the optimal shrinking factor $\eta(M,L)$ have already been proposed [11,13]. It is explicit that the input $|M\Psi\rangle$ belongs to symmetric subspace because we use the fidelity between input and output reduced density operators at a single qubit to define the quality of cloning for both UQCM and g-UQCM. The g-UQCM will reduce to UQCM if the input are M identical pure states $|M\Psi\rangle$. We can also study the concatenation of two quantum cloners [14]. The first one is the UQCM which acts on N identical pure qubits $|N\Psi\rangle$ and produces M copies, and the second cloner uses the output of first cloner as input and generates L copies. The output of a UQCM which is generally an entangled and/or mixed state belongs to symmetric subspace. Thus the second cloner can be formulated by a g-UQCM.

We propose the unitary cloning transformation of the g-UQCM as follows:

$$U(M,L)|(M-j)\Psi, j\Psi^\perp\rangle \otimes R$$
$$= \sum_{k=0}^{L-M} \alpha_{jk}(M,L)|(L-j-k)\Psi, (j+k)\Psi^\perp\rangle \otimes R_k, \tag{20}$$

where

$$\alpha_{jk}(M,L) = \sqrt{\frac{(L-M)!(M+1)!(L-j-k)!(j+k)!}{(L+1)!(L-M-k)!(M-j)!j!k!}},$$
$$j = 0,\ldots,M; \quad k = 0,\ldots,L-M. \tag{21}$$

R denotes the initial state of the UQCM and $M - N$ blank copies, and R_j are the orthonormalized internal states of the UQCM (ancilla states). In case $j = 0$, it reduces to the original UQCM with M identical pure input states and L copies [13]. This g-UQCM allows the input to be not only identical pure states but also mixed and/or entangled states in symmetric subspace. We now show that this g-UQCM is still optimal in the sense that the shrinking factor between input and output reduced density operators at each qubit achieves the upper bound. Substituting $|\Psi\rangle = |\uparrow\rangle, |\Psi^\perp\rangle = |\downarrow\rangle$ into (20), and applying this cloning transformation on the input density operator (17), $U(M,L)\rho^{\text{in}}(M)U^\dagger(M,L)$, taking trace over ancilla states, we can obtain the output density operator with L qubits. The reduced output density operator of each qubit is derived as

$$\rho_{\text{red.}}^{\text{out}}(M,L)$$

$$= |\uparrow\rangle\langle\uparrow| \sum_{j=0}^{M} \sum_{k=0}^{L-M} x_{jj} \alpha_{jk}^2(M,L) \frac{L-j-k}{L}$$

$$+ |\downarrow\rangle\langle\downarrow| \sum_{j=0}^{M} \sum_{k=0}^{L-M} x_{jj} \alpha_{jk}^2(M,L) \frac{j+k}{L}$$

$$+ |\uparrow\rangle\langle\downarrow| \sum_{j=0}^{M-1} \sum_{k=0}^{L-M} x_{jj+1} \alpha_{jk}(M,L) \alpha_{j+1k}(M,L) \frac{\sqrt{(L-j-k)(j+k+1)}}{L}$$

$$+ |\downarrow\rangle\langle\uparrow| \sum_{j=0}^{M-1} \sum_{k=0}^{L-M} x_{j+1j} \alpha_{jk}(M,L) \alpha_{j+1k}(M,L) \frac{\sqrt{(L-j-k)(j+k+1)}}{L}.$$

(22)

Comparing (22) with the reduced input density operator $\rho_{\text{red.}}^{\text{in}}(M)$ in (18) at each qubit of input state (17), and after some calculations, we have

$$\rho_{\text{red.}}^{\text{out}}(M,L) = \frac{M(L+2)}{L(M+2)} \rho_{\text{red.}}^{\text{in}}(M) + \frac{L-M}{L(M+2)} \cdot 1. \tag{23}$$

In the calculations, only the trace condition of the input density operator is used; the positivity condition of the density operator is not used. That means we even do not need (17) as a density operator, but the scaling form of cloning (23) is still holds. Thus we see that the shrinking factor characterizing the g-UQCM (20) achieves the upper bound and is independent from the arbitrary input density operators (17) in symmetric subspace. The unitary cloning transformation ((20) and (21)) is a universal and optimal cloner which allows the input to be arbitrary states in symmetric subspace.

As an example, we study the concatenation of a UQCM and a g-UQCM. Taking $|N\Psi\rangle$ as input, using cloning transformation (20), tracing over the ancilla states R_j, we can obtain the output density operator of M copies as

$$\rho^{\text{out}}(N,M) = \sum_{j=0}^{M-N} \alpha_{0j}^2(N,M) |(M-j)\Psi, j\Psi^\perp\rangle\langle j\Psi^\perp, (M-j)\Psi|. \tag{24}$$

We remark that (24) is the output density operator of a UQCM proposed by *Gisin* and *Massar* [13]. We now concatenate a g-UQCM to the N to M UQCM with (24) as input and produce L copies. Using the cloning transformation ((20) and (21)), the output density operator of the g-UQCM takes the form

$$\rho^{\text{out}}(N, M, L) = \sum_{j=0}^{M-N} \sum_{k=0}^{L-M} \alpha_{0j}^2(N, M) \alpha_{jk}^2(M, L) |(L-j-k)\Psi, (j+k)\Psi^\perp\rangle$$

$$\langle (j+k)\Psi^\perp, (L-j-k)\Psi |,$$

$$= \sum_{p=0}^{L-N} \alpha_{0p}^2(N, L) |(L-p)\Psi, p\Psi^\perp\rangle\langle p\Psi^\perp, (L-p)\Psi |,$$

(25)

where we have used a simple relation which can be derived from $(x+y)^{M-N}(x+y)^{L-M} = (x+y)^{L-N}$ to obtain the last equation. We can find the output density operator of the sequence of the concatenated cloners is the same as the output density operator of N to L UQCM; $\rho^{\text{out}}(N, M, L) = \rho^{\text{out}}(N, L)$. We already know that two UQCM are optimal. It is straightforward that the g-UQCM (20,21) is optimal, otherwise it would lead to a contradiction. We have shown here another method to prove the optimum of the g-UQCM in the case when inputs are identical pure states or the output density operator produced by a UQCM.

6.2 UQCM for *d*-Level System

Next, we study the *d*-level quantum system. Quantum cloning with N identical pure input states and M copies in arbitrary *d*-dimensional Hilbert spaces is formulated by CP map in [16, 17], and the optimal fidelity is given as $F(d: N, M) = \frac{N(M+d)+M-N}{M(N+d)}$. With the result of the optimal fidelity for *d*-level quantum cloning, the optimal fidelity of state estimation for finite and identical *d*-level quantum states can be obtained [24]. Similar to the two-level (qubit) case, the density matrix of the *d*-level state can be expressed by generalized Bloch vector $\boldsymbol{s} = (s_1, \cdots, s_{d^2-1})$ and the generators $\tau_i, i = 1, \cdots, d^2-1$ of the group $SU(d)$, $\rho = \frac{1}{d} + \frac{1}{2}\sum_{i=1}^{d^2-1} s_i \tau_i$, where the generators of $SU(d)$ are defined as $\text{Tr}\,\tau_i = 0$, $\text{Tr}(\tau_i \tau_j) = 2\delta_{ij}$. With N identical pure states as input, the reduced output density operator at each *d*-level state of N to M UQCM takes the form $\rho^{\text{out}} = \frac{1}{d} + \frac{1}{2}\eta(d: N, M)\sum_{i=1}^{d^2-1} s_i \tau_i$. Corresponding to optimal fidelity, the upper bound of the shrinking factor for both UQCM and g-UQCM is $\eta(d: N, M) = \frac{N(M+d)}{M(N+d)}$.

The 1 to 2 unitary cloning transformation of the *d*-level system was formulated in [15]. The 1 to M and a special case of N to M cloning transformations were given in [20], and the general unitary N to M UQCM was given in [18], where the form is different from this paper (in [16,17] the CP map of the general cloning transformation was derived). Similar to Gisin and Massar cloner, we present here the UQCM for the *d*-level system. Let $|\Psi\rangle$ be an arbitrary state in the d-level system, and let $|\Psi_1^\perp\rangle,\ldots,|\Psi_{d-1}^\perp\rangle$ be orthonormal states. The d-level N to M UQCM takes the following form:

$$U(d: N, M)|N\Psi\rangle \otimes R = \sum_{j}^{M-N} \alpha_{j}(N, M)|(N + j_0)\Psi, j_1\Psi_1^\perp,\ldots,$$

$$\ldots, j_{d-1}\Psi_{d-1}^\perp\rangle \otimes R_{j},$$

$$\alpha_{j}(N, M) = \sqrt{\frac{(M-N)!(N+d-1)}{(M+d-1)!}}\sqrt{\frac{(N+j_0)!}{N!j_0!}}, \quad (26)$$

where $\boldsymbol{j} = (j_0, j_1, \ldots, j_{d-1})$, state $|(N+j_0)\Psi, j_1\Psi_1^\perp, \ldots, j_{d-1}\Psi_{d-1}^\perp\rangle$ is a completely symmetric and normalized state with $N + j_0$ states in Ψ, j_i states in $\Psi_i^\perp, i = 1, \ldots, d-1$, and summation $\sum_{j}^{M-N}$ means sum over all variables under the condition $\sum_{i=0}^{d-1} j_i = M - N$. R_{k} are orthonormal internal states of the cloner, $\langle R_{k}|R_{k'}\rangle = \delta_{kk'}$. We next prove that this cloning transformation is the optimal UQCM. Since the optimal fidelity is already available [16], we just need to prove the fidelity of the cloning transformation (26) achieves this upper bound. As in the qubits case [13], the fidelity of the d-level UQCM can be calculated as

$$F(d: N, M) = \sum_{j}^{M-N} \alpha_j^2(N, M) \frac{(N+j_0)}{M} = \frac{N(M+d) + M - N}{M(N+d)}, \quad (27)$$

where $\alpha_j^2(N, M)$ is the probability of state $|(N+j_0)\Psi, j_1\Psi_1^\perp, \ldots, j_{d-1}\Psi_{d-1}^\perp\rangle \otimes R_{j}$ in the output, and $\frac{(N+j_0)}{M}$ is the ratio of the number of ways to choose $(N+j_0-1)\Psi, j_1\Psi_1^\perp, \ldots, j_{d-1}\Psi_{d-1}^\perp$ among $M-1$ d-level states over the number of ways to choose $(N+j_0)\Psi, j_1\Psi_1^\perp, \ldots, j_{d-1}\Psi_{d-1}^\perp$ among M d-level states. The fidelity (27) of d-level UQCM (26) is optimal; thus (26) is the optimal UQCM, and the shrinking factor achieves its upper bound. Tracing out the ancilla, we have the output density operator

$$\rho^{\text{out}}(d: N, M) = \sum_{j}^{M-N} \alpha_j^2(N, M)|(N+j_0)\Psi, j_1\Psi_1^\perp, \ldots,$$

$$\ldots, j_{d-1}\Psi_{d-1}^\perp\rangle\langle j_{d-1}\Psi_{d-1}^\perp, \ldots, j_1\Psi_1^\perp, (N+j_0)\Psi|. \quad (28)$$

We finally propose a g-UQCM which allows arbitrary states with M d-level states belonging to the symmetric subspace as input, and produces L copies. The cloning transformation takes the form,

$$U(d: M, L)|j_0\Psi, j_1\Psi_1^\perp, \ldots, j_{d-1}\Psi_{d-1}^\perp\rangle \otimes R$$

$$= \sum_{\boldsymbol{k}}^{L-M} \alpha_{\boldsymbol{jk}}(M, L)|(j_0 + k_0)\Psi, (j_1 + k_1)\Psi_1^\perp, \ldots,$$

$$\ldots, (j_{d-1} + k_{d-1})\Psi_{d-1}^\perp\rangle \otimes R_{\boldsymbol{k}}, \quad (29)$$

$$\alpha_{j,k}(M,L) = \sqrt{\frac{(L-M)!(M+d-1)!}{(L+d-1)!}} \sqrt{\prod_{i=0}^{d-1} \frac{(j_i+k_i)!}{j_i! k_i!}}, \tag{30}$$

where $\sum_{i=0}^{d-1} j_i = M, \sum_{i=0}^{d-1} k_i = L-M$ are assumed. The optimum of this g-UQCM for the d-level system can be proved by a similar method as for a two-level system. Using the output density operator (28) as input, applying the cloning transformation (29), one can prove the output of $\rho^{\mathrm{out}}(d\colon N, M, L)$ is the same as the output $\rho^{\mathrm{out}}(d\colon N, L)$ of one N to L UQCM. In the calculations, a relation derived from an identity $(\sum_{i=0}^{d-1} x_i)^{M-N}(\sum_{i=0}^{d-1} x_i)^{L-M} = (\sum_{i=0}^{d-1} x_i)^{L-N}$ is useful. We already know that the N to M UQCM and N to L UQCM are optimal. The g-UQCM which uses the output density operator (28) as input and generates L copies is thus optimal. The g-UQCM (29) allows the input to be mixed and/or entangled states supported in symmetric subspace, and the shrinking factor achieves the upper bound $\eta(d\colon M, L) = \frac{M(L+d)}{L(M+d)}$. We remark that the dimension of the internal state of the cloner (ancilla) is $\frac{(L-M+d-1)!}{(L-M)!(d-1)!}$ which is useful in POVM (positive operator valued measurement), see, for example [25,26]. Note the ancilla states R_k should be expressed more precisely as $R_k(\Psi)$ and can be realized in symmetric subspace $R_k(\Psi) = |k_0 \Psi, k_1 \Psi_1^\perp, \ldots, k_{d-1} \Psi_{d-1}^\perp\rangle$.

Suppose d-level quantum system is spanned by the orthonormal basis $|i\rangle, i = 0, \ldots, d-1$, an arbitrary pure state is written as the form $|\Phi\rangle = \sum_{i=0}^{d-1} c_i |i\rangle$ with $\sum_{i=0}^{d-1} |c_i|^2 = 1$. Then any M d-level states in symmetric subspace can be expressed as $|\boldsymbol{j}\rangle$, where j_i states are in $|i\rangle, i = 0, \ldots, d-1$, and $\sum_{i=0}^{d-1} j_i = M$. We take a special case; let $|\Psi\rangle = |0\rangle, |\Psi_i^\perp\rangle = |i\rangle, i = 1, \ldots, d-1$. The M to L quantum cloning transformation (29) can be rewritten as,

$$U(d\colon M,L)|\boldsymbol{j}\rangle \otimes R = \sum_{\boldsymbol{k}}^{L-M} \alpha_{\boldsymbol{jk}}(M,L)|\boldsymbol{j}+\boldsymbol{k}\rangle \otimes R_{\boldsymbol{k}}, \tag{31}$$

where we still denote the internal states of the cloner by $R_{\boldsymbol{k}}$ in this special case for convenience. These results coincide with the formulae in [18]. Because $|\boldsymbol{j}\rangle, \sum_{i=0}^{d-1} j_i = M$, can be the orthonormal basis for M states d-level system in symmetric subspace, this cloning transformation (31) is another independent and complete set of cloning transformation equivalent to (29).

As the qubits case, the g-UQCM can be used as a concatenated cloner, and the input can be arbitrary states in symmetric subspace. The input state consisting of M d-dimensional states in symmetric subspace is written as $\rho^{\mathrm{in}}(d\colon M) = \sum_{\boldsymbol{jj}'}^M x_{\boldsymbol{jj}'} |\boldsymbol{j}\rangle\langle \boldsymbol{j}'|$. The dimension of matrix $x_{\boldsymbol{jj}'}$ is $\frac{(M+d-1)!}{M!(d-1)!}$, and we let $\sum_{\boldsymbol{j}}^M x_{\boldsymbol{jj}} = 1$. Using the cloning transformation (31), the reduced density operator of output can still have an optimal shrinking factor $\eta(d\colon M, L) = \frac{M(L+d)}{L(M+d)}$ compared with the reduced density operator of input, $\rho^{\mathrm{out}}_{\mathrm{red.}}(d\colon M, L) = \eta(d\colon M, L)\rho^{\mathrm{in}}_{\mathrm{red.}}(M) + \frac{1}{d}(1 - \eta(d\colon M, L))$.

7 UQCM Realized in Real Physical Systems

We next review the results presented in [27]. Generally, it is believed that the quantum cloning transformation can be realized by quantum networks [28]. It is a little bit surprising that the photon stimulated emission can realize the UQCM automatically [29,30], and the corresponding fidelity is optimal. This is shown successfully in experiments [31]. In this scheme, it has been shown that certain types of three-level atoms can be used to optimally clone quantum information that is encoded as an arbitrary superposition of excitations in the photonic modes that correspond to the atomic transitions. The universality of the cloning is ensured if the cloning system is symmetric since all kinds od input state can be realized, see [29,30] for detailed arguments. Next, we shall first review the qubit case. We introduce extended initial states to be cloned which include different kinds of oscillaters corresponding to different polarizations of photons. So, we show that the UQCM allows an arbitrary input state in Bose subspace. And also with this extended initial state, we provide another way to prove that the cloning scheme in [29, 30] is universal. Then, we shall study the cloning of state in d-dimensional Hilbert space. A generalized Hamiltonian with $d+1$-level quantum system is studied, and the realization of the optimal cloning transformation for arbitrary symmetric states in d-dimensional Hilbert space is obtained. We shall show that the process of quantum cloning is actually governed by the Hamiltonian. We remark that if the cloning system can be represented by Bosonic operators, we can clone arbitrary states by this UQCM with fidelity achieving its upper bound.

We first briefly review the quantum cloning scheme proposed in [29, 30]. The cloning device is an inverted medium that can spontaneously emit photons of any polarization with the same probability. This property will ensure that the cloning transformation induced by the inverted medium is universal. For the qubit case, the inverted medium should consist of an ensemble of Λ atoms. The three-level system has two degenerate ground states $|g_1\rangle$ and $|g_2\rangle$ and an excited level $|e\rangle$. Quantum cloning with the V type of three-level system is similar to the Λ-type system. The ground states are coupled to the excited state by two modes of the electromagnetic field, a_1 and a_2, respectively. The interaction between field and inverted medium is described by the Hamiltonian

$$H = \gamma \left(a_1 \sum_{k=1}^{N} |e^k\rangle\langle g_1^k| + a_2 \sum_{k=1}^{N} |e^k\rangle\langle g_2^k| \right) + \text{H.c.} \tag{32}$$

The general superposition state of a qubit is expressed by the form $(\alpha a_1^\dagger + \beta a_2^\dagger)|0,0\rangle = \alpha|1,0\rangle + \beta|0,1\rangle$. The initial state considered in [29, 30] takes the following form:

$$|\Psi_{in}\rangle = \bigotimes_{k=1}^{N} |e^k\rangle \frac{(a_1^\dagger)^m}{\sqrt{m!}}|0,0\rangle. \tag{33}$$

Suppose we want to clone M identical pure states $|\Phi\rangle^{\otimes M} = (\alpha a_1^\dagger + \beta a_2^\dagger)^{\otimes N}|0,0\rangle$; it is argued that we only need to consider the cloning of initial state (33) with the Hamiltonian (32) [29, 30]. In this paper, we present another method. If we know how to clone the state

$$|\Psi_{in}, j\rangle = \bigotimes_{k=1}^{M} |e^k\rangle \frac{(a_1^\dagger)^{M-j}(a_2^\dagger)^j}{\sqrt{(M-j)!j!}}|0,0\rangle, \quad j = 0, 1, \ldots, M, \tag{34}$$

it will be straightforward to clone the general M identical pure states $|\Phi\rangle^{\otimes M} = (\alpha a_1^\dagger + \beta a_2^\dagger)^{\otimes M}|0,0\rangle$. And what is more interesting, we can extend the input of the UQCM to arbitrary states in Bose subspace because $\frac{(a_1^\dagger)^i(a_2^\dagger)^j}{\sqrt{i!j!}}|0,0\rangle, i + j = M$ constitutes a complete set of orthonormal bases of Bose subspace with M qubits. We remark that here the arbitrary input states in Bose subspace also include the mixed states.

For convenience, we use the same notations as used in [29, 30]. We denote by Schwinger representation the total angular momentum operator as $b_r c^\dagger \equiv \sum_{k=1}^{N} |e^k\rangle\langle g_r^k|, r = 1, 2$, where $c^\dagger$ is a creation operator of "e-type" excitation, b_r is a annihilation operator of g_r ground states, $r = 1, 2$. Now the Hamiltonian (32) becomes as follows in terms of harmonic-oscillator operators:

$$\mathcal{H} = \gamma(a_1 b_1 + a_2 b_2) c^\dagger + \text{H.c.} \tag{35}$$

Now, we study the case of initial states containing both kinds of oscillators $a_1^\dagger$ and $a_2^\dagger$ of $i + j$ qubits,

$$|\Psi_{in}, (i,j)\rangle = \frac{(a_1^\dagger)^i(a_2^\dagger)^j(c^\dagger)^N}{\sqrt{i!j!N!}}|0\rangle = |i_{a_1}, j_{a_2}\rangle|0_{b_1}, 0_{b_2}\rangle|N_c\rangle$$

$$\equiv |i,j\rangle_a |0,0\rangle_b |N\rangle_c. \tag{36}$$

With the initial state (36), the time evolution of the state acts as follows [30]:

$$|\Psi(t), (i,j)\rangle = e^{-iHt}|\Psi_{in}, (i,j)\rangle \sum_p (-iHt)^p/p!|\Psi_{in},(i,j)\rangle,$$

$$= \sum_{l=0}^{N} f_l(t)|F_l, (i,j)\rangle, \tag{37}$$

where $|\Psi_{in},(i,j)\rangle = |F_0,(i,j)\rangle$, and state $|F_l,(i,j)\rangle$ express that $i+j+l$ copies of the initial state (36) are obtained, and l is the additional photons that have been emitted. So, the output state of this cloning machine contains from 0 to N additional copies of the initial state (36). It is a superposition of $|F_l,(i,j)\rangle$. The probability of finding l additional copies is determined by its amplitude $|f_l(t)|^2$. After some calculations, we can find for the initial state (36), the output with l additional copies is

$$|F_l,(i,j)\rangle = \sum_{k=0}^{l} \sqrt{\frac{l!(i+j+1)!}{(i+j+l+1)!}} \sqrt{\frac{(i+l-k)!(j+k)!}{i!j!k!(l-k)!}}$$
$$|i+l-k,j+k\rangle_a |l-k,k\rangle_b |N-l\rangle_c. \quad (38)$$

So, the cloning transformation takes $i+j$ qubits in the form (36) as input, and produces $i+j+l$ output qubits in the form (38). And the action of Hamiltonian (35) on the state $|F_l,(i,j)\rangle$ is as follows:

$$\mathcal{H}|F_l,(i,j)\rangle = \gamma(\sqrt{(l+1)(N-l)(i+j+l+2)}|F_{l+1},(i,j)\rangle$$
$$+ \sqrt{l(N-l+1)(i+j+l+1)}|F_{l-1},(i,j)\rangle),$$
$$1 \le l < N,$$
$$\mathcal{H}|F_0,(i,j)\rangle = \gamma\sqrt{N(i+j+2)}|F_1,(i,j)\rangle,$$
$$\mathcal{H}|F_N,(i,j)\rangle = \gamma\sqrt{N(i+j+N+1)}|F_{N-1},(i,j)\rangle. \quad (39)$$

We remark that in case $i = m, j = 0$, we recover the previous results in [30].

We now consider the cloning of M identical pure input states to L copies. We have

$$|\Phi\rangle^{\otimes M} = (\alpha a_1^\dagger + \beta a_2^\dagger)^{\otimes M}|0,0\rangle$$
$$= \sum_{j=0}^{M} \frac{M!}{\sqrt{(M-j)!j!}} \alpha^{M-j}\beta^j \frac{(a_1^\dagger)^{M-j}(a_2^\dagger)^j}{\sqrt{(M-j)!j!}}|0,0\rangle. \quad (40)$$

We already know the cloning of the bases $\frac{(a_1^\dagger)^{M-j}(a_2^\dagger)^j}{\sqrt{(M-j)!j!}}|0,0\rangle$. Using the cloning transformation (38), we can obtain the output of L copies as

$$|\Phi\rangle^{out} = \sum_{j=0}^{M}\sum_{k=0}^{L-M} \frac{M!}{\sqrt{(M-j)!j!}} \alpha^{M-j}\beta^j$$
$$\sqrt{\frac{(L-M)!(M+1)!}{(L+1)!}} \sqrt{\frac{(L-j-k)!(j+k)!}{(M-j)!j!k!(L-M-k)!}}$$
$$|L-j-k,k+j\rangle_a |L-M-k,k\rangle_b |N-(M-L)\rangle_c. \quad (41)$$

The density operator of output can be obtained by taking the trace over b and c states. Tracing out all but one qubit (a type states), we can obtain the output reduced density operator. And the fidelity can be calculated to be optimal $F = M(L+2) + L - M/L(M+2)$. So, we know the cloning transformation is universal and optimal. A different criterion will also show the cloning transformation (38) is optimal and universal in the d-level case.

We know from (36), the cloning machine allows arbitrary mixed states in Bose subspace. We consider a input state of M qubits of the form

$$\rho = \sum_{jj'}^{M} \alpha_{jj'} \frac{(a_1^\dagger)^{M-j}(a_2^\dagger)^j}{\sqrt{(M-j)!j!}} |00\rangle\langle 00| \frac{(a_1)^{M-j'}(a_2)^{j'}}{\sqrt{(M-j')!j'!}}, \tag{42}$$

where $\alpha_{jj'}$ are arbitrary parameters, certainly we need here ρ to be a density operator. Using the cloning transformation (38), we can obtain l additional copies. And it can be proved that with input (42), the transformation (38) is still optimal [21]. We remark that (36) even allows the input states to have different qubits if they can be expressed by Bosonic operators.

We will study the photon optimal quantum cloning of states in d-dimensional Hilbert space (qudits). The atoms of the inverted medium of cloning have one excited state $|e\rangle$ and d ($d \geq 2$) ground states $|g_n\rangle, n = 1, 2, \ldots, d$, and each is coupled to a different degree of freedom of photons a_n. Similar to the qubit case, we denote by $b_r c^\dagger = \sum_{k=1}^{N} |e^k\rangle\langle g_r^k|, r = 1, \ldots, d$ for qudit systems. The Hamiltonian of the cloning system in terms of harmonic-oscillator operators is written as [30]

$$\mathcal{H}_d = \gamma(a_1 b_1 + \cdots + a_d b_d) + \text{H.c.} \tag{43}$$

We consider the general initial states in Bose subspace

$$|\Psi_{in}, \boldsymbol{j}\rangle = \prod_{i=1}^{d} \frac{(a_i^\dagger)^{j_i}}{\sqrt{j_i!}} \frac{(c^\dagger)^N}{\sqrt{N!}} |0\rangle \equiv |\boldsymbol{j}\rangle_a |\mathbf{0}\rangle_b |N\rangle_c, \tag{44}$$

where $\boldsymbol{j} = (j_1, j_2, \ldots, j_d)$, and we denote $\mathbf{0} = (0, 0, \ldots, 0)$. There are still N excited states in the initial state, so the number of additional copies is restricted by N. We remark that the initial state (44) to be cloned spans arbitrary states in Bose subspace and constitutes a set of orthonormal bases. One can see easily that the time evolution of states for qudits is the same as the qubits as presented in (37). That means the probability to obtain additional l copies is still $|f_l(t)|^2$. We still denote $|F_0, \boldsymbol{j}\rangle \equiv |\Psi_{in}, \boldsymbol{j}\rangle, \sum_i j_i = M$. The output of cloning with l additional copies from the initial state (44) can be calculated as

$$|F_l, \boldsymbol{j}\rangle = \sum_{k_i}^{l} \sqrt{\frac{(M+d-1)!l!}{(M+l+d-1)!}} \prod_{i=1}^{d} \sqrt{\frac{(k_i+j_i)!}{k_i!j_i!}} |\boldsymbol{j}+\boldsymbol{k}\rangle_a |\boldsymbol{k}\rangle_b |N-l\rangle_c, \tag{45}$$

where summation $\sum_{k_i}^{l}$ means taking the sum over all variables under the condition $\sum_i^d k_i = l$.

It is very interesting that the cloning transformation (45) with input (44) is completely determined by the interaction Hamiltonian (43). Given different input states to be cloned, the action of the Hamiltonian on the initial states will produce the corresponding cloning output. That means the procedure of quantum cloning is completely controlled by the Hamiltonian. And the detailed calculation shows the following results:

$$\mathcal{H}_d|F_l,\boldsymbol{j}\rangle = \gamma(\sqrt{(l+1)(N-l)(M+l+d)}|F_{l+1},\boldsymbol{j}\rangle$$
$$+\sqrt{l(N-l+1)(M+l+d-1)}|F_{l-1},\boldsymbol{j}\rangle), \quad l \le l < N,$$
$$\mathcal{H}_d|F_0,\boldsymbol{j}\rangle = \gamma\sqrt{N(M+d)}|F_1,\boldsymbol{j}\rangle,$$
$$\mathcal{H}_d|F_N,\boldsymbol{j}\rangle = \gamma\sqrt{N(M+N+d-1)}|F_{N-1},\boldsymbol{j}\rangle. \tag{46}$$

Now, we see how to clone M identical qudits to $M+l \equiv L$ copies. An arbitrary qudit take the form $|\Psi\rangle = \sum_{i=1}^{d} x_i a_i^\dagger|0\rangle$, with $\sum_{i=1}^{d} |x_i|^2 = 1$. The M identical qudits to be cloned can be expressed as follows:

$$|\Psi\rangle^{\otimes M} = (\sum_{i=1}^{d} x_i a_i^\dagger)^{\otimes M}|0\rangle = M!\sum_{j_i}\prod_{i=1}^{d}\frac{x_i^{j_i}}{\sqrt{j_i!}}\frac{(a_i^\dagger)^{j_i}}{\sqrt{j_i!}}|0\rangle. \tag{47}$$

Consider that we intend to clone this state in the system with N atoms in the excited state $|e\rangle$, that means the number of additional copies is restricted by N. With the help of cloning transformation (45), we can find the output of L copies of M identical qudits has the form:

$$|\Psi\rangle^{out} = M!\sum_{j_i}^{M}\sum_{k_i}^{l}\sqrt{\frac{(M+d-1)!l!}{(L+d-1)!}}\prod_{i=1}^{d}\frac{x_i^{j_i}}{j_i!}\sqrt{\frac{(k_i+j_i)!}{k_i!}}|\boldsymbol{j}+\boldsymbol{k}\rangle_a|\boldsymbol{k}\rangle_b, \tag{48}$$

where we omit the type-c state which counts the number of cloning the system produced. We can calculate the fidelity of cloning transformation is

$$F = \langle\Psi|\rho_{\text{red.}}^{out}|\Psi\rangle = \frac{M(L+d)+L-M}{L(M+d)}, \tag{49}$$

where $\rho_{\text{red.}}^{out}$ means taking trace over ancilla states, i.e., b-type states, and over all but one a-type states of $\rho^{out} = |\Psi\rangle^{out\ out}\langle\Psi|$. This fidelity is the optimal fidelity for identical pure input states in d-dimensional Hilbert space [16, 17]. And the cloning transformation is universal.

Next, instead of the fidelity between input and output reduced density operators of a single qudit, we use the fidelity between the output of L qudits and L identical pure qudits as measure of quality of cloning transformation (45). With the help of the result in (48), and considering the normalized factors, we can find that

$$F_{global} = \frac{L!(M+d-1)!}{M!(L+d-1)!}. \tag{50}$$

This is the optimal fidelity of cloning identical pure states [16]. We remark that optimal cloning of pure states was studied by *Werner* et al. [16, 17] by complete positive (CP) map realized by symmetric projection operators. In this paper, quantum cloning (45) obtained from the Hamiltonian is realized by unitary transformation. Thus we show that for both density operator and reduced density operator, the fidelities of cloning transformation in (45) are optimal for identical pure input states.

8 UQCM for Identical Mixed States

The result of this section was presented in a recent paper by *Fan* [32].

8.1 A 2 to 3 Universal Quantum Cloning for Mixed States

To copy two identical mixed qubits, we not only need the cloning transformations for triplet states in symmetric subspace, but we also need a cloning transformation for the singlet state. We consider the universal quantum cloning machine in the sense that the quality of the copies is independent of the input states. Since we consider arbitrary mixed qubits as input, each output state $\rho_{\text{red.}}^{(\text{out})}$ and the input ρ should satisfy the scalar form to satisfy the universality condition [14],

$$\rho_{\text{red.}}^{(\text{out})} = f\rho + \frac{1-f}{2}I, \tag{51}$$

where f is the shrinking factor, and I is the identity. The relationship between each input and output state is just like the input state goes through a depolarizing channel. We can find that the shrinking factor f can describe the quality of the copies. If $f = 1$, the output state is exactly the input state. If it is zero, the input state is completely destroyed, i.e., the output state contains no information of the input state. The optimal shrinking factor has already be obtained in [14] for identical pure input states. And it is showed that this shrinking factor is also the tight bound for arbitrary mixed states in symmetric subspace. It is obvious that the optimal shrinking factor for identical pure states is also an upper bound for identical mixed states. The problem is whether this bound can be saturated or not for the case of two identical mixed qubits.

To express our result explicitly, we first give the result for 2 to 3 cloning machine, we have 2 input state and 3 copies which may be entangled. We consider ρ to be an arbitrary mixed state

$$\rho = x_{00}|\uparrow\rangle\langle\uparrow| + x_{01}|\uparrow\rangle\langle\downarrow| + x_{10}|\downarrow\rangle\langle\uparrow| + x_{11}|\downarrow\rangle\langle\downarrow|, \tag{52}$$

with the restriction that this is a density operator.

We propose the following quantum cloning transformations:

$$U|2\uparrow\rangle|\uparrow\rangle \otimes R = \sqrt{\frac{3}{4}}|3\uparrow\rangle \otimes R_\uparrow + \sqrt{\frac{1}{4}}|2\uparrow,\downarrow\rangle \otimes R_\downarrow,$$

$$U|2\downarrow\rangle|\uparrow\rangle \otimes R = \sqrt{\frac{1}{4}}|\uparrow,2\downarrow\rangle \otimes R_\uparrow + \sqrt{\frac{3}{4}}|3\downarrow\rangle \otimes R_\downarrow,$$

$$U|\Psi^+\rangle|\uparrow\rangle \otimes R = \sqrt{\frac{1}{2}}|2\uparrow,\downarrow\rangle \otimes R_\uparrow + \sqrt{\frac{1}{2}}|\uparrow,2\downarrow\rangle \otimes R_\downarrow,$$

$$U|\Psi^-\rangle|\uparrow\rangle \otimes R = \sqrt{\frac{1}{2}}\widetilde{|2\uparrow,\downarrow\rangle} \otimes R_\uparrow + \sqrt{\frac{1}{2}}\widetilde{|\uparrow,2\downarrow\rangle} \otimes R_\downarrow. \tag{53}$$

Here let us introduce the notations. $|\Psi^+\rangle = (|\uparrow\downarrow\rangle + |\downarrow\uparrow\rangle)/\sqrt{2}$, $|\Psi^-\rangle = (|\uparrow\downarrow\rangle - |\downarrow\uparrow\rangle)/\sqrt{2}$, $|2\uparrow,\downarrow\rangle = (|\uparrow\uparrow\downarrow\rangle + |\uparrow\downarrow\uparrow\rangle + |\downarrow\uparrow\uparrow\rangle)/\sqrt{3}$ is a symmetric state with 2 spin up and 1 spin down, similarly for $|\uparrow,2\downarrow\rangle$. The state $\widetilde{|2\uparrow,\downarrow\rangle} = (|\uparrow\uparrow\downarrow\rangle + \omega|\uparrow\downarrow\uparrow\rangle + \omega^2|\downarrow\uparrow\uparrow\rangle)/\sqrt{3}$ is almost the same as the symmetric state $|2\uparrow,\downarrow\rangle$ but with the phase of $\omega = e^{2\pi i/3}$. R are ancilla state, and $R_\uparrow, R_\downarrow$ are orthogonal to each other. It can be checked easily that the above relations satisfy the unitary condition. We then show that this quantum cloning machine is universal and optimal in the sense the relation (51) is satisfied and the shrinking factor saturates the optimal bound. We expand the input state $\rho \otimes \rho$ in terms of the four bases $|2\uparrow\rangle, |2\downarrow\rangle, |\Psi^+\rangle, |\Psi^-\rangle$. By using the cloning transformations (53), tracing out the ancillary states $R_\uparrow, R_\downarrow$, we obtain the output state of three qubits. This state is a mixed state and may be entangled. What we are interested is the reduced density operator of each output qubit. One can see that each output qubit is the same from the cloning transformation (53). By some calculations, we find the following relation,

$$\rho_{\text{red.}}^{(\text{out})} = \frac{5}{6}\rho + \frac{1}{12}I. \tag{54}$$

Really, our cloning transformation (53) is universal and optimal since the shrinking factor $\frac{5}{6}$ is optimal. This is the first nontrivial quantum cloning of identical mixed qubits. We remark that two identical pure qubits can be expanded in the symmetric subspace, so the first three quantum cloning transformations are enough for this case. For the general identical mixed states, the cloning transformation for singlet state is necessary.

8.2 General 2 to $M(M > 2)$ UQCM

Next, we shall present our general result of 2 to M cloning, in which the cloning machine creates M copies out of 2 identical mixed qubits. The quantum cloning transformation is presented as follows:

$$U|2\uparrow\rangle|\uparrow\rangle \otimes R = \sum_{k=0}^{M-2} \alpha_{0k}|(M-k)\uparrow, k\downarrow\rangle \otimes R_k,$$

$$U|2\downarrow\rangle|\uparrow\rangle \otimes R = \sum_{k=0}^{M-2} \alpha_{2k}|(M-2-k)\uparrow, (2+k)\downarrow\rangle \otimes R_k,$$

$$U|\Psi^+\rangle|\uparrow\rangle \otimes R = \sum_{k=0}^{M-2} \alpha_{1k}|(M-1-k)\uparrow, (1+k)\downarrow\rangle \otimes R_k,$$

$$U|\Psi^-\rangle|\uparrow\rangle \otimes R = \sum_{k=0}^{M-2} \alpha_{1k}|\widetilde{(M-1-k)\uparrow, (1+k)\downarrow}\rangle \otimes R_k, \qquad (55)$$

where

$$\alpha_{jk} = \sqrt{\frac{6(M-2)!(M-j-k)!(j+k)!}{(2-j)!(M+1)!(M-2-k)!j!k!}}, \quad j = 0, 1, 2. \qquad (56)$$

As previously, the state $|i\uparrow, j\downarrow\rangle$ is a completely symmetrical state with i states spin up and j states spin down, the state $|\widetilde{i\uparrow, j\downarrow}\rangle$ is almost the same as $|i\uparrow, j\downarrow\rangle$, but each term has a different phase of $\binom{i+j}{i}$-th root of unity so that $|i\uparrow, j\downarrow\rangle$ and $|\widetilde{i\uparrow, j\downarrow}\rangle$ are orthogonal to each other. R_k are ancillary states and are orthogonal for different k. By tedious calculations, we can find that this quantum cloning machine is universal and optimal,

$$\rho_{\text{red.}}^{(\text{out})} = \frac{M+2}{2M}\rho + \frac{M-2}{4M}I, \qquad (57)$$

where the shrinking factor $(M+2)/2M$ achieve the optimal bound [14]. Thus we show that we can copy two identical mixed qubits as well as we copy two identical pure states.

9 Phase-Covariant Quantum Cloning Machine

We next review the results of a phase-covariant quantum cloning machine. We almost repeat the original calculations by *Buzěk* and *Hillery* in their seminal paper [11]. But the calculations are useful not only for UQCM but also for phase-covariant cloning machine. Though some advanced methods are available in studying those problems, we find the original method is still powerful and explicit. That is the main reason that we present some detailed calculations here.

In case of UQCM, the input states are arbitrary pure states. Next, we review the QCM for a restricted set of pure input states. The Bloch vector is restricted to the intersection of x–z (x–y and y–z) plane with the Bloch sphere. These kinds of qubits are the so-called equatorial qubits [33], and the corresponding QCM is called phase-covariant quantum cloning. We study the 1 to 2 cloning at first. Applying the method by *Bužek* and *Hillery* [11], we propose a possible extension of the original transformation. We demand that (I) the density matrices of the two output states are the same, and that (II) the distance between input density operator and the output density operators is input-state independent. To evaluate the distance of two states, we use both Hilbert–Schmidt norm and Bures fidelity. There is a family of transformations which satisfy the above two conditions. In a special point, we can obtain an optimal fidelity. The correspondent transformation for $x-z$ equator agrees with the results of *Bruß* et al. [33], who studied the optimal quantum cloning for equatorial qubits by taking BB84 states as input. The fidelity of quantum cloning for the equatorial qubits is higher than the original *Bužek* and *Hillery* UQCM [11]. This is expected as the more information about the input is given, the better one can clone each of its states. We also obtain by a simple transformation the quantum cloning transformations for equatorial qubits in the x–y plane. Using the approach presented in [28], we show that the optimal phase-covariant quantum cloning machines can be realized by networks consisting of quantum rotation gates and controlled NOT gates. The copied equatorial qubits are shown to be separable by using the Peres–Horodecki criterion. We then present the 1 to M phase-covariant quantum cloning transformations and prove that the fidelity is optimal. The next results are presented in [34].

10 Transformation

Instead of arbitrary input states, we consider the input state which we intend to clone to be a restricted set of states. It is a pure superposition state:

$$|\Psi\rangle = \alpha|0\rangle + \beta|1\rangle, \tag{58}$$

with $\alpha^2 + \beta^2 = 1$. Here, we use an assumption that α and β are real, in contrast to complex, when we consider the case of UQCM. That means the y component of the Bloch vector of the input qubits is zero. Because that there is just one unknown parameter in the input state under consideration, we expect that we can achieve a better quality in quantum cloning if we can find an appropriate phase-covariant QCM.

In order to have a better quality in phase-covariant quantum cloning than the UQCM, we need a different cloning transformation. We propose the following transformation:

$$|0\rangle_{a_1}|Q\rangle_{a_2 a_3} \to (|0\rangle_{a_1}|0\rangle_{a_2} + \lambda|1\rangle_{a_1}|1\rangle_{a_2})|Q_0\rangle_{a_3}$$
$$+ (|1\rangle_{a_1}|0\rangle_{a_2} + |0\rangle_{a_1}|1\rangle_{a_2})|Y_0\rangle_{a_3},$$
$$|1\rangle_{a_1}|Q\rangle_{a_2 a_3} \to (|1\rangle_{a_1}|1\rangle_{a_2} + \lambda|0\rangle_{a_1}|0\rangle_{a_2})|Q_1\rangle_{a_3}$$
$$+ (|1\rangle_{a_1}|0\rangle_{a_2} + |0\rangle_{a_1}|1\rangle_{a_2})|Y_1\rangle_{a_3}, \tag{59}$$

where the states $|Q_j\rangle_{a_3}, |Y_j\rangle_{a_3}, j = 0, 1$ are not necessarily orthonormal. We will sometimes drop the subscript a_3 for convenience. Explicitly, this transformation is a generalization of the original one proposed by *Bužek* and *Hillery* [11]. When $\lambda = 0$, this transformation is reduced to the original transformation. Here, we remark that it is yet unclear whether the cloning transformation presented above can achieve the optimal point if we choose appropriate parameters, because the supposed cloning transformation (59) is not the most general one. We shall show in the next sections that this cloning transformation indeed can achieve the optimal point. For convenience, we restrict λ to be real and $\lambda \neq \pm 1$. We also assume

$$\langle Q_0 | Q_1 \rangle = \langle Q_1 | Q_0 \rangle = 0. \tag{60}$$

Considering the unitarity of the transformation, we have the following relations:

$$(1 + \lambda^2)\langle Q_j | Q_j \rangle + 2\langle Y_j | Y_j \rangle = 1, \quad j = 0, 1, \tag{61}$$
$$\langle Y_0 | Y_1 \rangle = \langle Y_1 | Y_0 \rangle = 0. \tag{62}$$

As proposed by Bužek and Hillery, we further assume the following relations to reduce the free parameters:

$$\langle Q_j | Y_j \rangle = 0, \quad j = 0, 1. \tag{63}$$
$$\langle Y_0 | Y_0 \rangle = \langle Y_1 | Y_1 \rangle \equiv \xi, \tag{64}$$
$$\langle Y_0 | Q_1 \rangle = \langle Q_0 | Y_1 \rangle = \langle Q_1 | Y_0 \rangle = \langle Y_1 | Q_0 \rangle \equiv \frac{\eta}{2}. \tag{65}$$

For simplicity, we shall use the following standard notations:

$$|jk\rangle = |j\rangle_{a_1}|k\rangle_{a_2}, \quad j, k = 0, 1, \tag{66}$$

and

$$|+\rangle = \frac{1}{\sqrt{2}}(|10\rangle + |01\rangle), \quad |-\rangle = \frac{1}{\sqrt{2}}(|10\rangle - |01\rangle). \tag{67}$$

Obviously, $|\pm\rangle$ and $|00\rangle, |11\rangle$ constitute an orthonormal basis.

The output density operator $\rho_{ab}^{(out)}$ describing output state after the copying procedure reads

$$\rho_{a_1 a_2}^{(out)} = |00\rangle\langle 00|\{\frac{1-2\xi}{1+\lambda^2}[\lambda^2 + \alpha^2(1-\lambda^2)]\}$$
$$+ (|00\rangle\langle 10| + |00\rangle\langle 01| + |11\rangle\langle 10| + |11\rangle\langle 01| + |01\rangle\langle 00| + |10\rangle\langle 00|$$
$$+ |01\rangle\langle 11| + |10\rangle\langle 11|)[\frac{\eta}{2}\alpha\beta(\lambda+1)]$$
$$+ (|00\rangle\langle 11| + |11\rangle\langle 00|)(\frac{1-2\xi}{1+\lambda^2}\lambda)$$
$$+ \xi(|01\rangle\langle 10| + |01\rangle\langle 01| + |10\rangle\langle 10| + |10\rangle\langle 01|)$$
$$+ |11\rangle\langle 11|\{\frac{1-2\xi}{1+\lambda^2}[\alpha^2(\lambda^2-1)+1]\}, \tag{68}$$

where $\rho_{a_1 a_2}^{(out)} = \mathrm{Tr}_{a_3}[\rho_{a_1 a_2 a_3}^{(out)}]$ with $\rho_{a_1 a_2 a_3}^{(out)} \equiv |\Psi\rangle_{a_1 a_2 a_3}^{(out)} \,{}_{a_1 a_2 a_3}^{(out)}\langle\Psi|$. Taking the trace over mode a_2 or mode a_1, we can get the reduced density operator for mode a_1 or mode a_2, $\rho_{a_1}^{(out)}$ or $\rho_{a_2}^{(out)}$,

$$\rho_{a_1}^{(out)} = \rho_{a_2}^{(out)} = |0\rangle\langle 0|\left((\alpha^2 + \lambda^2\beta^2)\frac{1-2\xi}{1+\lambda^2} + \xi\right)$$
$$\cdot (|0\rangle\langle 1| + |1\rangle\langle 0|)\alpha\beta\eta(1+\lambda) + |1\rangle\langle 1|\left(\xi + (\beta^2 + \lambda^2\alpha^2)\frac{1-2\xi}{1+\lambda^2}\right). \tag{69}$$

We see that the output density operators $\rho_{a_1}^{(out)}$ and $\rho_{a_2}^{(out)}$ are exactly the same. However, it is well known that they are not equal to the original input density operator. Next, we first use the Hilbert–Schmidt norm to evaluate the distance between input density operator and output density operators.

11 Hilbert–Schmidt Norm

For two-dimensional space, the Hilbert–Schmidt norm is believed to give a reasonable result in comparing density matrices, though it becomes less good for finite-dimensional spaces as the dimension increases. The Hilbert–Schmidt norm defines the distance between the input and output density operators as

$$D_a \equiv \mathrm{Tr}[\rho_a^{(out)} - \rho_a^{(in)}]^2, \tag{70}$$

where $\rho_a^{(in)}$ is the input density operator. The distance between the two-mode density operators $\rho_{a_1 a_2}^{(out)}$ and $\rho_{a_1 a_2}^{(in)} = \rho_{a_1}^{(in)} \otimes \rho_{a_1}^{(in)}$, which corresponds to the ideal copy, is defined as:

$$D_{a_1 a_2}^{(2)} = \mathrm{Tr}[\rho_{a_1 a_2}^{(out)} - \rho_{a_1 a_2}^{(in)}]^2. \tag{71}$$

With the help of relation (69), we find

$$D_a = \{\xi + \frac{1-2\xi}{1+\lambda^2}[\alpha^2(1-\lambda^2) + \lambda^2] - \alpha^2\}^2 + 2\alpha^2(1-\alpha^2)(\lambda\eta + \eta - 1)^2$$
$$+ \{\xi - 1 + \frac{1-2\xi}{1+\lambda^2}[1 + \alpha^2(\lambda^2 - 1)] + \alpha^2\}^2. \quad (72)$$

We demand that this distance is independent of the parameter α^2. That means the quality of the copies it makes is independent of the input state.

$$\frac{\partial}{\partial \alpha^2} D_a = 0. \quad (73)$$

We can choose the following solution:

$$\eta = \frac{1-\lambda}{1+\lambda^2}(1 - 2\xi). \quad (74)$$

Thus we get

$$D_a = 2\left(\xi \frac{1-\lambda^2}{1+\lambda^2} + \frac{\lambda^2}{1+\lambda^2}\right)^2. \quad (75)$$

In case $\lambda = 0$, we find $\eta = 1 - 2\xi$ and $D_a = 2\xi^2$. These are exactly the original results obtained by *Bužek* and *Hillery* [11].

In order to calculate $D_{a_1 a_2}^{(2)}$, we can rewrite the output density operator $\rho_{a_1 a_2}^{(out)}$ by choosing the basis in (67). Substituting the relation (74) into the two-mode output density operator, we can obtain

$$\rho_{a_1 a_2}^{(out)} = |00\rangle\langle 00| \{\frac{1-2\xi}{1+\lambda^2}[\lambda^2 + \alpha^2(1-\lambda^2)]\}$$
$$+ (|00\rangle\langle +| + |+\rangle\langle 00| + |11\rangle\langle +| + |+\rangle\langle 11|) \{\sqrt{2}\alpha\beta \frac{1-\lambda^2}{2(1+\lambda^2)}(1-2\xi)\}$$
$$+ (|00\rangle\langle 11| + |11\rangle\langle 00|) \{\frac{1-2\xi}{1+\lambda^2}\lambda\}$$
$$+ 2\xi|+\rangle\langle +| + |11\rangle\langle 11| \{\frac{1-2\xi}{1+\lambda^2}[\alpha^2(\lambda^2-1)+1]\}. \quad (76)$$

By straightforward calculations, we also have

$$\rho_{a_1 a_2}^{(in)} = \alpha^4|00\rangle\langle 00| + \sqrt{2}\alpha^3\beta(|00\rangle\langle +| + |+\rangle\langle 00|) + \alpha^2\beta^2(|00\rangle\langle 11| + |11\rangle\langle 00|)$$
$$+ 2\alpha^2\beta^2|+\rangle\langle +| + \sqrt{2}\alpha\beta^3(|+\rangle\langle 11| + |11\rangle\langle +|) + \beta^4|11\rangle\langle 11|. \quad (77)$$

And with the definition (71), we obtain

$$D_{a_1 a_2}^{(2)} = (U_{11})^2 + (U_{22})^2 + (U_{33})^2 + 2(U_{12})^2 + 2(U_{13})^2 + 2(U_{23})^2, \quad (78)$$

where
$$U_{11} = \alpha^4 - \frac{1-2\xi}{1+\lambda^2}[\lambda^2 + \alpha^2(1-\lambda^2)],$$
$$U_{22} = 2\xi - 2\alpha^2 + 2\alpha^4,$$
$$U_{33} = \alpha^4 - 2\alpha^2 + 1 - \frac{1-2\xi}{1+\lambda^2}[\alpha^2(\lambda^2 - 1) + 1],$$
$$U_{12} = \sqrt{2}\alpha\beta[\alpha^2 - \frac{1-\lambda^2}{1+\lambda^2}(\frac{1}{2}-\xi)],$$
$$U_{13} = \alpha^2\beta^2 - \frac{1-2\xi}{1+\lambda^2}\lambda,$$
$$U_{23} = \sqrt{2}\alpha\beta[\beta^2 - \frac{1-\lambda^2}{1+\lambda^2}(\frac{1}{2}-\xi)]. \tag{79}$$

We still impose the condition
$$\frac{\partial}{\partial \alpha^2} D^{(2)}_{a_1 a_2} = 0, \tag{80}$$

and then
$$\xi = \frac{(1-\lambda)^2}{2(3-2\lambda+3\lambda^2)}. \tag{81}$$

Substitution of these results into D_a and $D^{(2)}_{a_1 a_2}$ gives
$$D_a = \frac{(1-2\lambda+5\lambda^2)^2}{2(3-2\lambda+3\lambda^2)^2}, \quad D^{(2)}_{a_1 a_2} = \frac{2(1-4\lambda+12\lambda^2-8\lambda^3+7\lambda^4)}{(3-2\lambda+3\lambda^2)^2}. \tag{82}$$

Therefore, we can have a family of transformations which satisfies the two conditions (I) and (II). In case $\lambda = 0$, we recover the Bužek and Hillery's result
$$D_a = \frac{1}{18} \approx 0.056, \quad D^{(2)}_{a_1 a_2} = \frac{2}{9} \approx 0.22. \tag{83}$$

Our aim is to find smaller D_a and $D^{(2)}_{a_1 a_2}$ for equatorial qubits, and to prove that the corresponding cloning transformation is the optimal QCM. We can show that in the region $0 < \lambda < 1/3$, both D_a and $D^{(2)}_{ab}$ take smaller values than the case $\lambda = 0$. When we choose
$$\lambda = 3 - 2\sqrt{2}, \tag{84}$$

both D_a and $D^{(2)}_{a_1 a_2}$ take their minimal values,
$$D_a = \frac{99-70\sqrt{2}}{68-48\sqrt{2}} \approx 0.043, \quad D_{a_1 a_2} = \frac{215-152\sqrt{2}}{8(3-2\sqrt{2})^2} \approx 0.17. \tag{85}$$

Thus for equatorial qubits, we can find smaller D_a and $D_{a_1 a_2}^{(2)}$, which means this QCM (59) has a higher fidelity than the original UQCM [11] in terms of the Hilbert–Schmidt norm. Actually, because we assume α and β are real, only a single unknown parameter is copied instead of two unknown parameters for the case of a general pure state. Thus a higher fidelity of quantum cloning can be achieved. The case of spin flip (universal NOT) has a similar phenomenon [28, 35].

Under the condition (84), we have

$$\xi = \frac{1}{8}, \quad \eta = \frac{\sqrt{2}-1}{12 - 8\sqrt{2}}. \tag{86}$$

We can realize vectors $|Q_j\rangle, |Y_j\rangle, j = 0, 1$ in two-dimensional space

$$|Q_0\rangle = (0, \frac{1}{4-2\sqrt{2}}), \qquad |Q_1\rangle = (\frac{1}{4-2\sqrt{2}}, 0),$$
$$|Y_0\rangle = (\frac{1}{2\sqrt{2}}, 0), \qquad |Y_1\rangle = (0, \frac{1}{2\sqrt{2}}). \tag{87}$$

The transformation (59) is rewritten as

$$|0\rangle_{a_1}|Q\rangle_{a_2 a_3} \to \frac{1}{4-2\sqrt{2}}[|00\rangle_{a_1 a_2} + (3 - 2\sqrt{2})|11\rangle_{a_1 a_2}]|\uparrow\rangle_{a_3}$$
$$+ \frac{1}{2}|+\rangle_{a_1 a_2}|\downarrow\rangle_{a_3}, \tag{88}$$

$$|1\rangle_{a_1}|Q\rangle_{a_2 a_3} \to \frac{1}{4-2\sqrt{2}}[|11\rangle_{a_1 a_2} + (3 - 2\sqrt{2})|00\rangle_{a_1 a_2}]|\downarrow\rangle_{a_3}$$
$$+ \frac{1}{2}|+\rangle_{a_1 a_2}|\uparrow\rangle_{a_3}. \tag{89}$$

This transformation agrees with the one obtained by *Bruß* et al. [33]. Using BB84 states as input, they showed that this transformation is the optimal cloning transformation for equatorial qubits. This means the proposed cloning transformation for the x–z equator (59) indeed realizes the optimal QCM within the Hilbert–Schmidt norm.

For an arbitrary λ with conditions (74) and (81) satisfied, we can still realize vectors $|Q_j\rangle, |Y_j\rangle, j = 0, 1$ in two-dimensional space,

$$|Q_0\rangle = q|\uparrow\rangle, \quad |Q_1\rangle = q|\downarrow\rangle,$$
$$|Y_0\rangle = y|\downarrow\rangle, \quad |Y_1\rangle = y|\uparrow\rangle, \tag{90}$$

where we use notations

$$q \equiv \sqrt{\frac{2}{3 - 2\lambda + 3\lambda^2}}, \quad y \equiv \frac{1-\lambda}{\sqrt{6 - 4\lambda + 6\lambda^2}}. \tag{91}$$

Thus all transformations (59) satisfy the conditions (I) and (II). Explicitly, the quantum cloning transformation for pure input states (58) can be written as:

$$|0\rangle_{a_1}|Q\rangle_{a_2a_3} \to (|00\rangle_{a_1a_2} + \lambda|11\rangle_{a_1a_2})q|\uparrow\rangle_{a_3} + (|10\rangle_{a_1a_2} + |01\rangle_{a_1a_2})y|\downarrow\rangle_{a_3},$$
$$|1\rangle_{a_1}|Q\rangle_{a_2a_3} \to (|11\rangle_{a_1a_2} + \lambda|00\rangle_{a_1a_2})q|\downarrow\rangle_{a_3} + (|10\rangle_{a_1a_2} + |01\rangle_{a_1a_2})y|\downarrow\rangle_{a_3}. \tag{92}$$

The distances defined by the Hilbert–Schmidt norm take the form (82).

12 Bures Fidelity

For finite-dimensional spaces, the Hilbert–Schmidt norm becomes less good when the dimension increases. Bures fidelity provides a more exact measurement of the distinguishability of two density matrices. In this section, we will use Bures fidelity to check the result in the previous section. The fidelity is defined as

$$F(\rho_1, \rho_2) = \mathrm{Tr}(\rho_1^{1/2} \rho_2 \rho_1^{1/2}). \tag{93}$$

The value of F ranges from 0 to 1. A larger F corresponds to a higher fidelity. $F = 1$ means two density matrices are equal. For a pure state, $\rho_1 = |\Psi\rangle\langle\Psi|$, the fidelity can be defined by an equivalent form $F = \langle\Psi|\rho_2|\Psi\rangle$. We shall use the definition (93) in this section.

It is known that a matrix

$$U = \begin{pmatrix} -\frac{\beta}{\alpha} & \frac{\alpha}{\beta} \\ 1 & 1 \end{pmatrix} \tag{94}$$

diagonalizes $\rho_a^{(in)}$ [36]

$$\rho_a^{(in)} = U \begin{pmatrix} 0 & 0 \\ 0 & 1 \end{pmatrix} U^{-1}. \tag{95}$$

We thus have

$$F(\rho_a^{(in)}, \rho_a^{(out)}) = \xi + \frac{(1-2\xi)[2\alpha^4(1-\lambda^2) + 2\alpha^2(\lambda^2-1) + 1]}{1+\lambda^2}$$
$$+ 2\alpha^2(1-\alpha^2)\eta(\lambda+1). \tag{96}$$

We demand that the fidelity be independent of the input state

$$\frac{\partial}{\partial \alpha^2} F(\rho_a^{(in)}, \rho_a^{(out)}) = 0. \tag{97}$$

This gives

$$\eta = \frac{1-\lambda}{1+\lambda^2}(1-2\xi), \tag{98}$$

and we obtain

$$F(\rho_a^{(in)}, \rho_a^{(out)}) = \frac{1 - \xi + \lambda^2 \xi}{1 + \lambda^2}. \tag{99}$$

Next, we use Bures fidelity to evaluate the distinguishability of density operators $\rho_{a_1 a_2}^{(out)}$ and $\rho_{a_1 a_2}^{(in)} = \rho_{a_1}^{(in)} \otimes \rho_{a_1}^{(in)}$. We have

$$F(\rho_{a_1 a_2}^{(in)}, \rho_{a_1 a_2}^{(out)}) = \frac{1 - 2\xi}{1 + \lambda^2}[\lambda^2 + \alpha^2(1 - \lambda^2)]\alpha^4 + 2\alpha^2(1 - \alpha^2)\lambda\frac{1 - 2\xi}{1 + \lambda^2}$$

$$+ 2\alpha^2(1 - \alpha^2)(1 - 2\xi)\frac{1 - \lambda^2}{1 + \lambda^2} + 4\alpha^2(1 - \alpha^2)\xi$$

$$+ \frac{1 - 2\xi}{1 + \lambda^2}[\alpha^2(\lambda^2 - 1) + 1](1 - \alpha^2)^2. \tag{100}$$

We again impose the condition

$$\frac{\partial}{\partial \alpha^2} F(\rho_{a_1 a_2}^{(in)}, \rho_{a_1 a_2}^{(out)}) = 0, \tag{101}$$

which gives

$$\xi = \frac{(1 - \lambda)^2}{2(3 - 2\lambda + 3\lambda^2)}. \tag{102}$$

Thus, we finally have two Bures fidelities for one- and two-mode density operators,

$$F(\rho_a^{(in)}, \rho_a^{(out)}) = \frac{5 - 2\lambda + \lambda^2}{2(3 - 2\lambda + 3\lambda^2)}, \tag{103}$$

$$F(\rho_{a_1 a_2}^{(in)}, \rho_{a_1 a_2}^{(out)}) = \frac{2}{3 - 2\lambda + 3\lambda^2}. \tag{104}$$

We find that Hilbert–Schmidt norm and Bures fidelity lead to the same relations ((74), (98)) and ((81),(102)). However, the fidelity (103) and (104) does not take the maximums simultaneously, which is different from the case of the Hilbert–Schmidt norm. In the region $0 < \lambda < 1/3$, for both $F(\rho_{a_1 a_2}^{(in)}, \rho_{a_1 a_2}^{(out)})$ and $F(\rho_a^{(in)}, \rho_a^{(out)})$, we can have a higher fidelity than the original UQCM which corresponds to $\lambda = 0$. This result agrees with the previous result by Hilbert–Schmidt norm. We use fidelity $F(\rho_a^{(in)}, \rho_a^{(out)})$ to define the quality of the copied equatorial qubits. When $\lambda = 3 - 2\sqrt{2}$, $F(\rho_a^{(in)}, \rho_a^{(out)})$ takes its maximum, which is the same as the case of Hilbert–Schmidt norm. Thus, we have shown that both Hilbert–Schmidt norm and Bures fidelity give the same result.

When $\lambda = 3 - 2\sqrt{2}$, $F(\rho_a^{(in)}, \rho_a^{(out)})$ takes its maximum

$$F(\rho_a^{(in)}, \rho_a^{(out)})|_{\lambda = 3 - 2\sqrt{2}} = \frac{1}{2} + \sqrt{\frac{1}{8}} \approx 0.8536, \tag{105}$$

which is larger than the original UQCM

$$F(\rho_a^{(in)}, \rho_a^{(out)})|_{\lambda=0} = \frac{5}{6} \approx 0.8333. \tag{106}$$

And we also have

$$F(\rho_{a_1a_2}^{(in)}, \rho_{a_1a_2}^{(out)})|_{\lambda=3-2\sqrt{2}} = \frac{1}{24 - 16\sqrt{2}} \approx 0.7286,$$

which is strictly larger than

$$F(\rho_{a_1a_2}^{(in)}, \rho_{a_1a_2}^{(out)})|_{\lambda=0} = \frac{2}{3} \approx 0.6667. \tag{107}$$

We remark that the optimal fidelity (105) also agrees with the result obtained by *Bruß* et al. [33].

In studying the optimal UQCM, the condition of orientation invariance of Bloch vector is generally imposed [12]. Under the symmetry condition (I), the condition of orientation invariance of Bloch vector is equivalent to the condition (II) that the distance between the input density operator and the output density operators is input state independent. We can check that for the case under consideration in this paper, the orientation invariance of Bloch vector means the relation (74) or (98), which is the subsequence of condition (II).

13 Quantum Cloning for $x - y$ Equatorial Qubits

We present in this section the cloning transformation for the x–y equator which can be obtained from the results of x–z equator by a transformation. The x–y plane equator takes the following form:

$$|\Psi\rangle = \frac{1}{\sqrt{2}}[|0\rangle + e^{i\phi}|1\rangle], \tag{108}$$

where $\phi \in [0, 2\pi)$. One can check that the y component of the Bloch vector of this state is zero. The cloning transformation takes the form

$$|0\rangle_{a_1}|00\rangle_{a_2a_3} \rightarrow \frac{2(1-\lambda)}{\sqrt{6-4\lambda+6\lambda^2}}|00\rangle_{a_1a_2}|0\rangle_{a_3}$$
$$+ \frac{1+\lambda}{\sqrt{6-4\lambda+6\lambda^2}}(|01\rangle_{a_1a_2} + |10\rangle_{a_1a_2})|1\rangle_{a_3},$$
$$|1\rangle_{a_1}|00\rangle_{a_2a_3} \rightarrow \frac{2(1-\lambda)}{\sqrt{6-4\lambda+6\lambda^2}}|11\rangle_{a_1a_2}|1\rangle_{a_3}$$
$$+ \frac{1+\lambda}{\sqrt{6-4\lambda+6\lambda^2}}(|01\rangle_{a_1a_2} + |10\rangle_{a_1a_2})|0\rangle_{a_3}. \tag{109}$$

The fidelity for this cloning transformation is calculated as

$$F(\rho^{(in)}, \rho^{(out)}) = \frac{5 - 2\lambda + \lambda^2}{6 - 4\lambda + 6\lambda^2}. \tag{110}$$

Corresponding to the fidelity (110), the reduced density operators of both copies at the output are equal and can be written as

$$\rho^{(out)} = \frac{5 - 2\lambda + \lambda^2}{6 - 4\lambda + 6\lambda^2}|\Psi\rangle\langle\Psi| + \frac{1 - 2\lambda + 5\lambda^2}{6 - 4\lambda + 6\lambda^2}|\Psi_\perp\rangle\langle\Psi_\perp|, \qquad (111)$$

where $|\Psi\rangle$ is the input state for equatorial qubits (108), and if we denote $|\Psi\rangle = \alpha|0\rangle + \beta|1\rangle$, we define $|\Psi_\perp\rangle \equiv \beta^*|0\rangle - \alpha^*|1\rangle$. We see that the copy contains $F(\rho^{(in)}, \rho^{(out)})$ of the state we want and $1 - F(\rho^{(in)}, \rho^{(out)})$ of that one we do not. The output density operator can be rewritten as

$$\rho^{(out)} = s|\Psi\rangle\langle\Psi| + \frac{1-s}{2} \cdot 1, \qquad (112)$$

where

$$s = 2F(\rho^{(in)}, \rho^{(out)}) - 1 = \frac{2(1 - \lambda^2)}{3 - 2\lambda + 3\lambda^2}. \qquad (113)$$

Actually, relations (110)–(113) are also correct for the case of x–z equator. We note that the output states of copies appear in a_1, a_2 qubits.

When $\lambda = 0$, we obtain the cloning transformations of UQCM with fidelity $\frac{5}{6}$. When $\lambda = 3 - 2\sqrt{2}$, we achieve the bound of the fidelity $\frac{1}{2} + \sqrt{\frac{1}{8}}$ and obtain the optimal quantum cloning transformations for equatorial qubits. Explicitly, we write here the optimal quantum cloning transformations for x–y equator (108),

$$|0\rangle_{a_1}|00\rangle_{a_2 a_3} \to \frac{1}{\sqrt{2}}|00\rangle_{a_1 a_2}|0\rangle_{a_3} + \frac{1}{2}(|01\rangle_{a_1 a_2} + |10\rangle_{a_1 a_2})|1\rangle_{a_3},$$

$$|1\rangle_{a_1}|00\rangle_{a_2 a_3} \to \frac{1}{\sqrt{2}}|11\rangle_{a_1 a_2}|1\rangle_{a_3} + \frac{1}{2}(|01\rangle_{a_1 a_2} + |10\rangle_{a_1 a_2})|0\rangle_{a_3}. \qquad (114)$$

Here we identify the internal states of the QCM as $|\uparrow\rangle \equiv |0\rangle$, $|\downarrow\rangle \equiv |1\rangle$, and for the x–z equator we also use these notations in the next sections.

14 Quantum Cloning Networks for Equatorial Qubits

In this section, following the method proposed by *Bužek* et al. [28], we show that the quantum cloning transformations for equatorial qubits can be realized by networks consisting of quantum logic gates. Let us first introduce the method proposed by *Bužek* et al. [28], and then analyze the case of phase-covariant cloning. The network is constructed by one- and two-qubit gates. The one-qubit gate is a single qubit rotation operator $\hat{R}_j(\vartheta)$, defined as

$$\hat{R}_j(\vartheta)|0\rangle_j = \cos\vartheta|0\rangle_j + \sin\vartheta|1\rangle_j, \quad \hat{R}_j(\vartheta)|1\rangle_j = -\sin\vartheta|0\rangle_j + \cos\vartheta|1\rangle_j. \qquad (115)$$

The two-qubit gate is the controlled NOT gate represented by the unitary matrix

$$\hat{P} = \begin{pmatrix} 1 & 0 & 0 & 0 \\ 0 & 1 & 0 & 0 \\ 0 & 0 & 0 & 1 \\ 0 & 0 & 1 & 0 \end{pmatrix}. \tag{116}$$

Explicitly, the controlled NOT gate $\hat{P}_{kl}$ acts on the basis vectors of the two qubits as follows:

$$\hat{P}_{kl}|0\rangle_k|0\rangle_l = |0\rangle_k|0\rangle_l, \quad \hat{P}_{kl}|0\rangle_k|1\rangle_l = |0\rangle_k|1\rangle_l,$$
$$\hat{P}_{kl}|1\rangle_k|0\rangle_l = |1\rangle_k|1\rangle_l, \quad \hat{P}_{kl}|1\rangle_k|1\rangle_l = |1\rangle_k|0\rangle_l. \tag{117}$$

Due to Bužek et al., the action of the copier is expressed as a sequence of two unitary transformations,

$$|\Psi\rangle_{a_1}^{(in)}|0\rangle_{a_2}|0\rangle_{a_3} \to |\Psi\rangle_{a_1}^{(in)}|\Psi\rangle_{a_1 a_2}^{(prep)} \to |\Psi\rangle_{a_1 a_2 a_3}^{(out)}. \tag{118}$$

This network can be described by a figure in [28]. The preparation state is constructed as

$$|\Psi\rangle_{a_2 a_3}^{(prep)} = \hat{R}_2(\vartheta_3)\hat{P}_{32}\hat{R}_3(\vartheta_2)\hat{P}_{23}\hat{R}_2(\vartheta_1)|0\rangle_{a_2}|0\rangle_{a_3}. \tag{119}$$

The quantum copying is performed by

$$|\Psi\rangle_{a_1 a_2 a_2}^{(out)} = \hat{P}_{a_3 a_1}\hat{P}_{a_2 a_1}\hat{P}_{a_1 a_3}\hat{P}_{a_1 a_2}|\Psi\rangle_{a_1}^{(in)}|\Psi\rangle_{a_2 a_3}^{(prep)}. \tag{120}$$

Note that the output copies appear in the a_2, a_3 qubits instead of a_1, a_2 qubits. For UQCM, we should choose [28]

$$\vartheta_1 = \vartheta_3 = \frac{\pi}{8}, \quad \vartheta_2 = -\arcsin\left(\frac{1}{2} - \frac{\sqrt{2}}{3}\right)^{1/2}. \tag{121}$$

We now consider the cloning transformations for equatorial qubits. The network proposed by Bužek et al. is rather general. We only need to take a different angles $\vartheta_j, j = 1, 2, 3$ to realize the phase-covariant cloning. In the case of cloning transformation for x–y equator (109), the preparation state takes the form

$$|\Psi\rangle_{a_2 a_3}^{(perp)} = \frac{2(1-\lambda)}{\sqrt{6 - 4\lambda + 6\lambda^2}}|00\rangle_{a_2 a_3}$$
$$+ \frac{1+\lambda}{\sqrt{6 - 4\lambda + 6\lambda^2}}\left(|01\rangle_{a_1 a_2} + |10\rangle_{a_2 a_3}\right). \tag{122}$$

The preparation state corresponding to cloning transformation (92) for x–z equator can be written as

$$|\Psi\rangle_{a_2 a_3}^{(perp)} = q|00\rangle_{a_2 a_3} + q\lambda|11\rangle_{a_2 a_3} + y|10\rangle_{a_2 a_3} + y|01\rangle_{a_2 a_3}. \tag{123}$$

We can check that for some angles $\vartheta_j, j = 1, 2, 3$, the above preparation states can be realized, Actually, we have several choices. When $\lambda = 0$, we obtain the result for UQCM. Here we present the result for the optimal case, i.e., $\lambda = 3 - 2\sqrt{2}$.

For x–y equator, let

$$\vartheta_1 = \vartheta_3 = \arcsin\left(\frac{1}{2} - \frac{1}{2\sqrt{3}}\right)^{\frac{1}{2}}, \quad \vartheta_2 = -\arcsin\left(\frac{1}{2} - \frac{\sqrt{3}}{4}\right)^{\frac{1}{2}}. \tag{124}$$

Then, the preparation state has the form

$$|\Psi\rangle_{a_2 a_3}^{(perp)} = \frac{1}{\sqrt{2}}|00\rangle_{a_2 a_3} + \frac{1}{2}(|01\rangle_{a_2 a_3} + |10\rangle_{a_2 a_3}). \tag{125}$$

For x–z equator, let

$$\vartheta_1 = \vartheta_3 = \arcsin\left(\frac{1}{2} - \sqrt{\frac{1}{8}}\right)^{\frac{1}{2}}, \quad \vartheta_2 = 0. \tag{126}$$

The preparation state is

$$|\Psi\rangle_{a_2 a_3}^{(perp)} = \left(\frac{1}{2} + \sqrt{\frac{1}{8}}\right)|00\rangle_{a_2 a_3} + \frac{1}{2\sqrt{2}}(|01\rangle_{a_2 a_3} + |10\rangle_{a_2 a_3})$$

$$+ \left(\frac{1}{2} - \sqrt{\frac{1}{8}}\right)|11\rangle_{a_2 a_3}. \tag{127}$$

After the preparation stage, perform the copying procedure (120), we obtain the output state. And the output copies appear in the a_2 and a_3 qubits. The optimal quantum cloning transformations for equatorial qubits can achieve the highest fidelity $\frac{1}{2} + \sqrt{\frac{1}{8}}$. The reduced density operator of both copies at the output in a_2 and a_3 qubits can be expressed as

$$\rho^{(out)} = \left(\frac{1}{2} + \sqrt{\frac{1}{8}}\right)|\Psi\rangle\langle\Psi| + \left(\frac{1}{2} - \sqrt{\frac{1}{8}}\right)|\Psi_\perp\rangle\langle\Psi_\perp|. \tag{128}$$

15 Separability of Copied Qubits and Quantum Triplicators

15.1 Separability

For the UQCM, the density matrix for the two copies $\rho_{a_2 a_3}^{(out)}$ is shown to be inseparable by use of Peres–Horodecki criterion [37,38]. That means it cannot be written as the convex sum

$$\rho_{a_2 a_3}^{(out)} = \sum_m w^{(m)} \rho_{a_2}^{(m)} \otimes \rho_{a_3}^{(m)}, \tag{129}$$

where the positive weights $w^{(m)}$ satisfy $\sum_m w^{(m)} = 1$. And there are correlations between the copies, i.e., the two qubits at the output of the quantum copier are nonclassically entangled [28]. We shall show in this section that, different from the UQCM, the copied qubits are separable for the case of optimal phase-covariant quantum cloning by Peres–Horodecki criterion.

Peres–Horodecki's positive partial transposition criterion states that the positivity of the partial transposition of a state is both a necessary and a sufficient condition for its separability [37, 38]. For x–z equator where the input state is $\alpha|0\rangle + \beta|1\rangle$, with $\alpha = \cos\theta, \beta = \sin\theta$, the partially transposed output density operator at a_2, a_3 qubits is expressed by a matrix,

$$[\rho_{a_2 a_3}^{(\text{out})}]^{T_2} = \frac{1}{3 - 2\lambda + 3\lambda^2} \begin{pmatrix} 2(\alpha^2 + \lambda^2\beta^2) & \alpha\beta(1-\lambda^2) & \alpha\beta(1-\lambda^2) & \frac{1}{2}(1-\lambda)^2 \\ \alpha\beta(1-\lambda^2) & \frac{1}{2}(1-\lambda)^2 & 2\lambda & \alpha\beta(1-\lambda^2) \\ \alpha\beta(1-\lambda^2) & 2\lambda & \frac{1}{2}(1-\lambda)^2 & \alpha\beta(1-\lambda^2) \\ \frac{1}{2}(1-\lambda)^2 & \alpha\beta(1-\lambda^2) & \alpha\beta(1-\lambda^2) & 2(\beta^2 + \alpha^2\lambda^2) \end{pmatrix}. \tag{130}$$

Here the cloning transformation corresponds to (92). Note that the output of copies appear in a_2, a_3 qubits. We have the following four eigenvalues:

$$\frac{1}{3 - 2\lambda + 3\lambda^2} \Big\{ \frac{1}{2}(1 - 6\lambda + \lambda^2), \quad 1 + \lambda^2 + \frac{1}{2}(1-\lambda)\sqrt{5 + 6\lambda + 5\lambda^2},$$
$$\frac{1}{2}(1 + 2\lambda + \lambda^2), \quad 1 + \lambda^2 - \frac{1}{2}(1-\lambda)\sqrt{5 + 6\lambda + 5\lambda^2} \Big\}. \tag{131}$$

For optimal phase-covariant quantum cloning, $\lambda = 3 - 2\sqrt{2}$, the four eigenvalues are

$$\{0, 0, \frac{1}{4}, \frac{3}{4}\}. \tag{132}$$

We see that none of the four eigenvalues is negative. This is different from the UQCM, where one negative eigenvalue exists for $\lambda = 0$. According to Peres–Horodecki criterion, the copied qubits in phase-covariant quantum cloning are separable. Analyzing the four eigenvalues (131), we find that the optimal point $\lambda = 3 - 2\sqrt{2}$ is the only separable point for the copied qubits. If we analyze the x–y equator, we obtain the same result.

15.2 Optimal Quantum Triplicators

The networks for equatorial qubits can realize the quantum copying. The copies at the output appear in a_2 and a_3 qubits. And the output density operator is written as

$$\rho^{(\text{out})} = \frac{2(1-\lambda^2)}{3 - 2\lambda + 3\lambda^2} \rho^{(\text{in})} + \frac{1 - 2\lambda + 5\lambda^2}{6 - 4\lambda + 6\lambda^2} \cdot 1. \tag{133}$$

Here, we are also interested in the output state in a_1 qubit. According to the cloning transformations or cloning networks for equatorial qubits, we find that the reduced density operator of the output state in a_1 qubit can be written as

$$\rho_{a_1}^{(out)} = \frac{(1+\lambda)^2}{3 - 2\lambda + 3\lambda^2}[\rho^{(in)}]^T + \frac{(1-\lambda)^2}{3 - 2\lambda + 3\lambda^2} \cdot 1, \tag{134}$$

where the superscript T means transposition. For x–z equator, the output density operator is invariant under the action of transposition. Comparing the output density operators in a_2 and a_3 qubits (133) and a_1 qubit (134), in case $\lambda = 1/3$, we have a triplicator,

$$\rho_{a_1}^{(out)} = \rho_{a_2}^{(out)} = \rho_{a_3}^{(out)} = \frac{2}{3}\rho^{(in)} + \frac{1}{6} \cdot 1, \tag{135}$$

with fidelity $\frac{5}{6}$ [28]. Explicitly, the triplicator cloning transformation for x–z equator has the form,

$$|0\rangle_{a_1}|00\rangle_{a_2a_3} \rightarrow \frac{1}{\sqrt{12}}[3|000\rangle_{a_1a_2a_3} + |011\rangle_{a_1a_2a_3}$$
$$+ |101\rangle_{a_1a_2a_3} + |110\rangle_{a_1a_2a_3}],$$

$$|1\rangle_{a_1}|00\rangle_{a_2a_3} \rightarrow \frac{1}{\sqrt{12}}[3|111\rangle_{a_1a_2a_3} + |100\rangle_{a_1a_2a_3}$$
$$+ |001\rangle_{a_1a_2a_3} + |010\rangle_{a_1a_2a_3}]. \tag{136}$$

For x–y equator, by applying a transformation $|0\rangle \leftrightarrow |1\rangle$ in a_1 qubit, and still let $\lambda = 1/3$, we find the output density operator in a_1 (134) equal to that of a_2 and a_3 (133). And the triplicator cloning for x–y equator takes the form,

$$|0\rangle_{a_1}|00\rangle_{a_2a_3} \rightarrow \frac{1}{\sqrt{3}}[|001\rangle_{a_1a_2a_3} + |100\rangle_{a_1a_2a_3} + |010\rangle_{a_1a_2a_3}],$$

$$|1\rangle_{a_1}|00\rangle_{a_2a_3} \rightarrow \frac{1}{\sqrt{3}}[|110\rangle_{a_1a_2a_3} + |011\rangle_{a_1a_2a_3} + |101\rangle_{a_1a_2a_3}]. \tag{137}$$

The fidelity for quantum triplicator is $\frac{5}{6}$. Actually, we can find the fidelity (110) takes the same value of $\frac{5}{6}$ when $\lambda = 0$ and $\lambda = 1/3$, corresponding to UQCM and quantum triplicator, respectively. *D'Ariano* and *Presti* [39] proved that the optimal fidelity for 1 to 3 phase-covariant quantum cloning is $\frac{5}{6}$, and presented the cloning transformation. The quantum triplicators presented above achieve the bound of the fidelity and agree with the results in [39].

16 Optimal 1 to M Phase-Covariant Quantum Cloning Machines

We have investigated the $1 \to 2$ and $1 \to 3$ optimal quantum cloning for equatorial qubits. In what follows, we shall study the general N to M ($M > N$) phase-covariant quantum cloning.

We first discuss $1 \to M$ phase-covariant quantum cloning. We start from the cloning transformations similar to the UQCM [13], then determine the parameters to give the highest fidelity, and finally prove that the determined cloning transformation is the optimal QCM for equatorial qubits. For x–y equator $|\Psi\rangle = (|\uparrow\rangle + e^{i\phi}|\downarrow\rangle)/\sqrt{2}$, we suppose the cloning transformations take the following form:

$$U_{1,M}|\uparrow\rangle \otimes R = \sum_{j=0}^{M-1} \alpha_j |(M-j)\uparrow, j\downarrow\rangle \otimes R_j,$$

$$U_{1,M}|\downarrow\rangle \otimes R = \sum_{j=0}^{M-1} \alpha_{M-1-j} |(M-1-j)\uparrow, (j+1)\downarrow\rangle \otimes R_j, \qquad (138)$$

where we use the same notations as those of [13], R denotes the initial state of the copy machine and $M-1$ blank copies, R_j are orthogonal normalized states of ancilla, and $|(M-j)\psi, j\psi_\perp\rangle$ denotes the symmetric and normalized state with $M-j$ qubits in state ψ and j qubits in state $\psi_\perp$. For an arbitrary input state, the case $\alpha_j = \sqrt{\frac{2(M-j)}{M(M+1)}}$ is the optimal $1 \to M$ quantum cloning [13]. Here we consider the case of x–y equator instead of an arbitrary input state. The quantum cloning transformations should satisfy the property of orientation invariance of the Bloch vector and that we have identical copies. The cloning transformation (138) already ensures that we have M identical copies. The unitarity of the cloning transformation demands the relation $\sum_{j=0}^{M-1} \alpha_j^2 = 1$. Under this condition, we can check that the cloning transformation has the property of orientation invariance of the Bloch vector. Thus, the relation (138) is the quantum cloning transformation for x–y equator. The fidelity of the cloning transformation (138) takes the form

$$F = \frac{1}{2}[1 + \eta(1, M)], \qquad (139)$$

where

$$\eta(1, M) = \sum_{j=0}^{M-1} \alpha_j \alpha_{M-1-j} \frac{C_{M-1}^j}{\sqrt{C_M^j C_M^{j+1}}}. \qquad (140)$$

We examine the cases of $M = 2, 3$. For $M = 2$, we have $\alpha_0^2 + \alpha_1^2 = 1$ and $\eta(1, M) = \sqrt{2}\alpha_0\alpha_1$. In case $\alpha_0 = \alpha_1 = 1/\sqrt{2}$, we have the optimal fidelity and recover the previous result (114). For $M = 3$, we have $\alpha_0^2 + \alpha_1^2 + \alpha_2^2 = 1$, and

$$\eta(1,3) = \frac{2}{3}\alpha_1^2 + \frac{2}{\sqrt{3}}\alpha_0\alpha_2. \tag{141}$$

For $\alpha_0 = \alpha_2 = 0, \alpha_1 = 1$, we have $\eta(1,3) = \frac{2}{3}$, which reproduces the case of quantum triplicator for x–y equator (137).

We present the result of 1 to M phase-covariant quantum cloning transformations. When M is even, we have $\alpha_j = \sqrt{2}/2, j = M/2 - 1, M/2$ and $\alpha_j = 0$, otherwise. When M is odd, we have $\alpha_j = 1, j = (M-1)/2$ and $\alpha_j = 0$, otherwise. The fidelity is $F = \frac{1}{2} + \frac{\sqrt{M(M+2)}}{4M}$ for M is even, and $F = \frac{1}{2} + \frac{(M+1)}{4M}$ for M is odd. The explicit cloning transformations have already been presented in (138).

Though the fidelity for $M = 2, 3$ are optimal, we need to prove that for general M, the fidelity achieves the bound as well. We apply the same method introduced by Gisin and Massar in [13]. In order to use some results later, we consider the general N to M cloning transformation. Generally, we write the N identical input state for equatorial qubits as

$$|\Psi\rangle^{\otimes N} = \sum_{j=0}^{N} e^{ij\phi}\sqrt{C_N^j}|(N-j)\uparrow, j\downarrow\rangle. \tag{142}$$

The most general N to M QCM for equatorial qubits is expressed as

$$|(N-j)\uparrow, j\downarrow\rangle \otimes R \to \sum_{k=0}^{M} |(M-k)\uparrow, k\downarrow\rangle \otimes |R_{jk}\rangle, \tag{143}$$

where R still denotes the $M - N$ blank copies and the initial state of the QCM, and $|R_{jk}\rangle$ are unnormalized final states of the ancilla. The unitarity relation is written as

$$\sum_{k=0}^{M} \langle R_{j'k}|R_{jk}\rangle = \delta_{jj'}. \tag{144}$$

The fidelity of the QCM takes the form

$$F = \langle \Psi|\rho^{\text{out}}|\Psi\rangle = \sum_{j',k',j,k} \langle R_{j'k'}|R_{jk}\rangle A_{j'k'jk}, \tag{145}$$

where ρ^{out} is the density operator of each output qubit by taking partial trace over all M but one output qubits. We impose the condition that the output density operator has the property of Bloch vector invariance, and find the following for $N = 1$,

$$A_{j'k'jk} = \frac{1}{4}\left\{\delta_{j'j}\delta_{k'k} + (1 - \delta_{j'j})\left[\delta_{k',(k+1)}\frac{\sqrt{(M-k)(k+1)}}{M}\right.\right.$$
$$\left.\left.+ \delta_{k,(k'+1)}\frac{\sqrt{(M-k')(k'+1)}}{M}\right]\right\}, \tag{146}$$

where $j, j' = 0, 1$ for case $N = 1$. The optimal fidelity of the QCM for equatorial qubits is related to the maximal eigenvalue λ_{max} of matrix A by $F = 2\lambda_{max}$ [13]. The matrix A (146) is a block diagonal matrix with block B given by

$$B = \frac{1}{4}\begin{pmatrix} 1 & \frac{\sqrt{(M-k)(k+1)}}{M} \\ \frac{\sqrt{(M-k)(k+1)}}{M} & 1 \end{pmatrix}. \tag{147}$$

Thus we have proved that the optimal fidelity of 1 to M QCM for equatorial qubits takes the form

$$F = 2\lambda_{max} = \begin{cases} \frac{1}{2} + \frac{\sqrt{M(M+2)}}{4M}, & \text{M is even,} \\ \frac{1}{2} + \frac{(M+1)}{4M}, & \text{M is odd.} \end{cases} \tag{148}$$

The experiment of the phase-covariant quantum cloning machine was performed by *Du* [40]. The general phase-covariant quantum cloning machine was studied by *D'Ariano* and *Macchiavello* [41]. Some related works can also be found in [42].

17 Some Known Results About Phase-Covariant Quantum Cloning Machine for Qubits and Qutrits

We first introduce the notations and review some known results for qubits [33, 34]. We consider the input state as

$$|\Psi\rangle^{(in)} = \frac{1}{\sqrt{2}}[|0\rangle + e^{i\phi}|1\rangle], \tag{149}$$

where $\phi \in [0, 2\pi)$. This state just has one arbitrary phase parameter ϕ instead of two free parameters for an arbitrary qubit. So, we already know partial information of this input state. One can check that the y component of the Bloch vector of this state is zero. This case is equivalent to the case that the input state is $\Psi\rangle = \cos\theta|0\rangle + \sin\theta|1\rangle$, in which the input state does not have arbitrary phase parameter. The optimal phase-covariant cloning transformation takes the form

$$U|0\rangle^{(in)}|Q\rangle = \frac{1}{\sqrt{2}}|00\rangle|0\rangle_a + \frac{1}{2}(|01\rangle + |10\rangle)|1\rangle_a,$$

$$U|1\rangle^{(in)}|Q\rangle = \frac{1}{\sqrt{2}}|11\rangle|1\rangle_a + \frac{1}{2}(|01\rangle + |10\rangle)|0\rangle_a, \qquad (150)$$

where $|Q\rangle$ is the blank state and the initial state of the cloning machine. The first states in the l.h.s. are input states. The states with sub-indices a are ancilla states of cloning machine which should be traced out to obtain the output state. The copies appear in the first two qubits in the r.h.s., actually the first two qubits are symmetric so that the reduced density matrices of copies are equal. The single qubit reduced density matrix of output can be calculated as

$$\rho_{\text{red.}}^{\text{out}} = \frac{1}{\sqrt{2}}\rho^{(in)} + \left(\frac{1}{2} - \sqrt{\frac{1}{8}}\right)I, \qquad (151)$$

where I is the identity matrix, and the input density matrix is $\rho^{(in)} = |\Psi\rangle\langle\Psi|$ defined in (149). We use fidelity to define the quality of the copies. The general definition of fidelity takes the form $F(\rho_1, \rho_2) = [\text{Tr}\sqrt{(\rho_1^{1/2}\rho_2\rho_1^{1/2})}]^2$ [43]. The value of F ranges from 0 to 1. A larger F corresponds to a higher fidelity. $F = 1$ means two density matrices are equal. We only consider about the pure input states, and the fidelity can be simplified as $F = {}^{(in)}\langle\Psi|\rho_{\text{red.}}^{(\text{out})}|\Psi\rangle^{(in)}$. The optimal fidelity of phase-covariant quantum cloning machine is obtained as

$$F_{optimal} = \frac{1}{2} + \sqrt{\frac{1}{8}}. \qquad (152)$$

As expected, this fidelity $F \approx 0.85$ is higher than the fidelity of UQCM $F \approx 0.83$.

In eavesdropping of the well-known BB84 quantum key distribution, because all four states $|0\rangle, |1\rangle, 1/\sqrt{2}(|0\rangle + |1\rangle), 1/\sqrt{2}(|0\rangle - |1\rangle)$ can be described by $|\Psi\rangle = \cos\theta|0\rangle + \sin\theta|1\rangle$. So, instead of the UQCM, we should at least use the cloning machine for equatorial qubits in eavesdropping. Actually in individual attack, we cannot do better than the cloning machine for equatorial qubits [33, 44]. The cloning machine presented in (150) can be used in analyzing the eavesdropping of other two mutually unbiased bases $1/\sqrt{2}(|0\rangle - |1\rangle), 1/\sqrt{2}(|0\rangle + |1\rangle), 1/\sqrt{2}(|0\rangle + i|1\rangle), 1/\sqrt{2}(|0\rangle - i|1\rangle)$.

The optimal fidelity of phase-covariant quantum cloning machine for qutrits was obtained by *D'Ariano* et al. [39] and *Cerf* et al. [45]:

$$F = \frac{5 + \sqrt{17}}{12}, \quad \text{for } d = 3. \qquad (153)$$

The more general case was studied by *Fan* et al. in [46]. Next we will review those results.

18 Phase-Covariant Cloning of Qudits

We study the quantum cloning of d-level states in the form

$$|\Psi\rangle^{(in)} = \frac{1}{\sqrt{d}} \sum_{j=0}^{d-1} e^{i\phi_j} |j\rangle, \tag{154}$$

where the arbitrary phase parameters $\phi_j \in [0, 2\pi), j = 0, \ldots, d-1$. A whole phase is not important, so we can assume $\phi_0 = 0$. The density operator of input state can be written as $\rho^{(in)} = \frac{1}{d} \sum_{jk} e^{i(\phi_j - \phi_k)} |j\rangle\langle k|$. In principle, for case 1 to M phase-covariant quantum cloning machine (with 1 input qudit and M output qudits), we can assume the most general cloning transformation take the following form:

$$U|j\rangle|Q\rangle = \sum_{\boldsymbol{k}}^{M} |\boldsymbol{k}\rangle|R_{j\boldsymbol{k}}\rangle, \tag{155}$$

where similar notations as in a 2-level quantum system are used, and $\boldsymbol{k} \equiv \{k_0, \ldots, k_{d-1}\}$. The summation $\sum_{\boldsymbol{k}}^{M}$ means take the summation over all possible values that satisfy the restriction $\sum_{j=0}^{d-1} k_j = M$. The quantum state $|\boldsymbol{k}\rangle$ is a normalized symmetric state with k_j states in $|j\rangle$. The ancilla states $|R_{j\boldsymbol{k}}\rangle$ are not necessarily orthogonal and normalized. The unitary relation means the restriction $\sum_{\boldsymbol{k}}^{M} \langle R_{j\boldsymbol{k}} | R_{j'\boldsymbol{k}} \rangle = \delta_{jj'}$. We remark here that as in UQCM, the output states are symmetrical so that every single qudit reduced density matrix of output is equal. Except the assumption that the output states are symmetric as in UQCM [11, 13, 16], the relation (160) is the most general cloning transformation. So, we can find the optimal phase-covariant cloning machine from (160). In UQCM, the property of Bloch vector invariance is often used. That is because we want the cloning machine to be universal, i.e., the quality of copies defined by fidelity between input state and the output states does not depend on the input states, for detailed argument see [12]. That means the output reduced density matrix can be written as a scalar form as in (151). However, we can find the optimal phase-covariant quantum cloning machine for a 2-level system also has the property of Bloch vector invariance. So, we still assume this property for a d-level phase cloning machine. This is a very useful relation. It implies the following relations considering the input state is in the form (154):

$$\langle R_{j\boldsymbol{k}'} | R_{j\boldsymbol{k}} \rangle \propto \delta_{\boldsymbol{k}\boldsymbol{k}'}, \tag{156}$$

$$\langle R_{j'\boldsymbol{k}'} | R_{j\boldsymbol{k}} \rangle \propto \delta_{k_0, k'_0} \cdots \delta_{k_{d-1}, k'_{d-1}} \delta_{k_j, k'_j + 1} \delta_{k_{j'}+1, k'_{j'}}, \tag{157}$$

where in $\cdots$, we do not have δ_{k_j,k'_j} and $\delta_{k_{j'},k'_{j'}}$, the same notations will be used later. The output single qudit reduced density matrix can be written as:

$$\rho_{\text{red}}^{(\text{out})} = \sum_{l=0}^{d-1} |l\rangle\langle l| \left[\frac{1}{d} \sum_{j=0}^{d-1} \sum_{k}^{M} \frac{k_l}{M} \langle R_{j\mathbf{k}}|R_{j\mathbf{k}}\rangle \right]$$

$$+ \sum_{j \neq j'} e^{i(\phi_j - \phi_{j'})} |j\rangle\langle j'| \left[\sum_{\mathbf{kk'}}^{M} \frac{\sqrt{k_j k'_{j'}}}{M} \langle R_{j'\mathbf{k'}}|R_{j\mathbf{k}}\rangle \delta_{k_0,k'_0} \cdots \right.$$

$$\left. \delta_{k_{d-1},k'_{d-1}} \delta_{k_j,k'_j+1} \delta_{k_{j'}+1,k'_{j'}} \right]. \quad (158)$$

The corresponding fidelity is written as

$$F = \frac{1}{d} + \frac{1}{d^2} \left[\sum_{\mathbf{kk'}}^{M} \frac{\sqrt{k_j k'_{j'}}}{M} \langle R_{j'\mathbf{k'}}|R_{j\mathbf{k}}\rangle \delta_{k_0,k'_0} \cdots \delta_{k_{d-1},k'_{d-1}} \delta_{k_j,k'_j+1} \delta_{k_{j'}+1,k'_{j'}} \right]. \quad (159)$$

Next, we shall turn our attention to 1 to 2 phase-covariant quantum cloning machine. Considering the restriction that the reduced density matrix of output should be written as a scalar form, and also considering the symmetric property of the input state (154) and the unitary restriction, we have the following phase-covariant quantum cloning transformation:

$$U|j\rangle|Q\rangle = \alpha|jj\rangle|R_j\rangle + \frac{\beta}{\sqrt{2(d-1)}} \sum_{l \neq j}^{d-1} (|jl\rangle + |lj\rangle)|R_k\rangle, \quad (160)$$

where α, β are real numbers, and $\alpha^2 + \beta^2 = 1$. Actually letting α, β be complex numbers does not improve the fidelity. $|R_j\rangle$ are orthonormal ancilla states. This is a simplified cloning transformation. Here we show this cloning transformation can be derived from (155) under the restrictions (156) and (157) for $M = 2$. The ancilla states $|R_j\rangle$ are orthogonal due to the relation (156). In the most general cloning transformation (155), the ancilla states should be denoted as $|R_{j\mathbf{k}}\rangle$. In the case of 1 to 2 cloning, for fixed j, j' if we choose $k_j = 2$, then $k_{j'} = 0$. According to relation (157), the ancilla state $|R_{j'\mathbf{k'}}\rangle$ can be identified with $|R_{j\mathbf{k}}\rangle$ when $k'_j = 1, k'_{j'} = 1$ with some normalization. So, we actually just need one ancilla state $|R_j\rangle$ to represent $|R_{j\mathbf{k}}\rangle$ and $|R_{j'\mathbf{k'}}\rangle$ if we have relations $k_j = 2, k_{j'} = 0; k'_j = 1, k'_{j'} = 1$. Without other states in (53), the cloning transformation (53) can achieve the optimal fidelity due to relation (159). In short, we can find the optimal cloning transformation from (53).

Substituting the input state (154) into the cloning transformation and tracing out the ancilla states, the output state takes the form

$$\rho^{(\text{out})} = \frac{\alpha^2}{d} \sum_j |jj\rangle\langle jj|$$
$$+ \frac{\alpha\beta}{d\sqrt{2(d-1)}} \sum_{j \neq l} e^{i(\phi_j - \phi_l)} [|jj\rangle(\langle jl| + \langle lj|) + (|jl\rangle + |lj\rangle)\langle ll|]$$
$$+ \frac{\beta^2}{2d(d-1)} \sum_{jj'} \sum_{l \neq j, j'} e^{i(\phi_j - \phi_{j'})}(|jl\rangle + |lj\rangle)(\langle lj'| + \langle j'l|). \quad (161)$$

Taking the trace over one qudit, we obtain the single qudit reduced density matrix of output:

$$\rho^{(\text{out})}_{\text{red.}} = \frac{1}{d}\sum_j |j\rangle\langle j| + \left(\frac{\alpha\beta}{d}\sqrt{\frac{2}{d-1}} + \frac{\beta^2(d-2)}{2d(d-1)}\right) \sum_{j \neq k} e^{i(\phi_j - \phi_k)}|j\rangle\langle k|. \quad (162)$$

The fidelity can be calculated as

$$F = \frac{1}{d} + \alpha\beta\frac{\sqrt{2(d-1)}}{d} + \beta^2\frac{d-2}{2d}. \quad (163)$$

Now, we need to optimize the fidelity under the restriction $\alpha^2 + \beta^2 = 1$. We find the optimal fidelity of 1 to 2 phase-covariant quantum cloning machine can be written as

$$F_{optimal} = \frac{1}{d} + \frac{1}{4d}(d - 2 + \sqrt{d^2 + 4d - 4}). \quad (164)$$

In case $d = 2, 3$, this result agrees with previous known results (152), (153), respectively. As expected, this optimal fidelity of phase-covariant quantum cloning machine is higher than the corresponding optimal fidelity of UQCM $F_{optimal} > F_{universal} = (d+3)/2(d+1)$. The optimal fidelity can be achieved when α, β take the following values,

$$\alpha = \left(\frac{1}{2} - \frac{d-2}{2\sqrt{d^2 + 4d - 4}}\right)^{\frac{1}{2}}, \quad \beta = \left(\frac{1}{2} + \frac{d-2}{2\sqrt{d^2 + 4d - 4}}\right)^{\frac{1}{2}}. \quad (165)$$

In case $d = 2$, the cloning transformation (53) recovers the previous result (150).

Thus we find the optimal phase-covariant quantum cloning machine for qudits (53), (165) and the corresponding optimal fidelity (164).

19 About Phase-Covariant Quantum Cloning Machines

Quantum measurements by mutually unbiased bases provide the optimal way of determining a quantum state. And the mutually unbiased bases have close relations with quantum cryptography. In d-dimensions, when d is prime, there are $d + 1$ mutually unbiased bases [47]. Except the standard basis $\{|0\rangle, |1\rangle, \ldots, |d-1\rangle\}$, the other d mutually unbiased bases take the form [47]

$$|\psi_t^l\rangle = \frac{1}{\sqrt{d}} \sum_{j=0}^{d-1} (\omega^t)^{d-j} (\omega^{-k})^{s_j} |j\rangle, \quad t = 0, \ldots, d-1, \tag{166}$$

where $s_j = j + \cdots + (d-1)$. And $l = 0, \ldots, d-1$ represent d mutually unbiased bases. The phase-covariant quantum cloning machine of qudits can clone all of these states equally well. So, we see if one uses d mutually unbiased bases (166) to perform quantum key distribution, the eavesdropper could use a phase-covariant quantum cloning machine to attack instead of the UQCM. If all $d+1$ mutually unbiased bases are used, we should use UQCM. However, there are no rigorous proofs about whether using a phase-covariant cloning machine in eavesdropping is optimal or not when d bases are used even though the cloning machine itself is optimal. We see the difference between d and $d+1$ mutually unbiased bases decreases when d becomes larger. Correspondingly, the gap between the fidelities of phase-covariant cloning machine and UQCM decreases when d becomes larger. When d is large enough, this gap becomes negligible.

In summary, we present in this paper the optimal phase-covariant quantum cloning machine for qudits (53), (165). The corresponding optimal fidelity (164) was found. In the $d = 2$ case, the results recover the previous result [18, 33]. In $d = 3$, the optimal fidelity agrees with the result obtained by D'Ariano et al. [39] and Cerf et al. [45].

20 Cerf's Asymmetric Quantum Cloning Machine

An arbitrary quantum pure state takes the form

$$|\psi\rangle = x_0|0\rangle + x_1|1\rangle, \quad \sum_j |x_j|^2 = 1. \tag{167}$$

A maximally entangled state is written as

$$|\Psi^+\rangle = \frac{1}{\sqrt{2}}(|00\rangle + |11\rangle). \tag{168}$$

We can write the complete quantum state of three particles as

$$|\psi\rangle_A |\Psi^+\rangle_{BC} = \frac{1}{2}\big[|\Psi^+\rangle_{AB}|\psi\rangle_C + (I \otimes X)|\Psi^+\rangle_{AB} X|\psi\rangle_C$$
$$+ (I \otimes Z)|\Psi^+\rangle_{AB} Z|\psi\rangle_C + (I \otimes XZ)|\Psi^+\rangle_{AB} XZ|\psi\rangle_C\big], \tag{169}$$

where I is the identity; X, Z are two Pauli matrices; and XZ is another Pauli matrix up to a whole factor i.

Denote the unitary transformation $U_{m,n} = X^m Z^n$, where $m, n = 0, 1$, and the relation (169) can be rewritten as

$$|\psi\rangle_A |\Psi^+\rangle_{BC} = \frac{1}{2} \sum_{m,n} (I \otimes U_{m,-n} \otimes U_{m,n}) |\Psi^+\rangle_{AB} |\psi\rangle_C. \tag{170}$$

Here we remark that $Z^{-1} = Z$ for a 2-level system. We write it in this form since this relation can be generalized directly to the general d-dimension system.

Now, suppose we do unitary transformation in the following form:

$$\sum_{\alpha,\beta} a_{\alpha,\beta} (U_{\alpha,\beta} \otimes U_{\alpha,-\beta} \otimes I) |\psi\rangle_A |\Psi^+\rangle_{BC}$$

$$= \frac{1}{2} \sum_{\alpha,\beta,m,n} (U_{\alpha,\beta} \otimes U_{\alpha,-\beta} U_{m,-n} \otimes U_{m,n}) |\Psi^+\rangle_{AB} |\psi\rangle_C$$

$$= \sum_{m,n} b_{m,n} (I \otimes U_{m,-n} \otimes U_{m,n}) |\Psi^+\rangle_{AB} |\psi\rangle_C, \tag{171}$$

where we defined

$$b_{m,n} = \frac{1}{2} \sum_{\alpha,\beta} (-1)^{\alpha n - \beta m} a_{\alpha,\beta}. \tag{172}$$

The amplitudes should be normalized $\sum_{\alpha,\beta} |a_{\alpha,\beta}|^2 = \sum_{m,n} |b_{m,n}|^2 = 1$. This is actually the asymmetric quantum cloning machine introduced by *Cerf* [48]. We find the quantum states of A and C now take the form

$$\rho_A = \sum_{\alpha,\beta} |a_{\alpha,\beta}|^2 U_{\alpha,\beta} |\psi\rangle \langle \psi| U^\dagger_{\alpha,\beta}, \tag{173}$$

$$\rho_C = \sum_{m,n} |b_{m,n}|^2 U_{m,n} |\psi\rangle \langle \psi| U^\dagger_{m,n}. \tag{174}$$

The quantum state of A is related to the quantum state C by the relationship between $a_{\alpha,\beta}$ and $b_{m,n}$.

The quantum state ρ_A is the original quantum state after the quantum cloning. The quantum state ρ_C is the copy.

Now, let's see a special case:

$$b_{0,0} = 1, \quad b_{0,1} = b_{1,0} = b_{1,1} = 0. \tag{175}$$

Correspondingly, we can choose

$$a_{0,0} = a_{0,1} = a_{1,0} = a_{1,1} = \frac{1}{2}. \tag{176}$$

So, we know the quantum states of A and C have the form

$$\rho_A = \frac{1}{2}I, \quad \rho_C = |\psi\rangle\langle\psi|. \tag{177}$$

As a quantum cloning machine, this means the original quantum state in A, $|\psi\rangle$, is completely destroyed,

This result can be generalized to a d-dimension system directly. Define the maximally entangled state as $|\Psi^+\rangle = \frac{1}{\sqrt{d}}\sum_j |jj\rangle$, and define also an arbitrary quantum state $|\psi\rangle = \sum_k x_k|k\rangle$ with normalization $\sum_j |x_j|^2 = 1$, and define operators $X|j\rangle = |j+1 \mathrm{mod} d\rangle$, $Z|j\rangle = \omega^j|j\rangle$, $\omega = e^{2\pi i/d}$, and $U_{\alpha,\beta} = X^\alpha Z^\beta$, $\alpha,\beta,=0,\ldots,d-1$. By straightforward calculations, we can find the following relation:

$$|\psi\rangle_A |\Psi^+\rangle_{BC} = \frac{1}{d}\sum_{m,n}(I \otimes U_{m,-n} \otimes U_{m,n}|\Psi^+\rangle_{AB}|\psi\rangle_C. \tag{178}$$

A general unitary transformation can be described as follows:

$$\sum_{\alpha,\beta} a_{\alpha,\beta}(U_{\alpha,\beta} \otimes U_{\alpha,-\beta} \otimes I)|\psi\rangle_A|\Psi^+\rangle_{BC}$$

$$= \frac{1}{d}\sum_{m,n,\alpha,\beta} a_{\alpha,\beta}\omega^{\alpha n - \beta m}(I \otimes U_{m,-n} \otimes U_{m,n})(U_{\alpha,\beta} \otimes U_{\alpha,-\beta})|\Psi^+\rangle_{AB}|\psi\rangle_C$$

$$= \sum_{m,n} b_{m,n}(I \otimes U_{m,-n} \otimes U_{m,n})|\Psi^+\rangle_{AB}|\psi\rangle_C. \tag{179}$$

where we define (see also [49]; we use a different method to obtain these results),

$$b_{m,n} = \frac{1}{d}\sum_{\alpha,\beta} \omega^{\alpha n - \beta m} a_{\alpha,\beta}, \tag{180}$$

and also the relations $U_{\alpha,-\beta}U_{m,-n} = \omega^{\alpha n - \beta m}U_{m,-n}U_{\alpha,-\beta}$ and $U_{\alpha,\beta} \otimes U_{\alpha,-\beta}|\Psi^+\rangle = |\Psi^+\rangle$ are used. As in a 2-level system, we still have the following relations:

$$\rho_A = \sum_{\alpha,\beta} |a_{\alpha,\beta}|^2 U_{\alpha,\beta}|\psi\rangle\langle\psi|U^\dagger_{\alpha,\beta}, \quad \rho_C = \sum_{m,n} |b_{m,n}|^2 U_{m,n}|\psi\rangle\langle\psi|U^\dagger_{m,n}, \tag{181}$$

but m,n,α,β take values between $0, d-1$. Since $b_{m,n}$ is completely determined by $a_{\alpha,\beta}$, by adjusting the parameters $a_{\alpha,\beta}$ of unitary transformations, we can control the quantum state ρ_C. This is Cerf's asymmetric quantum cloning machine [49]. If we know nothing about the quantum state $|\psi\rangle$, we can assume $a_{\alpha,\beta} = \frac{\eta}{d}$, $\alpha,\beta \neq 0$. Because of the normalization, we know $a_{0,0} = 1 - (d^2 - $

1)$\frac{\eta^2}{d^2}$. Similarly, we can assume $b_{m,n} = \frac{\lambda}{d}, m, n \neq 0, b_{0,0} = 1 - (d^2 - 1)\frac{\lambda^2}{d^2}$, so now we have the density operators of A and C as follows:

$$\rho_A = (1 - \eta^2)|\psi\rangle\langle\psi| + \frac{\eta^2}{d}I, \quad \rho_C = (1 - \lambda^2)|\psi\rangle\langle\psi| + \frac{\lambda^2}{d}I. \tag{182}$$

The relationship between $a_{\alpha,\beta}$ and $b_{m,n}$ shows that we should have $\lambda^2 + \eta^2 + 2\lambda\eta/d = 1$. Considering the cloning machine is optimal, we have Cerf's no-cloning theorem:

$$\lambda^2 + \eta^2 + 2\lambda\eta/d \geq 1. \tag{183}$$

The experimental realization of the asymmetric quantum cloning machine was recently made by Pan's group [50].

21 Duan and Guo Probabilistic Quantum Cloning Machine

While the previously mentioned quantum cloning machines can always succeed, at the same time, the copies cannot be perfect. *Duan* and *Guo* [51, 52] proposed a different quantum cloning machine: while the copying task can succeed with probability, but if it is successful, we can always obtain perfect copies. This kind of quantum cloning machine is called a probabilistic quantum cloning machine.

The simplest case for probabilistic quantum cloning machine is to copy two linearly independent states $S = \{|\Psi_0\rangle, |\Psi_1\rangle\}$ [51]. The cloning transformation can be proposed as:

$$U(|\Psi_0\rangle|\Sigma\rangle|m_p\rangle) = \sqrt{\eta_0}|\Psi_0\rangle|\Psi_0\rangle|m_0\rangle + \sqrt{1-\eta_0}|\Phi^0_{ABP}\rangle,$$
$$U(|\Psi_1\rangle|\Sigma\rangle|m_p\rangle) = \sqrt{\eta_1}|\Psi_1\rangle|\Psi_1\rangle|m_1\rangle + \sqrt{1-\eta_1}|\Phi^1_{ABP}\rangle, \tag{184}$$

where $|m_p\rangle, |m_0\rangle), |m_1\rangle$ are ancillary states. The measurements are performed in these states. When the measurements are $|m_0\rangle$ or $|m_1\rangle$, we know the states $S = \{|\Psi_0\rangle, |\Psi_1\rangle\}$ are copied perfectly. Otherwise, the cloning task fails. The probabilities of success are η_0 and η_1 for states $|\Psi_0\rangle$ and $|\Psi_1\rangle$, respectively. If we let $\eta_0 = \eta_1 = \eta$, we know that

$$\eta \leq \frac{1}{1 + |\langle\Psi_0|\Psi_1\rangle|}. \tag{185}$$

This is also a no-cloning theorem: Only orthogonal states can be cloned perfectly. And the optimal probabilistic quantum cloning is to let $\eta = 1/(1 + |\langle\Psi_0|\Psi_1\rangle|)$. It is also related with the problem of how to distinguish nonorthogonal quantum states.

The more complicated case is to copy a set of linearly independent states $S = \{|\Psi_0\rangle, |\Psi_1\rangle \ldots, |\Psi_n\rangle\}$. For optimal case, we need to analyze the matrix $X_{ij} = \langle\Psi_i|\Psi_j\rangle$ [52]. The result is the Duan–Guo bound to distinguish linearly independent quantum states.

22 A Brief Summary

Quantum cloning is an important subject in quantum information processing. It is closely related to quantum key distributions, quantum state estimation, quantum states distinguishability, etc. And on the other hand, quantum cloning is also an independent topic, and has its own aim and motivation. The author reviews several topics in quantum cloning machines and mainly reviews the results which were obtained by the author himself and his colleagues.

Acknowledgements

This work is supported in part by the "Bairen" program and a QuIP grant of CAS. The author would like to thank V. Bužek, G. M. D'Ariano, M. Hayashi, T. Hiroshima, H. Imai, H. Kobayashi, K. Matsumoto, T. Shimono, Y. Tsuda, M. Wadati, X. B. Wang and G. Weihs for encouragement and useful discussions in studying these topics. He also would like to thank T. Sakuragi for help when he was on the Imai ERATO project.

References

[1] C. H. Bennett, P. W. Shor: Quantum information theory, IEEE Trans. Inform. Theory **44**, 2724 (1998)
[2] M. A. Nielsen, I. C. Chuang: *Quantum Computation and Quantum Information* (Cambridge Univ. Press, Cambridge 2000)
[3] W. K. Wootters, W. H. Zurek: A single quantum cannot be cloned, Nature **299**, 802 (1982)
[4] D. Dieks: Communication by EPR devices, Phys. Lett. A **92**, 271 (1982)
[5] H. Barnum, C. M. Caves, C. A. Fuchs, R. Jozsa, B. Schumacher: Noncommuting mixed states cannot be broadcast, Phys. Rev. Lett. **76**, 2818 (1996)
[6] T. Mor: No cloning of orthogonal states in composite systems, Phys. Rev. Lett. **80**, 3137 (1998)
[7] M. Koashi, N. Imoto: No-clonig theorem of entangled states, Phys. Rev. Lett. **81**, 4264 (1998)
[8] M. Hayashi, K. Iwama, H. Nishimura, R. Raymond, S. Yamashita: Quantum network coding quant-ph/0601088
[9] V. Scarani, S. Iblisdir, N. Gisin, A. Acin: Quantum cloning, Rev. Mod. Phys. **77**, 1225 (2005)
[10] N. J. Cerf, J. Fiurasek: Optical quantum cloning – a review quant-ph/0512172
[11] V. Bužek, M. Hillery: Quantum copying: Beyond the no-cloning theorem, Phys. Rev. A **54**, 1844 (1996)
[12] D. Bruß, D. P. DiVincenzo, A. Ekert, C. A. Fuchs, C. Macchiavello, J. A. Smolin: Optimal universal and state-depending quantum cloning, Phys. Rev. A **57**, 2368 (1998)
[13] N. Gisin, S. Massar: Optimal quantum cloning machines, Phys. Rev. Lett. **79**, 2153 (1997)

[14] D. Bruß, A. Ekert, C. Macchiavello: Optimal universal quantum cloning and state estimation, Phys. Rev. Lett. **81**, 2598 (1998)
[15] V. Bužek, H. Hillery: Universal optimal cloning of arbitrary quantum states: From qubits to quantum registers, Phys. Rev. Lett. **81**, 5003 (1998)
[16] R. Werner: Quantum cloning of pure states, Phys. Rev. A **58**, 1827 (1998)
[17] M. Keyl, R. Werner: Optimal cloning of pure states, testing single clones, J. Math. Phys. **40**, 3283 (1999)
[18] H. Fan, K. Matsumoto, M. Wadati: Quantum cloning machines of a d-level system, Phys. Rev. A **64**, 064301 (2001)
[19] P. Zanardi: Quantum cloning in d dimensions, Phys. Rev. A **58**, 3484 (1998)
[20] S. Albeverio, S. M. Fei: A remark on the optimal cloning of an n-level quantum system, Eur. Phys. J. B **14**, 669 (2000)
[21] H. Fan, K. Matsumoto, X. B. Wang, H. Imai, M. Wadati: A universal cloner allowing the input to be arbitrary states in symmetric subspaces quant-ph/0107113 (unpublished)
[22] N. Gisin, S. Popescu: Spin flips and quantum information for antiparallel spins, Phys. Rev. Lett. **83**, 432 (1999)
[23] R. Derka, V. Bužek, A. Ekert: Universal algorithm for optimal estimation of quantum states from finite ensembles via realizable generalized measurment, Phys. Rev. Lett. **80**, 1571 (1998)
[24] D. Bruß, C. Macchiavello: Optimal state estimation for d-dimensional quantum systems, Phys. Lett. A **253**, 249 (1999)
[25] A. Peres, W. K. Wootters: Optimal detection of quantum information, Phys. Rev. Lett. **66**, 1119 (1991)
[26] C. A. Fuchs, A. Peres: Quantum state disturbance versus information gain: Uncertainty relations for quantum information, Phys. Rev. A **53**, 2038 (1996)
[27] H. Fan, G. Weihs, K. Matsumoto, H. Imai: Cloning of symmetric d-level photonic states in physical systems, Phys. Rev. A **66**, 024307 (2002)
[28] V. Bužek, S. Braunstein, M. Hillery, D. Bruß: Quantum copying: A network, Phys. Rev. A **56**, 3446 (1997)
[29] C. Simon, G. Weihs, A. Zeilinger: Optimal quantum cloning via stimulated emmision, Phys. Rev. Lett. **84**, 2993 (2000)
[30] J. Kempe, C. Simon, G. Weihs: Optimal photon cloning, Phys. Rev. A **62**, 032302 (2000)
[31] A. Lamas-Linaresm, C. Simon, J. C. Howell, D. Bouwmeester: Experimental quantum cloning of single photons, Science **296**, 712 (2002)
[32] H. Fan, B. Y. Liu, K. J. Shi: Quantum cloning of identical mixed states quant-ph/0601017
[33] D. Bruß, M. Cinchetti, G. M. D'Ariano, C. Macchiavello: Phase-covariant quantum cloning, Phys. Rev. A **62**, 012302 (2000)
[34] H. Fan, K. Matsumoto, X. B. Wang, M. Wadati: Quantum cloning machines for equatorial qubits, Phys. Rev. A **65**, 012304 (2002)
[35] V. Bužek, M. Hillery, R. F. Werner: Optimal manipulations with qubits: Universal-not gate, Phys. Rev. A **60**, R2626 (1999)
[36] L. C. Kwek, C. H. Oh, X. B. Wang, Y. Yeo: Buzek-Hillery cloning revisited using bures metric and trace norm, Phys. Rev. A **62**, 052313 (2000)
[37] A. Peres: Separability criterion for density matrices, Phys. Rev. Lett. **77**, 1413 (1996)

[38] M. Horodecki, P. Horodecki, R. Horodecki: Separability of mixed states: Necessary and sufficient conditions, Phys. Lett. A **223**, 1 (1996)
[39] G. M. D'Ariano, P. L. Presti: Optimal nonuniversally covariant cloning, Phys. Rev. A **64**, 042308 (2001)
[40] J. F. Du, D. Durt, P. Zou, H. Li, L. C. Kwek, C. H. Lai, C. H. Oh, A. Ekert: Experimental quantum cloning with prior partial information, Phys. Rev. Lett. **94**, 040505 (2005)
[41] G. M. D'Ariano, C. Macchiavello: Optimal phase-covariant cloning with qubits and qutrits, Phys. Rev. A **67**, 042306 (2003)
[42] T. Durt, J. F. Du: Characterization of low-cost one-to-two qubit cloning, Phys. Rev. A **69**, 062316 (2004)
[43] R. Josza: Fidelity for mixed quantum states, J. Mod. Opt. **41**, 2314 (1994)
[44] C. A. Fuchs, N. Gisin, R. B. Griffiths, C. S. Niu, A. Peres: Optimal eavesdropping in quantum cryptography: Information bound and optimal strategy, Phys. Rev. A **56**, 1163 (1997)
[45] N. J. Cerf, T. Durt, N. Gisin: Cloning a qutrit, J. Mod. Opt. **49**, 1355 (2002)
[46] H. Fan: Quantum cloning of mixed states in symmetric subspaces, Phys. Rev. A **67**, 022317 (2003)
[47] S. Bandyopadhyay, P. O. Boykin, V. Roychowdhury, F. Vatan: A new proof for the existence of mutually unbiased bases, Algorithmica **34**, 512 (2002)
[48] N. J. Cerf: Pauli cloning of a quantum bit, Phys. Rev. Lett. **84**, 4497 (2000)
[49] N. J. Cerf: Asymmetric quantum cloning in any dimension, J. Mod. Opt. **47**, 187 (2000)
[50] Z. Zhao, A. N. Zhang, X. Q. Zhou, Y. A. Chen, C. Y. Lu, A. Karlsson, J. W. Pan: Experimental realization of optimal asymmetric cloning and telecloning via partial teleportation, Phys. Rev. Lett. **94**, 030501 (2005)
[51] L. M. Duan, G. C. Guo: A probabilistic cloning machine for replicating two non-orthogonal states, Phys. Lett. A **243**, 261 (1998)
[52] L. M. Duan, G. C. Guo: Probabilistic cloning and identification of linearly independent quantum states, Phys. Rev. Lett. **80**, 4999 (1998)

Index

asymmetric quantum cloning machine, 105

Bloch vector, 69

cloning machine, 65

fidelity, 64

no-cloning theorem, 63

Peres–Horodecki criterion, 93
phase covariant, 81
probabilistic quantum cloning machine, 106

Schwinger representation, 75
shrinking factor, 67

UQCM, 63, 65

Entanglement and Quantum Error Correction

Tohya Hiroshima and Masahito Hayashi

ERATO Quantum Computation and Information Project, Japan Science and Technology Agency 201 Daini Hongo White Bldg. 5-28-3, Hongo, Bunkyo-ku, Tokyo 113-0033, Japan
{tohya,masahito}@qci.jst.go.jp

Abstract. Quantum entanglement is a fundamental topic in quantum information that has various aspects. In order to discover its essence, we studied this topic from various viewpoints.

1 Introduction

Quantum entanglement is well acknowledged to be a physical resource in various types of quantum information processing such as quantum dense coding [1] and quantum teleportation [2]. The latter is one of the basic building blocks of the quantum repeater that is a key to long-distance quantum communication. Entanglement also has significance in computer science. Error-correcting codes provide the fundamental framework of fault-tolerant quantum computation [3]. In the quantum version of zero-knowledge proof, quantum entanglement is indispensable. In a nutshell, quantum entanglement is fundamental in quantum information science and technology. Therefore, in this project we have launched the theoretical study of quantum entanglement from various viewpoints. We focus on the following topics:

1. entanglement distillation
2. error correction
3. basic characteristics of bipartite entanglement
4. SLOCC convertibility
5. protocol assisted by multipartite entangled state

Each component of the following sections was first written by researchers who were responsible to the corresponding work and were edited subsequently by Hiroshima and Hayashi.

2 Entanglement Distillation

To obtain the seemingly magical powers of quantum information processing, it is desirable to share maximally entangled states, which makes worthwhile the study of entanglement distillation or the production of a maximally entangled state from given partially entangled states through local operations and classical communications (LOCC). In particular, if the initial state is pure, this protocol is called entanglement concentration. Our results on this topic are classified into three types:

1. exponents of optimal concentration
2. universal entanglement concentration
3. entanglement in Boson–Fock space
4. computation of distillable entanglement of a certain class of bipartite mixed states

2.1 Background of Concentration

If the initial state is pure, the known results are summarized as follows. As proven by *Bennett* et al. [4], when $n\,(\gg 1)$ copies of the pure state $|\phi\rangle$ are shared by Alice and Bob, whose respective Hilbert spaces are denoted by $\mathcal{H}_A$ and $\mathcal{H}_B$, respectively, they can produce, through local operations, $2^{nH(\mathbf{p}_\phi)}$-dimensional maximally entangled states with the probability 1 asymptotically. Here, $\mathbf{p}_\phi = (p_{1,\phi}, \ldots, p_{d,\phi})$ are the Schmidt coefficients of $|\phi\rangle$, (i.e., $|\phi\rangle = \sum_i \sqrt{p_{i,\phi}} |e_{i,A}\rangle|e_{i,B}\rangle$), with $p_{1,\phi} \geq p_{2,\phi} \geq \ldots \geq p_{d,\phi}$, $H(\mathbf{p})$ is the Shannon entropy of $\mathbf{p}$, and k-dimensional maximally entangle state means the state such that $\frac{1}{\sqrt{k}} \sum_i |f_{i,A}\rangle|f_{i,B}\rangle$. (Without loss of generality, $\mathcal{H}_A = \mathcal{H}_B = d$ is assumed.)

2.2 Exponents of Optimal Concentration

We treated the case where the initial state $|\phi\rangle$ is known. By using the method of types in information theory (Fig. 1), we analyzed this problem in terms of the error rate in the following two settings [5, 6]:

(i) **Probabilistic Setting.** We gave the number of Bell states distilled per copy, as a function of an error exponent, which represents the rate of decrease in failure probability as n tends to infinity. The formula fills the gap between the least upper bound of distillable entanglement in probabilistic concentration, which is the well-known entropy of entanglement, and the maximum attained in deterministic concentration.

(ii) **Deterministic Setting.** In addition to the probabilistic argument, we considered another type of entanglement concentration scheme, where the initial state is deterministically transformed into a (possibly mixed) final state whose fidelity to a maximally entangled state of a desired size converges to 1 in the asymptotic limit. We showed that the same formula as in the probabilistic argument is valid for the argument on fidelity by replacing the success probability with the fidelity.

Furthermore, we also discussed entanglement yield when optimal success probability or optimal fidelity converges to zero in the asymptotic limit (strong converse), and gave the explicit formulae for those cases.

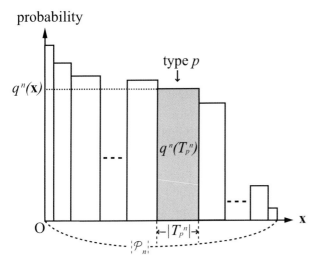

Fig. 1. Analysis of entanglement concentration by the method of types

2.3 Universal Entanglement Concentration

We treated the case where $|\phi\rangle$ is unknown, and the *perfect* (not approximate) entangled state is needed. We proposed a protocol $\{C_*^n\}$ that produces a $2^{nH(\mathbf{p})_\phi}$-dimensional maximally entangled state asymptotically with the probability 1 even in this difficult setting. This kind of protocol is called a *universal distortion-free entanglement concentration* [5].

(i) **Advantage Over the Previous Protocol.** Performing the entanglement concentration protocol by *Bennett* et al. [4] after the estimation of the Schmidt basis of $|\phi\rangle$ by measuring m ($n \gg m \gg 1$) copies of $|\phi\rangle$, we can produce the *approximate* $2^{nH(\mathbf{p}_\phi)}$-dimensional maximally entangled state, and call this type protocol *universal approximate entanglement concentration*. In this way, however, the final state is not quite a maximally entangled state, because of the errors in the estimation. The difficulty of construction of a universal distortion-free concentration mainly comes from the lack of knowledge about the Schmidt basis, and our protocol overcomes this difficulty. Indeed, if the Schmidt basis is known, the protocol by *Bennett* et al. [4] is successfully applied to produce perfect maximally entangled states. This difficulty is overcome by focusing on the symmetry of the n-tensored pure state $|\phi\rangle^{\otimes n}$.

(ii) **Optimality.** It was also proven that our protocol is nonasymptotically optimal for all universal distortion-free concentrations, as well as asymptotically optimal for all universal approximate concentrations, in terms of failure probability and the average dimension of the outcoming maximally entangled state. Remarkably, our protocol uses only local operations and no classical communication, and still achieves optimality in such strong senses.

(iii) **Relation to State Estimation.** We also studied universal concentration from the theory of state estimation, as the logarithm of the dimension of output maximally entangled state gives a natural estimate of $H(\mathbf{p}_\phi)$. It turns out that our universal concentration protocol gives a better estimate of the entanglement measure $H(\mathbf{p}_\phi)$ than any other *global* measurements. This argument gives another proof of optimality of our universal concentration protocol.

2.4 Entanglement in Boson–Fock Space

Quantum states in photon number states are useful resources for various tasks of quantum information. In the following, we focus on entanglement distillation in the Boson–Fock space, whose mathematical framework was built up many years ago. However, the entanglement properties of states in such a space was not throughly investigated until recently, when the papers by *Duan* and *Simon* [7, 8] appeared.

We mainly did the following studies under such a topic:

(i) **Detecting the Inseparability and Distillability.** A number of criteria on the inseparability and distillability for the multi-mode Gaussian states were naturally drawn.[1] We showed for the first time that the 2-mode squeezed states in dephasing channel [9] are always inseparable. We finally gave an explicit formula for the state in a subspace of a global Gaussian state. This formula, together with the known results for Gaussian states, gave the criteria for the inseparability and distillability in a subspace of the global Gaussian state [10].

(ii) **A Theorem for Beam Splitter Entangler.** It had been conjectured that the entanglement output state from a beam splitter (Fig. 2) requires the nonclassicality in the input state [11, 12]. Here we gave a proof for this conjecture. The proof is very simple: Given a classical state, it must be positive in certain P-representation. Then after an arbitrary rotation transformation, i.e., after it passes through any linear optical devices, it must be still positive in another P-representation. So, after any linear optical devices, it must be still classical therefore un-entangled.

(iii) **Properties of Beam Splitter Entangler.** An explicit formula was given for quantifying entanglement in the output state of a beam splitter, given the squeezed vacuum states input in each mode. For the general Gaussian states input, an explicit formula is given as the necessary

---

[1] In order to obtain these criteria, we gave an explicit formula for the partial transposition (PT) operation for the continuous variable states in Fock space, and gave the necessary and sufficient condition for the positivity of Gaussian operators.

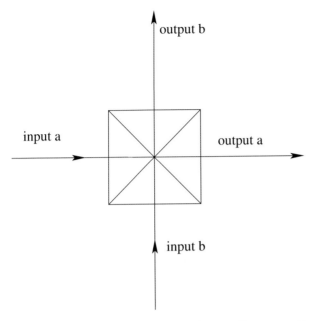

Fig. 2. A schematic diagram for the beam splitter operation. Both the input and the output are two mode states. The different mode is distinguished by the propagating direction of the field

and sufficient condition for the inseparability of the output state from a beam splitter [13].

2.5 Computation of Distillable Entanglement of a Certain Class of Bipartite Mixed States

A maximally correlated state on the composite Hilbert space $\mathcal{H}_A \otimes \mathcal{H}_B$ of the form

$$\rho_{MC} = \sum_{j,k=1}^{\min\{d_A,d_B\}} \alpha_{jk} |jj\rangle \langle kk| \qquad (1)$$

has significant entanglement properties. Here, $d_{A(B)} = \dim \mathcal{H}_{A(B)}$, and $|jj\rangle$ denotes $|j_A\rangle \otimes |j_B\rangle$ with $|j_{A(B)}\rangle$ being an orthonormal basis in $\mathcal{H}_{A(B)}$. A salient feature is that the distillable entanglement [14] E_D of the maximally correlated state is given by the following simple formula [15]:

$$E_D(\rho_{MC}) = I_A(\rho_{MC}) = I_B(\rho_{MC}), \qquad (2)$$

where $I_{A(B)}(\rho) = S(\rho_{A(B)}) - S(\rho)$, $\rho_{A(B)} = \text{Tr}_{B(A)} \rho$, and $S(\rho) = -\text{Tr} \rho \log_2 \rho$ denotes the von Neumann entropy of ρ.

We showed that a class of bipartite mixed states composed of *simultaneously Schmidt decomposable* vectors $|\psi_\alpha\rangle$,

$$\rho = \sum_{\alpha,\beta=1}^{l} a_{\alpha\beta} |\psi_\alpha\rangle \langle\psi_\beta|, \tag{3}$$

can be cast in the maximally correlated states by local unitary transformation [14].

Any two independent generalized Bell states in a $d \otimes d$ system are simultaneously Schmidt decomposable. Therefore, the generalized Bell diagonal states of rank 2,

$$\rho = \lambda \left|\psi_{nm}^{(d)}\right\rangle \left\langle\psi_{nm}^{(d)}\right| + (1-\lambda) \left|\psi_{n'm'}^{(d)}\right\rangle \left\langle\psi_{n'm'}^{(d)}\right|, \tag{4}$$

with $0 < \lambda < 1$, take the form of maximally correlated states by local unitary transformation. Here, the generalized Bell states in a $d \otimes d$ system are defined as

$$\left|\psi_{nm}^{(d)}\right\rangle = \left(Z^n \otimes X^{-m}\right) \left|\psi_+^{(d)}\right\rangle, \tag{5}$$

for $n, m = 0, 1, \cdots, d-1$ with $\left|\psi_+^{(d)}\right\rangle = d^{-1/2} \sum_{k=0}^{d-1} |k\rangle \otimes |k\rangle$. In (5), unitary matrices X and Z are given by $X|k\rangle = |k-1 \pmod d\rangle$ and $Z|k\rangle = \omega_d^k |k\rangle$ for $k = 0, 1, \cdots, d-1$ with $\omega_d = \exp\left(2\pi\sqrt{-1}/d\right)$. The distillable entanglement of the state (4) is given by $E_D(\rho) = \log_2 d - S(\rho)$. This is the generalization of the known result that the distillable entanglement of Bell diagonal states of rank 2 is given by $1 - S(\rho)$ [16]. More generally, the mixed state $\rho = \sum_{\alpha,\beta=1}^{2} a_{\alpha\beta} \left|\psi_{n_\alpha m_\alpha}^{(d)}\right\rangle \left\langle\psi_{n_\beta m_\beta}^{(d)}\right|$ also takes the form of maximally correlated states by local unitary transformation and the distillable entanglement is given by the formula (2) [14].

3 Quantum Error Correction

In a quantum system, any noisy process is described by a quantum channel which gives the state evolution. In information theory, "coding" is known as a method to protect information from noise. In this method, by choosing a suitable subset (which is called a code), we can recover our information from the signal blurred by the noise. In particular, when we apply this method to protecting the quantum state, the method is called quantum error correction.

In fact, quantum error correction is closely related to entanglement distillation because noise of the entangled state can be regarded as noise of the channel. Using this correspondence, we studied entanglement distillation from the viewpoint of quantum error correction.

3.1 Mathematical Formulation of Quantum Channel

As a mathematical aspect, any quantum channel is described by trace-preserving completely positive (TP-CP) maps. As a simple noise model, we usually treat the depolarizing channel in the two-dimensional space, which is given as follows:

$$\rho \mapsto \mathcal{A}(\rho) = (1-3p)\rho + p(\sigma_1\rho\sigma_1 + \sigma_2\rho\sigma_2 + \sigma_3\rho\sigma_3), \tag{6}$$

where σ_i is the Pauli matrix. We also investigate QECC under the following Pauli channel known as a more general model in the two-dimensional space:

$$\mathcal{A}(\rho) = \sum_{i=0}^{3} p(i)\sigma_i\rho\sigma_i, \tag{7}$$

where σ_0 is the identity matrix I and $p(i)$ is a probability distribution. However, a TP-CP map does not necessarily have the above form, and we need to take into account a more general TP-CP map. We also discussed the problem of find a TP-CP map satisfying a given condition. We focused on a kind of approximation problem via a quantum channel, which is called quantum channel resolvability.

3.2 Background of Information Theory and Coding Theory

In the classical information theory (Shannon theory), it is known that in Shannon's channel coding theorem there exists a code satisfying:

(i) The transmission rate is close to a certain number called the channel capacity.
(ii) The error probability, i.e., the probability that the recovered message is different from the true message, is close to 0.

However, it is very hard to find a code satisfying the conditions **(i)** and **(ii)** and the following:

(iii) The decoding time is small.

In coding systems, efficient decoding methods are needed. Therefore, researchers in coding theory usually limit their code to a code having an algebraic structure. Such a code is called an algebraic code and has a simpler structure than other codes. Roughly speaking, its complexity also increases with the length n of the code, i.e., the number of bits used for the code. However, when we fix the the transmission rate R of our code, there is the following relation between the error probability and the code-length n. Let $P_{n,k}^*$ be the minimum of the error probability over all possible choices of a code $\mathcal{C}$ with $\log|\mathcal{C}| \geq k$ and decode, where $|\mathcal{C}|$ is the number of possible messages to be sent with $\mathcal{C}$. Shannon's channel coding theorem says that if the transmission rate R is less than the capacity $C(W)$ of the channel W, then $P^*_{n,Rn} \to 0$. A stronger result has been long known in information theory. There exists a function $E_{\rm r}(R,W)$, which is called the reliability function of W, such that

$$\liminf_{n\to\infty} -\frac{1}{n}\log P^*_{n,Rn} = E_{\rm r}(R,W), \tag{8}$$

i.e., $P^*_{n,Rn} \approx \exp[-nE_{\rm r}(R,W)]$, and

$$E_{\rm r}(R,W) > 0 \quad \text{if} \quad R < C(W), \tag{9}$$

which gives the above-mentioned relation between the error probability and the code length. Determination of the limit of $-\frac{1}{n}\log P^*_{n,Rn}$, say, $E_{\rm cl}(R,W)$, is one of the central issues in classical information theory, which remains unsolved for low rates. We can say that there is no reason to employ codes of rates near the capacity exclusively, because the lower the value of R, the greater the value of $E_{\rm cl}(R,W)$, and hence the less $P^*_{n,Rn} \approx \exp[-nE_{\rm cl}(R,W)]$ is exponentially. The primary motivation of this work is the natural problem of establishing the function corresponding to $E_{\rm cl}(R,W)$ or a suitable lower bound of $E_{\rm cl}(R,W)$ in quantum settings.

3.3 Exponential Evaluation of Quantum Error Correcting Codes

In the present work, we assumed the memoryless operation of the Pauli channel $\mathcal{A}$ (7) in the sense that the channel $\mathcal{A}$ acts as $\mathcal{A}^{\otimes n}(\rho)$ on a state ρ on the n-qubit system. We proved that there exists a quantum error correcting code (QECC) with the length n and the rate R whose fidelity[2] is greater than $1 - \exp[-nE(R,\mathcal{A}) + o(n)]$ for some function $E(R,\mathcal{A})$ when they are used on quantum channels $\mathcal{A}$, i.e., the highest fidelity of QECCs of length n and rate R was proven to be lower-bounded by $1 - \exp[-nE(R,\mathcal{A}) + o(n)]$ [17]. The $E(R,\mathcal{A})$ is positive below some threshold R_0, which implies R_0 is a lower bound on the quantum capacity. We also obtained the same results in the d-dimensional case, i.e., proved the existence such a code for generalized Pauli channels [17].

3.3.1 Extensions

The above result was strengthened in the following three directions:

(i) **Fidelity of QECCs on General Memoryless Quantum Channel.** We extended the above results to the memoryless operation of the general channel $\mathcal{A}$. For this purpose, we defined the function $E(R,\mathcal{A})$ to be the function $E(R,\mathcal{A}')$, where $\mathcal{A}'(\rho) = \sum_{i=0}^{3} P_{\mathcal{A}}(i)\sigma_i\rho\sigma_i$ and we associate a probability distribution $P_{\mathcal{A}}$ with $\mathcal{A}$ in a certain manner. This function $E(R,\mathcal{A}')$ plays the same role as the above [18].

---

[2] Fidelity is a measure between two quantum states. If they coincide, it equals 1. If they are completely different, it equals 0.

(ii) **Fidelity of QECCs on Channels with Correlation.** An extension of the treated channel class to channels with classical Markovian memory is done [19]. This result gives evidence, from an information-theoretic viewpoint, that the standard quantum error correction schemes work reliably even in the presence of correlated errors.

(iii) **Rates Achievable with Algebraic QECCs.** As a quantum analogue of algebraic codes, symplectic (stabilizer) codes are known. An improvement on the rate R_0 was also obtained, i.e., was shown to be attained by symplectic codes [20]. In other words, the highest information rate at which quantum error-correction schemes work reliably on a channel, which is called the quantum capacity, was proven to be lower-bounded by the limit of the quantity termed coherent information[3] maximized over the set of input density operators which are proportional to the projections onto the code spaces of symplectic codes. Quantum channels considered in [20] are those subject to independent errors and modeled as tensor products of copies of a completely positive linear map on a Hilbert space of finite dimension, and the codes that were proven to have the desired performance are symplectic codes. On the depolarizing channel, this work's bound is actually the highest possible rate at which symplectic codes work reliably.

Yet other results on QECCs were obtained as follows:

(iv) **Teleportation, Entanglement Distillation and QECCs.** We quantitatively discussed relations among teleportation, entanglement distillation and error-correcting codes [22]. This is explained in Sect. 2.

(v) **Formula for Fidelity of QECCs.** We gave a refined formula for the fidelity of symplectic quantum error-correcting codes. Namely, we showed that the fidelity of a symplectic (stabilizer) code, if properly defined, exactly equals the "probability" of the correctable errors for general quantum channels [23]. In [23], we also observed that exponential convergence of the fidelity of quantum codes to unity is always possible for any transmission rates below the quantum capacity.

3.4 Relation Between Teleportation and Entanglement Distillation

Next, we treat entanglement distillation from mixed states based on quantum error correction, which have been discussed by *Bennett* et al. [16]. Especially, they argue that achievable information rates for quantum error correction, i.e., those at which quantum error-correcting codes (quantum codes) reliably are also achievable as rates for one-way entanglement distillation. More precisely, they associate with an arbitrary bipartite mixed state a map called a

---

[3] Coherent information is a quantum information quantity defined by *Schumacher* et al. [21].

teleportation channel, which represents the change suffered by a teleported state when the bipartite mixed state is used for teleportation [2] in place of the ideal maximally entangled state. Then, they argue that an achievable rate for quantum codes on the teleportation channel is also achievable as the asymptotic yield of distillation schemes for the bipartite state.

In [22], we did the next three things:

1. **Formula for Teleportation Channel.** To deal with correlated states, the formula for the teleportation channel using $(\mathbb{Z}/d\mathbb{Z})^2$ was generalized to that for teleportation using Weyl's projective unitary representation of $(\mathbb{Z}/d\mathbb{Z})^n$, which allows any correlation among the n bipartite systems shared by two parties, and proved in such a way that the role of (characters of) the underlying group $(\mathbb{Z}/d\mathbb{Z})^n$ becomes clear.
2. **Entanglement Distillation by QECCs.** We refined *Bennett* et al.'s observation [16]. Namely, while they had discussed only asymptotically achievable rates, we directly worked with fidelity, and showed that tradeoffs between the fidelity and rates of quantum codes can be transformed into those between the fidelity and rates of one-way distillation protocols.
3. **Application to Known QECCs.** We applied these arguments to the known results on quantum codes. Namely, we presented exponential lower bounds on the largest fidelity that can be attained by one-way distillation protocols using the generalized formula in (1), and transformations in (2).

For example, reliable distillation with a positive asymptotic rate and exponential decay of unity minus fidelity was shown to be possible of a sequence of Bell states $|00\rangle \pm |11\rangle$, $|01\rangle \pm |10\rangle$, which occur according to the probability measure of a Markov chain.

3.5 Application to Quantum Key Distribution

Applying our study on quantum error correction code to quantum key distribution (QKD) [24], we gave a sufficient condition for a CSS code to achieve the Shannon rate $1 - h((p_x + p_z)/2)$ mentioned in [25], where p_x is the bit error rate and p_z is the phase error rate. That is, we showed that codes of "balanced weight spectra (distributions)" achieve it. The weight spectra are known as important characteristics of error-correcting codes in coding theory. We also showed the existence of codes of "balanced weight spectra", to prove the achievability of the rate $1 - h((p_x + p_z)/2)$ in BB84 protocol. Though our result is an existence theorem, as usual in information theory, this would show the direction to designers of codes for QKD. We also argued that $1 - h(p_x) - h(p_z)$ is achievable if we use codes of a similar balanced property. We also proved the security of the BB84 protocol using the codes of balanced weight spectra against any joint (coherent) attacks [24]. From these discussions, we can check that the Eve's information and error probability goes to 0 exponentially.

4 Basic Characteristics of Bipartite Entanglement

We characterized bipartite entanglement by the following methods:

1. **Concurrence Hierarchy.** We generalized concurrence, which is a useful entanglement measure for the two-dimensional case.
2. **Entanglement of Formation.** We discussed the relation between and the channel capacity. We also proved the additivity of EoF in several examples in the chapter by Matsumoto.
3. **Entanglement of Purification and Mixed States Compression.** We showed that the optimal compression rate of visible compression of mixed states.
4. **Simultaneous Schmidt Decomposition.** The notion of simultaneous Schmidt decomposition (SSD) was introduced. The necessary and sufficient condition for the simultaneous Schmidt decomposability of the set of bipartite state vectors was given.
5. **Bell-Type Inequalities via Combinatorial Approach.** The set of Bell inequalities is closely related to the set of entanglement states. In this approach, we analyzed the latter by discussing the former.
6. **Pseudo-Telepathy Game**. The pseudo-telepathy game is an approach that deals with Bell's inequality without inequalities from the point of computer science. Several pseudo-telepathy games are known, and graph coloring game is one of them. We propose a quantum protocol to win the graph coloring game on all Hadamard graphs.

4.1 Concurrence Hierarchy

We treated entanglement measures. It is acceptable that we use several quantities simultaneously as entanglement measures. It is also true that even for pure states, several quantities are necessary to quantify entanglement. Concurrence is one of the widely accepted entanglement measures for a two qubit system, which is directly related with EoF. There are several proposals to define concurrence for a higher-dimensional system. And all were shown to be essentially the same. Considering that one quantity may be not enough to quantify entanglement, we did the next three things [26]:

(i) We proposed to use $d-1$ quantities to quantify entanglement, called the concurrence hierarchy.
(ii) We found some formulae for this concurrence hierarchy.
(iii) We studied its relationship with the majorization scheme in entanglement transformation.

The first nontrivial quantity in our proposal is the concurrence, which has already been proposed by several groups. We showed that this concurrence hierarchy is useful in the entanglement measure.

4.2 Optimal Compression Rates and Entanglement of Purification

Quantum data compression was initiated by *Schumacher* [27]. As the quantum information source, he focused on the quantum states ensemble $(p_x, W_x)_{x \in \mathcal{X}}$, in which the quantum state W_x generates with the probability p_x. He showed that the asymptotic optimal compression rate $R(W, p)$ is equal to the entropy $H(W_p)$ of the average state $W_p \equiv \sum_x p_x W_x$ of this ensemble. In the original version his problem, the encoder is restricted to the quantum operation. However, *Horodecki* [28] considered another problem, in which the encoder is defined as any map from $\mathcal{X}$ to the quantum states. This formulation is called visible, while the former is called blind. He also showed that even in the visible setting if any state W_x is pure, the optimal rate $R(W, p)$ is equal to the entropy rate $H(W_p)$. However, it had been an open problem to characterize the rate $R(W, p)$ in the mixed states case. *Horodecki* [29] studied this problem and succeeded in its characterization. However, his characterization contains a limiting expression. Hence, it is an open problem whether it can be characterized without any limiting expression.

On the other hand, *Terhal* et al. [30] introduced entanglement of purification $E_p(\rho)$ for any partially entangled state ρ as the minimum value of $H(\mathbf{p}_\phi)$ among purification ϕ of ρ. They also consider the generation of the tensor product of any partially entangled state ρ on the composite system $\mathcal{H}_A \otimes \mathcal{H}_B$ from maximal entangled states in the asymptotic form with the restriction of the rate of the classical communication to be zero asymptotically. Indeed, when the target state ρ is pure, this optimal rate is $H(\text{Tr}_A \rho)$, which is equal to the optimal rate without any restriction for the rate of the classical communication [31]. Their main result is that the optimal rate with this restriction is equal to $\lim \frac{E_p(\rho^{\otimes n})}{n}$. Of course, if the entanglement of purification satisfies the additivity, i.e., $E_p(\rho) + E_p(\sigma) = E_p(\rho \otimes \sigma)$, this optimal rate is equal to the entanglement purification. However, this additivity is still open.

In [32], we gave another formula for the optimal visible compression rate $R(W, p)$ as

$$R(W, p) = \lim \frac{1}{n} E_p(\tilde{W}_p^{\otimes n}), \quad \tilde{W}_p \equiv \sum_x p_x |e_x^A\rangle\langle e_x^A| \otimes W_x.$$

Using this relation, we clarify the relation between the two problems, the mixed state compression and the state generation from maximally entangled state with zero-rate classical communication. Hence, if the additivity of entanglement of purification is proved, the optimal rate of visible compression is equal to entanglement purification.

4.3 Simultaneous Schmidt Decomposition and Maximally Correlated States

We introduced a notion of simultaneous Schmidt decomposition of a set of bipartite pure state vectors $\{|\psi_\alpha\rangle\}_{\alpha=1}^{l}$. If all $|\psi_\alpha\rangle$ are written as the following form,

$$|\psi_\alpha\rangle = \sum_{k=1}^{\min\{d_A, d_B\}} b_k^{(\alpha)} |k_A\rangle \otimes |k_B\rangle, \tag{10}$$

with common biorthogonal bases $|k_A\rangle \otimes |k_B\rangle$, we call $|\psi_\alpha\rangle$ simultaneously Schmidt decomposable. In (10), coefficients $b_k^{(\alpha)}$ are complex numbers. A necessary and sufficient condition for the simultaneous Schmidt decomposability was given. We explored the properties of several bipartite mixed states in light of the condition of simultaneous Schmidt decomposability. In particular, for generalized Bell states in a $d \otimes d$ system, the condition for the simultaneous Schmidt decomposability was shown to be a simple algebraic relation between indices (n, m) of the states. We also discussed the local distinguishability of the generalized Bell states that are simultaneously Schmidt decomposable [14].

4.4 Bell-Type Inequalities Via Combinatorial Approach

In this research, we considered the explicit representation of tight Bell-type inequalities for bipartite systems with many 0/1 valued observables, especially the enumeration of new Bell-type inequalities.

In 1986, *Pitowsky* [33] pointed out that tight Bell-type inequalities for bipartite systems are facet inequalities of correlation polytope of the complete bipartite graph, which is the projection of well-known correlation polytope of complete graph. However, because the projection of facet produces many non tight faces in the case of general operation (Fourier–Motzkin elimination), the explicit representation of tight Bell-type inequalities which are projected in this manner is not known, except for the smallest case, namely the bipartite system with two 0/1 valued observables.

In [34], we investigated the relationship of the projection operation of correlation polytope of graph and elimination of edge of graph. As a result, we found the following:

1. **Interactability of Enumeration.** We showed that the membership of the correlation polytope of the complete bipartite graph is NP-complete. This means that we cannot hope for the existence of efficient algorithms which computes the list of all Bell-type inequalities from the number of observables as input. However, we also showed that in bipartite system case, if we obtain a tight Bell-type inequality for small number of observables, its simple extension (0-lifting) is always tight for larger numbers.

2. **New Efficient Enumeration Algorithm Based on Combinatorics.**
We constructed an efficient algorithm which enumerates tight Bell-type inequalities from known facets of the correlation polytope of the complete graph. For this algorithm, we showed that it is sound, i.e., the output is always tight and mutually inequivalent in the sense of permutation and switching of the coefficients. As the output, we obtained 16 236 representations of tight general Bell-type inequalities, except 5 of them are previously unknown.

4.5 Quantum Graph Coloring Game

We deal with graph coloring games, an example of pseudo-telepathy, in which two players can convince a verifier that a graph G is c-colorable where c is less than the chromatic number of the graph. They win the game if they convince the verifier. It is known that the players cannot win if they share only classical information, but they can win in some cases by sharing entanglement. The smallest known graph where the players win in the quantum setting, but not in the classical setting, was found by *Galliard* et al. [35] and has 32 768 vertices. It is a connected component of the Hadamard graph G_N with $N = c = 16$. Their protocol applies only to Hadamard graphs where N is a power of 2. We propose a protocol that applies to all Hadamard graphs [36]. Combined with a result of *Frankl* [37], this shows that the players can win on any induced subgraph of G_{12} having 1609 vertices, with $c = 12$. Moreover, combined with a result of *Godsil* and *Newman* [38], our result shows that all Hadamard graphs G_N ($N \geq 12$) and $c = N$ yield pseudo-telepathy games.

5 SLOCC Convertibility

Next, we focus on stochastic local operation and classical communication (SLOCC) as a wider class of operations than LOCC. In the following we discuss multipartite entanglement and bipartite entanglement in infinite-dimensional space by a viewpoint of SLOCC convertibility.

5.1 Multipartite Entanglement

We studied basic characteristics of quantum correlation entanglement in quantum multipartite systems. Entanglement plays significant roles in applications to quantum information, where quantum theory can broaden and improve our information processing, compared with current classical methods. Since this innovative resource has strange nonlocal (nonseparable) properties of entanglement, the characterization of entanglement in terms of LOCC is of great interest. In particular, entanglement is expected to be intriguing and valuable in the multiparty situation, since network nature (i.e., interactions of many elements) is essential for high information processing. The rich properties of multipartite entanglement also would offer us renewed insight into mysteries in the fundamental aspects of quantum theory.

However, the studies of multipartite entanglement turned out to be challenging, due to the fact that many useful techniques, such as the Schmidt decomposition, utilized in the two-party situation cannot be generalized straightforwardly to the multiparty situation. Our results brought breakthroughs as follows:

(i) First, a guiding principle for the characterization of multipartite entanglement, applicable to arbitrary n-partite systems, was introduced. Central ideas were a duality between entangled classes, and multidimensional generalized determinants, i.e., hyperdeterminants, associated with the duality.
(ii) Second, in virtue of these ideas, systematic characterizations of entanglement in several multiparty situations were obtained.

(i) Duality and Hyperdeterminant. Focusing on a duality, a generalization of the Legendre transformation, between the set of separable states and that of entangled states, we showed that entanglement is classified in a unified manner for both two-party and multi party situations. The key entanglement measure associated with the duality is a multidimensional determinant called the hyperdeterminant, describing the basic nature of multipartite entanglement. The hyperdeterminant for the 3-qubit system has been already known as the so-called residual entanglement "3-tangle", but the importance of hyperdeterminants (of several formats) for the characterization of arbitrary n-partite entanglement, along with their valid definition associated with the duality, was clarified through our research.

The basic convertibility properties of multipartite entanglement, i.e., the equivalence (reversible) classes of entanglement and monotonic (irreversible) properties of entanglement, are captured as partial orders of multipartite entanglement under SLOCC. It was found in general that partially ordered structures of multipartite entanglement appear in terms of the hyperdeterminant and its singularities, where different entangled classes correspond to different types of degeneracy [39, 40]. Moreover, the hyperdeterminant of a given format distinguishes the generic kind of multipartite entanglement from the other. So, as is the case in the 3-qubit system, hyperdeterminants are expected to be key ingredients in limited shareability of multipartite entanglement, which is a fundamental phenomenon of quantum multipartite systems in contrast with the classical counterparts.

(ii) Characterizations of Multipartite Entanglement. We illustrated the systematic characterizations of multipartite entanglement, by addressing the 3-qubit ($2 \times 2 \times 2$) case, the 2-qubit and 1-qutrit ($2 \times 2 \times 3$) case, the 2-qubit and the rest ($2 \times 2 \times l$, $l \geq 4$) case, and partially the n-qubit (2^n) case [41]. Since the 3-qubit case and partially the 4-qubit case had been studied so far, our studies presented valuable examples. It is known through these studies that, in the multiparty situations, there are various inequivalent entangled classes which cannot be converted to each other even probabilistically. Two representative states of the 3-qubit GHZ and W classes are famous in their different physical properties and applications to quantum information processing. The GHZ state $|000\rangle + |111\rangle$ has the maximal amount of generic 3-qubit entanglement measured by the hyperdeterminant of format $2 \times 2 \times 2$, called the 3-tangle. The GHZ state violates the Bell's inequality maximally, and enable us to extract one Bell state between any two parties out of three with probability 1. On the other hand, the W state $|001\rangle + |010\rangle + |100\rangle$ has the maximal amount of average pairwise entanglement distributed over three parties. So, the states in the W class can be utilized in the optimal quantum cloning.

The complete entanglement structure of 2-qubit and the rest ($2 \times 2 \times l$) system not only includes the results of the 3-qubit system, but is also practically important because of, for example, its implications to 2-qubit mixed states. We showed that there exist nine essentially different entangled classes, and they constitute a five-graded partially ordered structure under SLOCC Fig.3. The five-graded partial order of nine entangled classes is found to cause various multipartite phenomena, which cover the aforementioned shareability of multipartite entanglement as fundamental and multiparty LOCC protocols, as described later, as applications. Remarkably, the generic (maximal dimensional) class is a unique "maximally entangled" class, lying on the top of the hierarchy. All $2 \times 2 \times l$ states were shown to be deterministically prepared from one maximally entangled state, which is indeed two Bell states shared over three parties, in the generic class. This makes a clear contrast with the n-qubit ($n \geq 3$) situation where there is no such a resource as to create any state even probabilistically. Also, it can be readily seen that downward (non-invertible) flows in this partially ordered structure correspond to multiparty LOCC protocols, such as entanglement swapping or the creation of 3-qubit GHZ and W entanglement.

5.2 Bipartite Entanglement in Infinite-Dimensional Space

Next, we focus on bipartite entanglement in arbitrary pure states in infinite-dimensional space like Boson–Fock space while we treated the multipartite entanglement in the above. The convertibility properties of two different entangled states under local operations are important for qualitative and quantitative understanding of entanglement. Though we now have a better understanding for finite dimensional bipartite systems [42], in infinite dimensional systems, LOCC and SLOCC convertibility are investigated for only a limited class of local operations [43] (Gaussian operations). In [44], we investigated SLOCC convertibility in infinite-dimensional systems as follows:

(i) **General Formulation of Convertibility-Monotones.** We developed a general formulation for constructing a pair of convertibility

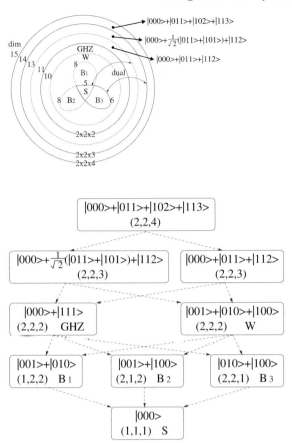

Fig. 3. (*Top*) The onion-like classification of multipartite entanglement in the 2-qubit and the rest ($2 \times 2 \times l$, $l \geq 4$) quantum system. Divided by "onion skins", there are nine different SLOCC entangled classes, each of which is a set of states interconvertible under invertible local operations. (*Bottom*) The five-graded partially ordered structure of nine entangled classes. Every class is labeled by its representative, its set of local ranks and its name. Noninvertible local operations, indicated by *dashed arrows*, degrade higher entangled classes into lower entangled ones. Both figures partly include the cases for $l = 3$ and 2

monotones using order properties. The monotones are considered as generalizations of distillable entanglement and entanglement cost. This formulation can be applied to many different situations to analyze entanglement convertibility.

(ii) **SLOCC Incomparable Pure States.** We applied the formulation on (i) to SLOCC convertibility for genuine infinite dimensional pure states in the single-copy situation. By constructing an example, we proved the existence of SLOCC incomparable pure bipartite states, a new property of entanglement in infinite dimensional systems. In contrast, incomparable pure states only exist for multipartite systems (such as GHZ and W states for three-qubit states) in finite dimensional systems. The ordering property under SLOCC convertibility is changed fundamentally, from total ordering to nontotal (partial) ordering, with the shift in dimensionality from finite to infinite.

It had been widely believed that the fundamental entanglement properties of finite- and infinite-dimensional systems are similar. However, we showed that there exists a significant difference in convertibility.

6 Protocols Assisted by Multipartite Entangled State

Many quantum information processing protocols were proposed. They are performed by using Bell states, i.e., bipartite maximally entangled states. We proposed two protocols based on multipartite entangled states as alternative protocols.

6.1 Teleportation by W State

Entanglement in three qubits is more complicated than that in two qubits. As is mentioned in Sect. 5, it is known that there are two inequivalent classes of tripartite entangled states, the GHZ class and W class. These two classes cannot be converted to each other even under SLOCC. W states have some interesting properties, and are more robust against the loss of one qubit. In [45], we showed that a W state can be used to probabilistically realize the teleportation from the sender to one of two receivers. In this process, a two-particle Bell state measurement (BSM) and a single-qubit projection measurement are needed. While the BSM depends on which receiver revives the teleported state, the probability of success is independent of the teleported state. Besides, we also considered the teleportation of a two-particle entangled state by a W state, and found that a W state cannot be used to do that, although a GHZ state can be used to do it.

6.2 Remote State Preparation of Entangled State

Remote state preparation (RSP) is called "teleportation of a known state", which means the sender-Alice knows the precise state that she will transmit to the receiver-Bob. Her task is to help Bob construct a state that is unknown to him by means of a prior shared entanglement and classical communication. It was pointed out that RSP is one of the examples of studying the classical communication cost in quantum information processing (CC-CIQIP). CCCIQIP is important for better understanding the fundamental laws of quantum information processing. It can also be regarded as the natural generalization of quantum communication complexity, and has received much interest recently. In [46], we proposed the following three schemes and obtained the following properties:

(i) **RSP of a Qubit State Using GHZ State.** In this setting, Alice knows the desired qubit state, and Bob wants to construct it with the help of Alice and Charlie. For this task, Bob needs only two classical bits, i.e., one bit from Alice and the other bit from Charlie, while the teleportation of a qubit state by using the GHZ state needs three classical bits.

(ii) **RSP of a Bipartite Entangled State Using GHZ State.** In this setting, Alice knows the desired special bipartite entangled state with two parameters, and Bob and Charlie want to construct it with the help of Alice. For this task, Alice needs to send only one classical bit to both, and Bob and Charlie need to perform local operations. The number of needed classical bits is less than that for the teleportation scheme.

(iii) **RSP of an n-partite Entangled State Using $n+1$-partite GHZ State.** In this setting, Alice knows the desired special n-partite entangled state with two parameters, and the other n persons want to construct it with the help of Alice. Similarly to (ii), Alice needs to send only one classical bit to them, and they need to perform local operations. However, in the teleportation, the number of classical bits increases with particle number n.

References

[1] C. H. Bennett, S. J. Wiesner: Communication via one- and two-particle operators on Einstein–Podolsky–Rosen states, Phys. Rev. Lett. **69**, 2881–2884 (1992)

[2] C. H. Bennett, G. Brassard, C. Crépeau, R. Jozsa, R. A. Peres, W. K. Wootters: Teleporting an unknown quantum state via dual classical and Einstein–Podolsky–Rosen channels, Phys. Rev. Lett. **70**, 1895–1899 (1993)

[3] P. W. Shor: Fault-tolerant quantum computation, in *Proceedings, 37th Annual Symposium on Fundations of Computer Science* (IEEE Press, Los Alamos, CA 1996) pp. 56–65

[4] C. H. Bennett, H. J. Bernstein, S. Popescu, B. Schumacher: Concentrating partial entanglement by local operations, Phys. Rev. A **53**, 2046 (1996) quant-ph/9511030

[5] M. Hayashi, M. Koashi, K. Matsumoto, F. Morikoshi, A. Winter: Error exponents for entanglement concentration, J. Phys. A: Math. Gen. **36**, 527–553 (2003)

[6] M. Hayashi, K. Matsumoto: Universal distortion-free entanglement concentration, in *Proceedings of the International Symposium on Information Theory 2004 (ISIT2004)* (2004) p. 323

[7] L. M. Duan, G. Giedke, J. I. Cirac, P. Zoller: Inseparability criterion for continuous variable systems, Phys. Rev. Lett. **84**, 2722–2725 (2000)
[8] R. Simon: Peres–Horodecki separability criterion for continuous variable systems, Phys. Rev. Lett. **84**, 2726–2729 (2000)
[9] T. Hiroshima: Decoherence and entanglement in two-mode squeezed vacuum states, Phys. Rev. A **63**, 022305 (2001)
[10] X. B. Wang, K. Matsumoto, A. Tomita: Detecting the inseparability and distillability of continuous variable state in fock space, Phys. Rev. Lett. **87**, 137903 (2001)
[11] X. B. Wang: Theorem for the beam splitter entangler, Phys. Rev. A **66**, 024303 (2002)
[12] M. S. Kim, W. Son, V. Buzek, P. L. Knight: Entanglement by a beam splitter: Nonclassicality as a prerequisite for entanglement, Phys. Rev. A **56**, 032323 (2002)
[13] X. B. Wang: Properties of a beam-splitter entangler with gaussian input states, Phys. Rev. A **66**, 064304 (2002)
[14] T. Hiroshima, M. Hayashi: Finding a maximally correlated state: Simultaneous Schmidt decomposition of bipartite pure states, Phys. Rev. A **70**, 30302(R) (2004)
[15] E. M. Rains: A semidefinite program for distillable entanglement, IEEE Trans. Inf. Theory **47**, 2921–2933 (2001)
[16] C. H. Bennett, D. P. DiVincenzo, J. A. Smolin, W. K. Wootters: Mixed-state entanglement and quantum error correction, Phys. Rev. A **54**, 3824–3851 (1996)
[17] M. Hamada: Exponential lower bound on the highest fidelity achievable by quantum error-correcting codes, Phys. Rev. A **65**, 052305 (2002)
[18] M. Hamada: Lower bounds on the quantum capacity and highest error exponent of general memoryless channels, IEEE Trans. Information Theory pp. 2547–2557 (2002)
[19] M. Hamada: A lower bound on the quantum capacity of channels with correlated errors, J. Math. Phys. **43**, 4382–4390 (2002)
[20] M. Hamada: Information rates on the quantum capacity and highest error exponent of general memoryless channels, IEEE Trans. Information Theory **51**, 4263–4277 (2005)
[21] B. Schumacher, M. A. Nielsen: Quantum data processing and error correction, Phys. Rev. A **54**, 2629 (1996)
[22] M. Hamada: Teleportation and entanglement distillation in the presence of correlation among bipartite mixed states, Phys. Rev. A **68**, 012301 (2003)
[23] M. Hamada: Notes on the fidelity of symplectic quantum error-correcting codes, Int. J. Quant. Inform. **1**, 443–463 (2003)
[24] M. Hamada: Reliability of Calderbank-Shor-Steane codes and security of quantum key distribution, J. Phys. A: Math. Gen. **37**, 8303–8328 (2004)
[25] P. Shor, J. Preskill: Simple proof of security of the BB84 quantum key distribution protocol, Phys. Rev. Lett. **85**, 441–444 (2000)
[26] H. Fan, K. Matsumoto, H. Imai: Quantify entanglement by concurrence hierarchy, J. Phys. A: Math. Gen. **36**, 4151–4158 (2003)
[27] B. Schumacher: Quantum coding, Phys. Rev. A pp. 2738–2747 (1995)
[28] M. Horodecki: Limits for compression of quantum information carried by ensembles of mixed states, Phys. Rev. A **57**, 3364–3369 (1997)

[29] M. Horodecki: Optimal compression for mixed signal states, Phys. Rev. A **61**, 052309 (2000)
[30] B. M. Terhal, M. Horodecki, D. W. Leung, D. P. DiVincenzo: The entanglement of purification, J. Math. Phys. **43**, 4286 (2002)
[31] H. -K. Lo, S. Popescu: Classical communication cost of entanglement manipulation: Is entanglement an interconvertible resource?, Phys. Rev. Lett. **83**, 1459 (1999)
[32] M. Hayashi: Optimal visible compression rate for mixed states is determined by entanglement purification, to appear in Phys. Rev. A., quant-ph/0511267
[33] D. Avis, H. Imai, T. Ito, Y. Sasaki: Two-party Bell inequalities derived from combinatorics via triangular elimination, J. Phys. A: Math. Gen. **38**, 10971–10987 (2005)
[34] I. Pitowsky: The range of quantum probability, J. Math. Phys. **27**, 1556–1565 (1986)
[35] The impossibility of pseudo-telepathy without quantum entanglement, in V. Galliard, A. Tapp, S. Wolf (Eds.): (Proceedings of ISIT 2003) p. 457, quant-ph/0211011
[36] D. Avis, J. Hasegawa, Y. Kikuchi, Y. Sasaki: A quantum protocol to win the graph colouring game on all hadamard graphs, IEICE Transactions on Fundamentals **E89-A**, 1318–1381 (2006)
[37] P. Frankl: Orthogonal vectors in the n-dimensional cube and codes with missing distances, Combinatorica **6(3)**, 279 (1986)
[38] C. D. Godsil, M. W. Newman: Colouring an orthogonality graph, e-Print math.CO/0509151
   URL arXiv.org
[39] A. Miyake, M. Wadati: Multipartite entanglement and hyperdeterminants, Quant. Inf. Comp. **2**, 540–555 (2002)
[40] A. Miyake: Classification of multipartite entangled states by multidimensional determinants, Phys. Rev. A **67**, 012108 (2003)
[41] A. Miyake, F. Verstraete: Multipartite entanglement in $2 \times 2 \times n$ quantum systems, Phys. Rev. A **69**, 012101 (2004)
[42] M. A. Nielsen: Conditions for a class of entanglement transformations, Phys. Rev. Lett. **83**, 436–439 (1999)
[43] G. Giedke, J. Eisert, J. I. Cirac, M. B. Plenio: Entanglement transformations of pure Gaussian states, Quant. Inf. Comp. **3**, 211 (2003)
[44] M. Owari, K. Matsumoto, M. Murao: Entanglement convertibility for infinite-dimensional pure bipartite states, Phys. Rev. A **70**, 050301(R) (2004)
[45] B. S. Shi, A. Tomita: Teleportation of an unknown state by W state, Phys. Lett. A **296**, 161–164 (2002)
[46] B. S. Shi, A. Tomita: Remote state preparation of an entangled state, J. Opt. B: Quantum Semiclass **4**, 380–382 (2002)

Index

n-partite, 125, 129

beam splitter (BS), 114
beam splitter entangler, 114
Bell state, 112, 120, 126
Bell state measurement (BSM), 128
Bell type inequality, 123
bipartite, 129
bipartite,entanglement,infinite-dimensional space, 126
Boson–Fock space, 114, 126

coherent information, 119
compression, 121
concurrence, 121

entangled state, 126, 128
entanglement, 114, 124
entanglement concentration, 112
entanglement distillation, 111, 119
entanglement of formation (EoF), 121

fidelity, 118

Gaussian state, 114
GHZ state, 126, 129

hyperdeterminant, 125

LOCC (local operation and classical communication), 112

maximally entangled state, 111
multipartite, 124
multipartite,entangled state, 128

QECC (quantum error correcting code), 118–120
quantum capacity, 119
quantum key distribution (QKD), 120

reliability function, 118
remote state preparation (RSP), 128

Simultaneous Schmidt decomposition, 123
SLOCC convertibility, 126
stochastic local operation and classical communication (SLOCC), 124
symplectic code, 119

TP-CP (trace-preserving completely positive) map, 117

W state, 126

On Additivity Questions

Keiji Matsumoto[1,2]

[1] ERATO Quantum Computation and Information Project, JST.
  Hongo White Building, 5-28-3 Hongo, Bunkyo-ku, Tokyo 113-0033, Japan
[2] Quantum Information Science Group, National Institute of Informatics,
  2-1-2 Hitotsubashi, Chiyoda-ku, Tokyo 101-8430, Japan
  keiji@nii.ac.jp

1 Introduction

In quantum information theory, there are several open problems which center around whether certain quantities are additive or not. The additivity of Holevo capacity is the oldest of these. If this conjecture is true, it follows that entangled signal states do not improve the capacity of quantum channels. Another additivity conjecture is about the minimum entropy of the output of the quantum channel.

Also, there are additivity conjectures about an entanglement measure, namely, the entanglement of formation (EoF). The thermodynamic limit of this quantity gives entanglement cost, which is defined as the number of maximally entangled pairs required to prepare ρ by LOCC in an asymptotic way. Additivity conjecture of EoF implies this thermodynamic limit is equal to the original quantity, simplifying the computation of entanglement cost to a large extent. Another implication of this conjecture is that making ρ and σ altogether requires the same amount of maximally entangled states, as they are produced separately. In other words, there is no catalytic effect in entanglement dilution, which is different from entanglement distillation. There is yet another additivity-like conjecture about EoF, called strong superadditivity [1]. The intuitive appeal of the strong superadditivity property is that by measuring the entanglement via EoF, a system can only appear less entangled if judged by looking at its subsystems individually.

Our project had studied these conjectures from various aspects. First, we found some relations between additivity conjectures about a channel and EoF of a state [2,3]. Our research is one of the earliest efforts toward this direction, and the concept of channel state, proposed by us, is a key main machinery of *Shor's* celebrated work [4] on equivalence of additivity questions.

Second, we proved additivity relations for some specific channels and states. Especially, after continuous commitment to the study of antisymmetric states, we finally proved the additivity of Holevo capacity of Werner–Holevo channels [5, 6, 7, 8, 9], which had been potentially counter examples to the additivity conjecture. Also, *Fan* [10] proved additivity of EoF of some other special states, and *Hiroshima* [11] treated additivity and multiplicativity of some Gaussian channels for Gaussian inputs.

Third, additivity questions are studied numerically [12, 13]. Especially in the study of a qubit 4-state channel in [12], we utilized results from theory convex optimizations.

The manuscript is organized as follows. After stating definitions of the problems with some comments in Sect. 2, we state our results on equivalence of additivity, proofs of additivity in specific examples, and numerical studies in Sect. 3, 4, and 5, respectively.

2 Additivity Questions: Definitions and Comments

2.1 Holevo Capacity, Output Minimum Entropy, and Maximum Output p-Norm

We consider coding of classical information via the quantum channel

$$T : \mathcal{B}(\mathcal{K}) \longrightarrow \mathcal{B}(\mathcal{H}_A),$$

where $\mathcal{K}$ and $\mathcal{H}_A$ are Hilbert spaces. If the encoding is restricted to separable states it is known [14, 15] that the capacity is given by Holevo capacity, defined by

$$C(T) = \sup_{\{p_i, \pi_i\}} \left\{ S\left(\sum_i p_i T(\rho_i)\right) - \sum_i p_i S(T(\rho_i)) \right\}, \quad (1)$$

where $\{p_i, \pi_i\}$ moves over all the pure state ensembles on $\mathcal{K}$ and $S(\rho) = -\text{Tr}\rho \log \rho$ is the von Neumann entropy of a state. It is a consequence of Carathéodory's theorem and the convex structure of this problem that the above supremum can be replaced with the maximum over $(\dim \mathcal{H}_A)^2$ pairs of $\{p_i, \pi_i\}$ [16, 17].

It is conjectured that a product of channels making use of entangled input states does not help to increase the capacity:

$$C(T_1 \otimes T_2) = C(T_1) + C(T_2). \quad (2)$$

This would imply that $C(T)$ is the classical capacity of T. Observe that here the inequality "$\geq$" follows immediately from the fact that the right-hand side can be achieved using product states. Without additivity, the general formula for this capacity reads

$$\lim_{n \to \infty} \frac{1}{n} C\left(T^{\otimes n}\right).$$

The question (2) is implicit in [18] and the above references, and made explicit in [19], where it was speculated that the answer may be negative. Another early reference to this conjecture is by *Osawa* and *Nagaoka* [20, 21], in which they checked (2) by careful numerical simulation for number of examples.

Despite much recent activity on the question [22, 23], and even proofs of the additivity conjecture in some cases [24, 25, 26, 27, 28, 29, 30], it is still an open problem. Also, there have been several numerical studies [12, 20, 21], which we will discuss later.

In showing these results, many of the authors first show that minimum output entropy,

$$S_{\min}(T) := \min_{\rho \in S(\mathcal{K})} S(T(\rho)),$$

is additive,

$$S_{\min}(T_1 \otimes T_2) = S_{\min}(T_1) + S_{\min}(T_2). \tag{3}$$

Many of authors, following the suggestion in [23], show this relation via the multiplicativity of maximum p-norm,

$$\nu_p(T_1 \otimes T_2) = \nu_p(T_1)\nu_p(T_2), \tag{4}$$

where

$$\nu_p(T) := \max_{\rho \in S(\mathcal{K})} \left(\text{Tr}\,(T(\rho))^p\right)^{\frac{1}{p}}$$

is a maximum p-norm of T. By differentiating with respect p and letting $p \to 1$, this leads to (3). For this technical tool to work, (4) has to be proved for for all p in the interval $[1, 1+\epsilon]$ with some $\epsilon > 0$. It is known that (4) is false for large p for some channels, such as Werner–Holevo channels [31].

Example 1 *Consider the generalized depolarizing channels of qubits:*

$$T^{\mathbf{p}} : \rho \longmapsto \sum_{s=0,x,y,z} p_s \sigma_s \rho \sigma_s^\dagger,$$

with $\sigma_0 = \mathbf{1}_2$, the familiar Pauli matrices

$$\sigma_x = \begin{pmatrix} 0 & 1 \\ 1 & 0 \end{pmatrix}, \ \sigma_y = \begin{pmatrix} 0 & -i \\ i & 0 \end{pmatrix}, \ \sigma_z = \begin{pmatrix} 1 & 0 \\ 0 & -1 \end{pmatrix},$$

and a probability distribution $\mathbf{p} = (p_s)_{s=0,x,y,z}$. For these channels additivity of the capacity under tensor product with an arbitrary channel was proved in [28]. Note that up to unitary transformations on input and output system each unital qubit channel has this form, by the classification of qubit maps of King and Ruskai [25], and Fujiwara and Algoet [32]. By this result we also can assume that

$$p_0 + p_z - p_x - p_y \geq |p_0 + p_y - p_x - p_z|, |p_0 + p_x - p_y - p_z|. \tag{5}$$

It is easy to see that for such a channel the capacity is given by

$$C(T) = 1 - S_{\min}(T),$$

with the minimal output entropy achieved at the eigenstates $|0\rangle, |1\rangle$ of σ_z: $S_{\min}(T) = S(T(|0\rangle\langle 0|)) = S(T(|1\rangle\langle 1|))$. An optimal ensemble is the uniform distribution on these states.

Example 2 Consider the d-dimensional depolarizing channel with parameter λ:

$$T_{depol}^\lambda : X \longmapsto \lambda X + (1-\lambda)\frac{\mathrm{Tr}X}{d}\mathbf{1}_d,$$

with $-\frac{1}{d^2-1} \leq \lambda \leq 1$ for complete positivity, to ensure that T can be represented as a mixture of generalized Pauli actions:

$$T_{depol}^\lambda(\rho) = \lambda\rho + (1-\lambda)\sum_{i=1}^{d^2-1}\frac{1}{d^2-1}\sigma_i\rho\sigma_i^\dagger,$$

with an orthogonal set of unitaries (a "nice error basis", see e.g., [33] for constructions) σ_i, i.e.,

$$\sigma_0 = \mathbf{1}_d, \quad \mathrm{Tr}(\sigma_i^\dagger\sigma_j) = d\delta_{ij},$$

and $p_0 = \lambda + (1-\lambda)/d^2$. It is quite obvious that

$$C(T_{depol}^\lambda) = \log d - S_{\min}(T_{depol}^\lambda) = \log d - S(T(|\psi\rangle\langle\psi|)),$$

for arbitrary $|\psi\rangle \in \mathbb{C}^d$, optimal input ensembles being those mixing to $\frac{1}{d}\mathbf{1}_d$. It is easy to evaluate this latter von Neumann entropy:

$$S\left(T_{depol}^\lambda(|\psi\rangle\langle\psi|)\right) = H\left(\lambda + \frac{1-\lambda}{d}, \frac{1-\lambda}{d}, \ldots, \frac{1-\lambda}{d}\right),$$

$$= H\left(\left(1-\frac{1}{d}\right)(1-\lambda), 1 - \left(1-\frac{1}{d}\right)(1-\lambda)\right)$$

$$+ \left(1-\frac{1}{d}\right)(1-\lambda)\log(d-1).$$

For this channel, first Bruss et al. [24] showed $C(T_{depol}^{\lambda \otimes 2}) = 2C(T_{depol}^\lambda)$. Later, Fujiwara and Hashizume [26] proved $C(T_{depol}^\lambda \otimes T_{depol}^{\lambda'}) = C(T_{depol}^\lambda) + C(T_{depol}^{\lambda'})$. $C(T_{depol}^\lambda \otimes T) = C(T_{depol}^\lambda) + C(T)$ is obtained in [30].

Example 3 Werner–Holevo channel $T_{WH}^d : \mathcal{S}(\mathbb{C}^d) \to \mathcal{S}(\mathbb{C}^d)$ is defined by

$$T_{WH}^d(\rho) = \frac{1}{d-1}(\mathbf{1}_d - \rho^T).$$

This is a family of channels used in [31] to disprove general multiplicativity conjecture of the maximal output p-norm of a channel. The additivity of Holevo capacity and multiplicativity of maximum p-norm for $1 \leq p \leq 2$ of WH channels are first shown in our project [8].

2.2 Entanglement of Formation

Let ρ be a state on $\mathcal{H}_A \otimes \mathcal{H}_B$. The entanglement of formation (EoF) of ρ is defined as

$$E_f(\rho) := \inf_{\{p_i, \pi_i\}} \sum_i p_i E(\pi_i), \qquad (6)$$

where $\{p_i, \pi_i\}$ moves over all the pure state ensembles with $\sum_i p_i \pi_i = \rho$, and the (entropy of) entanglement for a pure state π on $\mathcal{H}_A \otimes \mathcal{H}_B$ is defined as

$$E(\pi) := S\left(\mathrm{Tr}_{\mathcal{H}_B} \pi\right) = S\left(\mathrm{Tr}_{\mathcal{H}_A} \pi\right).$$

If the rank of ρ is finite the inf function is in fact a min, achieved for an ensemble of at most $(\mathrm{rank}\,\rho)^2$ elements. This quantity was proposed in [34] as a measure of how costly in terms of entanglement the creation of ρ is.

It is conjectured (but only in a few cases proved; the only published examples are in [35]) that E_f is an additive function with respect to tensor products:

$$E_f(\rho_1 \otimes \rho_2) = E_f(\rho_1) + E_f(\rho_2). \qquad (7)$$

Observe that, as in the case of the Holevo capacity, "$\leq$" follows easily from the fact that the right-hand side is achieved by product state ensembles. If this turned out to be true, the entanglement cost $E_c(\rho)$ of ρ, i.e., the asymptotic rate of EPR pairs to approximately create n copies of ρ, is given by $E_f(\rho)$. In [36] it was proved rigorously that

$$E_c(\rho) = \lim_{n \to \infty} \frac{1}{n} E_f\left(\rho^{\otimes n}\right).$$

Note that the function E_f has the property of being a convex roof:

$$E_f(\rho) = \inf_{\{p_i, \rho_i\}} \sum_i p_i E_f(\rho_i), \qquad (8)$$

where $\{p_i, \rho_i\}$ moves over all the (not necessarily pure state) ensembles with $\sum_i p_i \rho_i = \rho$. The cases in which E_f is known are arbitrary states of 2×2 systems [37], isotropic states in arbitrary dimension [38], Werner and OO-symmetric states [1], and some other highly symmetric states [35].

Strong superadditivity of EoF, first considered in [1], is defined as follows. Let ρ be a state on $\mathcal{H}_1 \otimes \mathcal{H}_2$, where $\mathcal{H}_i = \mathcal{H}_{Ai} \otimes \mathcal{H}_{Bi}$ ($i = 1, 2$). Then superadditivity means that

$$E_f(\rho) \geq E_f(\mathrm{Tr}_{\mathcal{H}_1}\rho) + E_f(\mathrm{Tr}_{\mathcal{H}_2}\rho), \qquad (9)$$

where all entanglements of formation are understood with respect to the A-B partition of the respective system. Note that this implies additivity of E_f when applied to $\rho_1 \otimes \rho_2$ since we remarked above that the other inequality is trivial.

The intuitive appeal of the strong superadditivity property is that by measuring the entanglement via E_f, a system can only appear less entangled if judged by looking at its subsystems individually. Observe that it is sufficient to prove superadditivity for a *pure* state $\rho = |\psi\rangle\langle\psi|$, as then we can apply it to an optimal decomposition of ρ, together with the convex roof property, (8) [39].

There are some cases where strong superadditivity is proved. *Vollbrecht* and *Werner* [1] noted that it is true if one of the marginal states, say $\mathrm{Tr}_{\mathcal{H}_1}\rho$, is separable: because then its E_f is 0, and (9) simply expresses the monotonicity of E_f under local operations (in this case, partial traces). In [35], (16), it is actually proved if the partial trace in one of the subsystems is entanglement breaking.

Also, if the reduced state $\mathrm{Tr}_{\mathcal{H}_{A1} \otimes \mathcal{H}_{A2}}\rho$ of a pure state ρ is a tensored state, (9) is proved straightforwardly: such ρ is generated by applying a unitary U_A over $\mathcal{H}_{A1} \otimes \mathcal{H}_{A2}$ to $\sigma_1 \otimes \sigma_2$, with $\sigma_i \in \mathcal{S}(\mathcal{H}_i)$ being a pure state. Observe also due to the monotonicity of E_f under local operations, we have

$$E_f(\sigma_1) \geq E_f(\mathrm{Tr}_{\mathcal{H}_2}(U_A \otimes \mathbf{1}_B (\sigma_1 \otimes \sigma_2) U_A \otimes \mathbf{1}_B)),$$

and the similar inequality for $E_f(\sigma_2)$. Therefore, due to $E_f(\rho) = E_f(\sigma_1 \otimes \sigma_2) = E_f(\sigma_1) + E_f(\sigma_2)$, we have (9).

As a whole, the additivity question about EoF had been less understood than the additivity question about quantum channels. As discussed later, our project clarified that we can translate additivity results about quantum channels to that of quantum states, and vice versa.

3 Linking Additivity Conjectures

3.1 Channel States

In this subsection, the concept of channel state is introduced based on [2]. Due to a theorem of *Stinespring* [40] the completely positive and trace-preserving map T can be presented as the composition of an isometric embedding of $\mathcal{K}$ into a bipartite system $\mathcal{H} = \mathcal{H}_A \otimes \mathcal{H}_B$ with a partial trace:

$$T : \mathcal{B}(\mathcal{K}) \xrightarrow{U} \mathcal{B}(\mathcal{H}_A \otimes \mathcal{H}_B) \xrightarrow{\mathrm{Tr}_{\mathcal{H}_B}} \mathcal{B}(\mathcal{H}_A). \qquad (10)$$

See [41] for a discussion on how to construct this from the so-called Kraus (operator sum) representation [42], $T(\rho) = \sum_i A_i \rho A_i^\dagger$ with $\sum_i A_i^\dagger A_i = \mathbf{1}_\mathcal{K}$, of T. We shall use this construction later on using the examples 1 and 2.

By embedding into larger spaces we can present U as a restriction of a unitary, which often we silently assume done. Denote isometric embedding $U\mathcal{K} \subset \mathcal{H}_1 \otimes \mathcal{H}_2$ also by $\mathcal{K}$, so long as no confusion is likely to arise. Then we can say that T is equivalent to the partial trace channel, with inputs restricted to states on $\mathcal{K}$. This entails:

Lemma 1
$$C(T) = \sup_{\rho \in \mathcal{S}(\mathcal{K})} \{S\left(\mathrm{Tr}_{\mathcal{H}_B} \rho\right) - E_f(\rho)\}. \tag{11}$$

Note that if we choose the dimension of $\mathcal{H}_B$ large enough, every channel from $\mathcal{K}$ to $\mathcal{H}_A$ corresponds to a subspace of $\mathcal{H}_A \otimes \mathcal{H}_B$ (though not uniquely), and vice versa.

This has interesting consequences: let ρ_T be a state which we maximize (11). ρ_T, called a channel state of T, is not unique, but $T(\rho_T) = \mathrm{Tr}_{\mathcal{H}_B} \rho_T$ is unique, for (11) is strictly concave due to strict concavity of von Neumann entropy.

For channel states, the additivity of E_f is implied by the additivity of C for the corresponding channels. Let ρ_{T_1}, ρ_{T_2} be a channel state of T_1, T_2, respectively. Then, assuming additivity, we get

$$\begin{aligned} S(\mathrm{Tr}_{\mathcal{H}_{B1}} \rho_{T_1}) &- E_f(\rho_{T_1}) + S(\mathrm{Tr}_{\mathcal{H}_{B2}} \rho_{T_2}) - E_f(\rho_{T_2}) \\ &= C(T_1) + C(T_2) = C(T_1 \otimes T_2) \\ &\geq S(\mathrm{Tr}_{\mathcal{H}_{B1}} \rho_{T_1} \otimes \mathrm{Tr}_{\mathcal{H}_{B2}} \rho_{T_2}) - E_f(\rho_{T_1} \otimes \rho_{T_2}), \end{aligned} \tag{12}$$

hence

$$E_f(\rho_{T_1} \otimes \rho_{T_2}) \geq E_f(\rho_{T_1}) + E_f(\rho_{T_2}),$$

which by our earlier remarks implies additivity. Thus we have proved:

Theorem 1 *If for any two channels T_1 and T_2, each with a Stinespring dilation chosen as in (10), $C(T_1 \otimes T_2) = C(T_1) + C(T_2)$, then for a channel state ρ_{T_1} and ρ_{T_2} of T_1 and T_2,*

$$E_f(\rho_{T_1} \otimes \rho_{T_2}) = E_f(\rho_{T_1}) + E_f(\rho_{T_2}).$$

□

Most interesting is the case when we know $C(T^{\otimes n}) = nC(T)$, because then we can conclude $E_f(\rho^{\otimes n}) = nE_f(\rho)$, thus determining the entanglement cost of ρ (see Sect. 2.2). For example, *King* [27, 28] proved this for unital qubit channels, *Shor* [29] for entanglement-breaking channels, and *King* [30] for arbitrary depolarizing channels, giving rise to a host of states for which we thus know that the entanglement cost equals E_f.

In these applications, the following theorem in [1] is useful:

Theorem 2 *If $E_f\left(\sum_i p_i \pi_i\right) = \sum_i p_i E(\pi_i)$ for some probability distribution $\{p_i\}$ with $p_i \neq 0$ for all i, $E_f\left(\sum_i q_i \pi_i\right) = \sum_i q_i E(\pi_i)$ for any probability distribution $\{q_i\}$.*

This immediately implies the following lemma:

Lemma 2 *Let $\rho_\alpha = \sum_i p_i^\alpha \pi_i^\alpha$ be an optimal decomposition of ρ_α ($\alpha = 1, 2$), i.e., $E_f(\rho_\alpha) = \sum_i p_i^\alpha E(\pi_i^\alpha)$, with $p_i^\alpha \neq 0$ for any i, α. Suppose*

$$E_f(\rho_1 \otimes \rho_2) = E_f(\rho_1) + E_f(\rho_2).$$

Then, for states σ_α ($\alpha = 1, 2$) with $\sigma_\alpha = \sum_i q_i^\alpha \pi_i^\alpha$, additivity of EoF holds,

$$E_f(\sigma_1 \otimes \sigma_2) = E_f(\sigma_1) + E_f(\sigma_2).$$

Also, for a state $\rho' = \sum_{i,j} p'_{ij} \pi_i^1 \otimes \pi_j^2$,

$$E_f(\rho') = E_f(\operatorname{Tr}_{\mathcal{H}_1} \rho') + E_f(\operatorname{Tr}_{\mathcal{H}_2} \rho').$$

3.2 Strong Superadditivity and Additivity of Holevo Capacity

In this subsection, we prove that strong superadditivity of EoF suggests additivity of Holevo capacity with linear cost constraints. Additivity of Holevo capacity *without* constraints is shown by (11) in Sect. 3.1. Let us consider Stinespring dilation of two channels $T_i : S(\mathcal{K}_i) \to S(\mathcal{H}_{Ai})$ in $\mathcal{H}_i = \mathcal{H}_{Ai} \otimes \mathcal{H}_{Bi}$ ($i = 1, 2$). Denote by ρ a supposedly optimal state on $\mathcal{K}_1 \otimes \mathcal{K}_2$. Then we have

$$C(T_1 \otimes T_2) = S(\rho) - E_f(\rho) \leq S(\operatorname{Tr}_{\mathcal{H}_1} \rho) + S(\operatorname{Tr}_{\mathcal{H}_1} \rho) - E_f(\rho),$$
$$\leq S(\operatorname{Tr}_{\mathcal{H}_1} \rho) + S(\operatorname{Tr}_{\mathcal{H}_1} \rho) - E_f(\operatorname{Tr}_{\mathcal{H}_1} \rho) - E_f(\operatorname{Tr}_{\mathcal{H}_1} \rho),$$

where the first and second inequality is due to subadditivity of von Neumann entropy and strong superadditivity of EoF.

Let us show that it even implies an additivity formula for the classical capacity under linear cost constraints (see [43]): in this problem, there is given a self-adjoint operator A on the input system, and a real number α, additional to the channel T. As signal states we allow only such states σ on $\mathcal{H}^{\otimes n}$ for which $\operatorname{Tr}(\sigma \widehat{A}) \leq n\alpha + o(n)$, with

$$\widehat{A} = \sum_{k=1}^n \mathbf{1}^{\otimes(k-1)} \otimes A \otimes \mathbf{1}^{\otimes(n-k)}.$$

(That is, their average cost is asymptotically bounded by α.) Then it can be shown [43, 44] that the capacity $C(T; A, \alpha)$ in the thus constrained system *and using product states* is given by a maximization as in (1), only that the ensembles $\{p_i, \pi_i\}$ are restricted by $\sum_i p_i \operatorname{Tr}(\pi_i A) \leq \alpha$. (The same treatment applies if there are several linear cost inequalities of this kind. It is only for simplicity of notation that we stick to the case of a single one.) Because of the linearity of this condition in the states this yields a formula for $C(T; A, \alpha)$ very similar to Theorem 1:

$$C(T; A, \alpha) = \sup\{S\left(\operatorname{Tr}_{\mathcal{H}_B} \rho\right) - E_f(\rho) : \rho \text{ state on } \mathcal{K}, \ \operatorname{Tr}(\rho A) \leq \alpha\}. \quad (13)$$

By the general arguments given in previous sections we can conclude that this function is concave in α. The question, of course, is again, if entangled inputs help to increase the capacity, or if

$$C\left(T^{\otimes n}; \widehat{A}, n\alpha\right) \stackrel{?}{=} nC(T; A, \alpha). \quad (14)$$

We shall show that this indeed follows from the strong superadditivity, by showing the following: for channels T_1, T_2, cost operators A_1, A_2, and cost threshold $\widetilde{\alpha}$:

$$C\left(T_1 \otimes T_2; A_1 \otimes \mathbf{1} + \mathbf{1} \otimes A_2; \widetilde{\alpha}\right) = \sup_{\alpha + \alpha' = \widetilde{\alpha}} \{C(T_1; A_1, \alpha) + C(T_2; A_2, \alpha')\}.$$

Then, by induction and using the concavity, the equality in (14) follows.

Indeed, "$\geq$" is obvious by choosing, for $\alpha + \alpha' = \widetilde{\alpha}$, optimal states ρ_1, ρ_2 in the sense of (13), and considering $\rho_1 \otimes \rho_2$. In the other direction, assume any optimal ω for the product system, with marginal states ρ_1 and ρ_2. By definition,

$$\operatorname{Tr}\{\rho_1 \otimes \rho_2(A_1 \otimes \mathbf{1} + \mathbf{1} \otimes A_2)\} = \operatorname{Tr}\{\omega(A_1 \otimes \mathbf{1} + \mathbf{1} \otimes A_2\} \leq \widetilde{\alpha},$$

so also the product $\rho_1 \otimes \rho_2$ is admissible, and since there exist α, α' summing to $\widetilde{\alpha}$ such that $\operatorname{Tr}(\rho_1 A_1) \leq \alpha$, $\operatorname{Tr}(\rho_2 A_2) \leq \alpha'$, the claim follows in exactly the same way as for the unconstrained capacity.

We have thus proved:

Theorem 3 *Strong superadditivity of E_f (9) for all the states implies additivity of entanglement of formation, of the Holevo capacity, and of the Holevo capacity with cost constraint under tensor products.*

3.3 Equivalence Theorem by Shor, and One More New Equivalent Additivity Question

Using the concept of a channel state, *Shor* [4] had shown all the additivity conjectures are equivalent. Combining the main theorem of [4] and theorem 3 above, we obtain the following theorem.

Theorem 4 *The following (i)–(v) are equivalent:*

(i) *(7) holds for all quantum states*
(ii) *(9) holds for all quantum states*
(iii) *(2) holds for all quantum channels*
(iv) *(14) holds for all quantum channels*

(v) (3) holds for all quantum channels

Remark 1 *The main theorem of [4] showed (i), (ii), (iii), and (v) are equivalent.*

Due to this theorem, we have to work on only one of the additivity questions. Among them, many of researchers are focusing on (3) for its simplicity. A natural question is whether there is a simple additivity question about entanglement or not. Let us consider

$$E_m(\rho) := \min_{\{p_i, \pi_i\}} \min_i E(\pi_i), \tag{15}$$

where $\{p_i, \pi_i\}$ runs all over the ensembles of pure bipartite states with $\sum_i p_i \pi_i = \rho$. The additivity and the strong superadditivity of this quantity means

$$E_m(\rho_1 \otimes \rho_2) = E_m(\rho_1) + E_m(\rho_2), \tag{16}$$

and

$$E_m(\rho) \geq E_m(\mathrm{Tr}_{\mathcal{H}_2}\rho) + E_m(\mathrm{Tr}_{\mathcal{H}_1}\rho), \tag{17}$$

respectively. In the latter, ρ is a state on $\mathcal{H}_1 \otimes \mathcal{H}_2$ with $\mathcal{H}_i = \mathcal{H}_{Ai} \otimes \mathcal{H}_{Bi}$ ($i = 1, 2$), and entanglement is defined in A-B split.

Theorem 5 *The following (i)–(v) are equivalent:*

(i) (17) for all the pure states
(ii) (17) for all the states
(iii) (16) for all the states
(iv) (3) for all the quantum channels
(v) (9) for all the states

Proof For (iv)⇔(v) due to [4], it suffices to show (v)⇒(i)⇔(ii)⇒(iii)⇒(iv). In the following, let $\rho \in \mathcal{S}(\mathcal{H}_1 \otimes \mathcal{H}_2)$.
(v)⇒ (i): Let ρ be a pure state. Then,

$$\begin{aligned}E_m(\rho) = E(\rho) = E_f(\rho) &\geq E_f(\mathrm{Tr}_{\mathcal{H}_2}\rho) + E_f(\mathrm{Tr}_{\mathcal{H}_1}\rho), \\ &\geq E_m(\mathrm{Tr}_{\mathcal{H}_2}\rho) + E_m(\mathrm{Tr}_{\mathcal{H}_1}\rho).\end{aligned}$$

(i)⇒ (ii): Let π_* be a pure state living in the support of ρ with $E_m(\rho) = E(\pi_*)$. Then,

$$\begin{aligned}E_m(\rho) = E(\pi_*) &\geq E_m(\mathrm{Tr}_{\mathcal{H}_2}\pi_*) + E_m(\mathrm{Tr}_{\mathcal{H}_1}\pi_*), \\ &\geq E_m(\mathrm{Tr}_{\mathcal{H}_2}\rho) + E_m(\mathrm{Tr}_{\mathcal{H}_1}\rho),\end{aligned}$$

in which the second inequality comes from the assumption, and the third inequality due to the fact that the support of $\mathrm{Tr}_{\mathcal{H}_i}\pi_*$ is a subset of the support of $\mathrm{Tr}_{\mathcal{H}_i}\rho$.

(ii)⇐ (i), (ii) ⇒ (iii): Trivial.
(iii)⇒: (iv) Let T_i be a CPTP map from $\mathcal{B}(\mathcal{K}_i)$ to $\mathcal{B}(\mathcal{H}_{Ai})$. Then, we have

$$E_m\left(\frac{\mathbf{1}_{\mathcal{K}_i}}{\dim \mathcal{K}_i}\right) = \min_{\phi \in \mathcal{K}_i} S\left(\mathrm{Tr}_{\mathcal{H}_{Bi}}|\phi\rangle\langle\phi|\right) = \min_{\phi \in \mathcal{K}_i} S\left(T_i(|\phi\rangle\langle\phi|)\right) = S_{\min}(T_i),$$

and

$$E_m\left(\frac{\mathbf{1}_{\mathcal{K}_1}}{\dim \mathcal{K}_1} \otimes \frac{\mathbf{1}_{\mathcal{K}_2}}{\dim \mathcal{K}_2}\right) = S_{\min}(T_1 \otimes T_2). \tag{18}$$

Combining there equations, we have the assertion. □

Combining this theorem with Theorem 4, we can conclude the additivity of the new entanglement quantity is equivalent to all the other additivity questions.

Among all the additivity conjectures which are equivalent with each other, many researchers are focusing on additivity of the minimum output entropy. However, in the existing proofs of this additivity conjecture for the special cases (e.g., [8, 28, 29, 30]), they first show the strong superadditivity of E_m for all the pure states living in $\mathcal{K}_1 \otimes \mathcal{K}_2$,

$$E(\rho) \geq E_m(\mathrm{Tr}_{\mathcal{H}_2}\rho) + E_m(\mathrm{Tr}_{\mathcal{H}_1}\rho),$$

which naturally leads to the additivity of the minimum output entropy. Also, in many states for which the additivity or the strong super additivity of EoF is shown, EoF is equal to E_m (e.g., [8]). Therefore, it seems to the author that one cannot bypass strong superadditivity relations about E_m to prove any of the additivity relations. Therefore, considering its simplicity, this quantity is a good alternative to minimum output entropy.

3.4 Group Symmetry

In proving Theorem 3, we in fact had assumed (9) only for states living in $\mathcal{K}_1 \otimes \mathcal{K}_2$. On the other hand, the additivity of EoF of a state cannot be implied by the additivity of Holevo capacity of channels corresponding to subspaces $\mathcal{K}_1$ and $\mathcal{K}_2$. Instead, we have to consider nontrivial extensions of these channels, which are defined on infinite dimensional Hilbert spaces.

However, imposing a group symmetry via representation on the involved (sub)spaces as follows, we can prove very strong equivalence between the additivity properties of channels corresponding to subspaces $\mathcal{K}_1$ and $\mathcal{K}_2$ and states living in these subspaces.

Assume that a compact group G (with Haar measure dg) acts irreducibly both on $\mathcal{K}$ and $\mathcal{H}_A$ by a unitary representation (which we denote by V_g and U_g), which commutes with the map T (partial trace):

$$T\left(V_g \sigma V_g^\dagger\right) = U_g\left(T\sigma\right) U_g^\dagger. \tag{19}$$

We also consider the contravariant condition

$$T\left(V_g \sigma V_g^\dagger\right) = \overline{U_g}\, (T\sigma)\, \overline{U_g^\dagger}. \tag{20}$$

In the general, nonproduct case of (19), we can show (P denotes the projection onto $\mathcal{K}$ in $\mathcal{H}_A \otimes \mathcal{H}_B$):

Lemma 3 *Suppose also (19) or (20) holds for any $g \in G$, where G is a compact group with Haar measure dg, and V_g and U_g are irreducible representations on $\mathcal{K}$ and $\mathcal{H}_A$, respectively. Then we have*

$$C(T) = S\left(\frac{1}{\dim \mathcal{H}_B}\mathbf{1}_{\mathcal{H}_B}\right) - E_f\left(\frac{1}{\dim \mathcal{K}}\mathbf{1}_{\mathcal{K}}\right),$$
$$= \log \dim \mathcal{H}_B - E_f\left(\frac{1}{\dim \mathcal{K}}\mathbf{1}_{\mathcal{K}}\right). \tag{21}$$

$$E_f\left(\frac{1}{\dim \mathcal{K}}\mathbf{1}_{\mathcal{K}}\right) = E_m\left(\frac{1}{\dim \mathcal{K}}\mathbf{1}_{\mathcal{K}}\right), \tag{22}$$
$$= \min\{E(\psi) : |\psi\rangle \in \mathcal{K}\}.$$

Also

$$E_f(\rho) = E_m(\rho) = \min\{E(\psi) : |\psi\rangle \in \mathcal{K}\}$$

for all ρ spanned by $\{V_g|\psi_0\rangle\langle\psi_0|V_g^\dagger : g \in G\}$, where $|\psi_0\rangle$ is a pure state with $E(|\psi_0\rangle) = \min\{E(|\psi\rangle) : |\psi\rangle \in \mathcal{K}\}$. In particular, if in addition the action of G in $\mathcal{K}$ is transitive, this is true for all ρ supported on $\mathcal{K}$.

Proof We prove the assertion under the assumption (19). The assertion under the assumption (20) can be proved similarly. Observe that

$$S\left(\sum_i p_i T(\rho_i)\right) - \sum_i p_i S(T(\rho_i))$$
$$= S\left(\sum_i p_i U_g T(\rho_i) U_g^\dagger\right) - \sum_i p_i S(U_g T(\rho_i) U_g^\dagger),\ \forall g,$$
$$= \int \left\{S\left(\sum_i p_i U_g T(\rho_i) U_g^\dagger\right) - \sum_i p_i S(U_g T(\rho_i) U_g^\dagger)\right\} dg,$$
$$\leq S\left(\sum_i p_i \int U_g T(\rho_i) U_g^\dagger dg\right) - \min_\rho S(T(\rho)),$$
$$= S\left(\frac{1}{\dim \mathcal{H}_B}\mathbf{1}_{\mathcal{H}_B}\right) - \min\{E(\psi) : |\psi\rangle \in \mathcal{K}\}, \tag{23}$$

where the third inequality is due to the concavity of von Neumann entropy. Consider the ensemble $\{V_g|\psi_0\rangle\langle\psi_0|V_g^\dagger, dg\}$ with dg being a Haar measure.

By Shur's lemma, this gives a decomposition of $\frac{1}{\dim \mathcal{K}} \mathbf{1}_{\mathcal{K}}$, and this ensemble achieves the upper bound (23). Therefore, if (22) is correct, we have (21).

In (22) "$\geq$" is trivially true, and for the opposite direction choose a minimum entanglement pure state $|\psi_0\rangle \in \mathcal{K}$, and consider the decomposition $\{V_g|\psi_0\rangle\langle\psi_0|V_g^\dagger, dg\}$ of $\frac{1}{\dim \mathcal{K}} \mathbf{1}_{\mathcal{K}}$. All these states $V_g|\psi_0\rangle\langle\psi_0|V_g^\dagger$ have the same entanglement,

$$\begin{aligned} E(V_g|\psi_0\rangle) &= S\left(\mathrm{Tr}_1\left(V_g|\psi_0\rangle\langle\psi_0|V_g^\dagger\right)\right), \\ &= S\left(U_g \mathrm{Tr}_1|\psi_0\rangle\langle\psi_0|U_g^\dagger\right), \\ &= S\left(\mathrm{Tr}_1|\psi_0\rangle\langle\psi_0|\right) = E(\psi_0), \end{aligned} \qquad (24)$$

using (19). As for the capacity, in the light of (11) and using (21), the "$\leq$" is trivial, and the argument just given proves equality.

Due to Theorem 2 and (22), for *all* states ρ spanned by $\{V_g|\psi_0\rangle\langle\psi_0|V_g^\dagger : g \in G\}$, we have (3). $\square$

Note that in [35] a similar assertion was argued by making use of being in the "product case", i.e., the case where $V_g = \widetilde{U}_g \otimes U_g$ for a unitary representation of G on $\mathcal{H}_B$, denoted $\widetilde{U}_g$. In their argument, the group action on $\mathcal{K}$ is performable by LOCC, and they utilize nonincrease of E_f under LOCC transformations.

Theorem 6 *Suppose also (19) holds for any $g_i \in G_i$, where G_i is a compact group with Haar measure dg_i, and $V_{g_i}^{(i)}$ and $U_{g_i}^{(i)}$ are irreducible representations on $\mathcal{K}_i$ and $\mathcal{H}_{A_i}$, respectively. Then we have*

(i) If $C\left(\bigotimes T_i\right) = \sum_i C(T_i)$, we have $S_{\min}\left(\bigotimes T_i\right) = \sum_i S_{\min}(T_i)$, and vice versa.
(ii) If $E_f\left(\bigotimes_i \frac{1}{\dim \mathcal{K}_i} \mathbf{1}_{\mathcal{K}_i}\right) = \sum_i E_f\left(\frac{1}{\dim \mathcal{K}_i} \mathbf{1}_{\mathcal{K}_i}\right)$, we have $C\left(\bigotimes T_i\right) = \sum_i C(T_i)$.
(iii) If $C\left(\bigotimes T_i\right) = \sum_i C(T_i)$, then we have

$$E_f\left(\bigotimes_i \rho_i\right) = \sum_i E_f(\rho_i)$$

for all ρ_i spanned by

$$\{V_{g_i}^{(i)}(|\psi_{0,i}\rangle\langle\psi_{0,i}|)V_{g_i}^{(i)\dagger} : g_i \in G_i\},$$

with $|\psi_{0,i}\rangle = \mathrm{argmin}\{E(|\psi\rangle) : |\psi\rangle \in \mathcal{K}_i\}$. (In particular, if in addition the action of G_i in $\mathcal{K}_i$ is transitive, this coincides with the totality of states supported on $\mathcal{K}_i$.)

(iv) Suppose the action of G_i in $\mathcal{K}_i$ is transitive. Then, if E_m satisfies strong superadditivity $E_m(\rho) \geq \sum_i E_m\left(\rho|_{\mathcal{H}_i}\right)$ for all the states in $\bigotimes_i \mathcal{H}_i$, so does EoF,

$$E_f(\rho) \geq \sum_i E_f\left(\rho|_{\mathcal{H}_i}\right).$$

Also, additivity of Holevo capacity and $S_{\min}$ are concluded.

Proof We prove the assertion under the assumption (19). The assertion under the assumption (20) can be proved in a parallel fashion. (i) Almost parallel with the derivation of (23), we have

$$C\left(\bigotimes_i T_i\right) \leq S\left(\bigotimes_i \frac{1}{\dim \mathcal{K}_i}\mathbf{1}_{\mathcal{K}_i}\right) - S_{\min}\left(\bigotimes_i T_i\right),$$

$$= \sum_i S\left(\frac{1}{\dim \mathcal{K}_i}\mathbf{1}_{\mathcal{K}_i}\right) - S_{\min}\left(\bigotimes_i T_i\right).$$

This implies our assertion.

(ii) Almost parallel with the derivation of (23), we have

$$C\left(\bigotimes_i T_i\right) \leq S\left(\bigotimes_i \frac{1}{\dim \mathcal{K}_i}\mathbf{1}_{\mathcal{K}_i}\right) - \min\left\{E(\psi) : |\psi\rangle \in \bigotimes_i \mathcal{K}_i\right\},$$

$$= \sum_i S\left(\frac{1}{\dim \mathcal{K}_i}\mathbf{1}_{\mathcal{K}_i}\right) - E_f\left(\bigotimes_i \frac{1}{\dim \mathcal{K}_i}\mathbf{1}_{\mathcal{K}_i}\right),$$

$$= \sum_i S\left(\frac{1}{\dim \mathcal{K}_i}\mathbf{1}_{\mathcal{K}_i}\right) - \sum_i E_f\left(\frac{1}{\dim \mathcal{K}_i}\mathbf{1}_{\mathcal{K}_i}\right),$$

$$\leq \sum_i C(T_i).$$

For "$\geq$" is trivial, we have the assertion.

(iii) This is a direct consequence of Theorem 1 and Lemma 2.

(iv) Strong superadditivity of EoF and E_m are derived as follows:

$$E_f(\rho) \geq E_m(\rho) \geq \sum_i E_m\left(\rho|_{\mathcal{H}_i}\right) = \sum_i E_f\left(\rho|_{\mathcal{H}_i}\right).$$

Additivity of Holevo capacity follows from this almost in parallel with the proof of Theorem 3. Additivity of $S_{\min}$ naturally follows due to (i). □

3.5 Analysis of Examples

3.5.1 Example 1

Note this channel is covariant with respect to the action of Pauli group

$$T^{\mathbf{P}}(\sigma_i \rho \sigma_i) = \sigma_i T^{\mathbf{P}}(\rho)\sigma_i,$$

and $\mathbb{C}^2$ is irreducible space with respect to this group action. Hence, we can make use of the results in the preceding section.

It is easy to construct a Stinespring dilation for this map, by an isometry $U : \mathbb{C}^2 \longrightarrow \mathbb{C}^2 \otimes \mathbb{C}^4$, in block form:

$$U = \begin{pmatrix} \sqrt{p_0}\sigma_0 \\ \sqrt{p_x}\sigma_x \\ \sqrt{p_y}\sigma_y \\ \sqrt{p_z}\sigma_z \end{pmatrix},$$

and the corresponding subspace $\mathcal{K} \subset \mathbb{C}^2 \otimes \mathbb{C}^4$ is spanned by

$$|\psi_{\mathbf{p}}\rangle = \sqrt{p_0}|0\rangle \otimes |0\rangle + \sqrt{p_x}|1\rangle \otimes |x\rangle + i\sqrt{p_y}|1\rangle \otimes |y\rangle + \sqrt{p_z}|0\rangle \otimes |z\rangle,$$
$$|\psi_{\mathbf{p}}^{\perp}\rangle = \sqrt{p_0}|1\rangle \otimes |0\rangle + \sqrt{p_x}|0\rangle \otimes |x\rangle - i\sqrt{p_y}|0\rangle \otimes |y\rangle - \sqrt{p_z}|1\rangle \otimes |z\rangle.$$

Pure states $|\psi_{\mathbf{p}}\rangle$ and $|\psi_{\mathbf{p}}^{\perp}\rangle$ achieve the output minimum entropy.

Recall the additivity of Holevo capacity of $T^{\mathbf{p}}$ is shown. From these observations, for any ρ_i spanned by $|\psi_{\mathbf{p}_i}\rangle\langle\psi_{\mathbf{p}_i}|$ and $|\psi_{\mathbf{p}_i}^{\perp}\rangle\langle\psi_{\mathbf{p}_i}^{\perp}|$,

$$E_f(\rho_i) = S_{\min}(T^{\mathbf{p}_i}) = H(p_{i,0} + p_{i,z}, 1 - p_{i,0} - p_{i,z}),$$

and

$$E_f\left(\bigotimes_i \rho_i\right) = \sum_i E_f(\rho_i) .$$

In particular, for all the states ρ spanned by $|\psi_{\mathbf{p}}\rangle\langle\psi_{\mathbf{p}}|$ and $|\psi_{\mathbf{p}}^{\perp}\rangle\langle\psi_{\mathbf{p}}^{\perp}|$,

$$E_c(\rho) = E_f(\rho) = H(p_{i,0} + p_{i,z}, 1 - p_{i,0} - p_{i,z}). \tag{25}$$

3.5.2 Example 2

This channel is covariant with respect to the action of $SU(d)$, $\rho \to U\rho U^{\dagger}$:

$$T_{depol}^{d,\lambda}(U\rho U^{\dagger}) = UT_{depol}^{d,\lambda}(\rho)U^{\dagger}.$$

In addition, this representation of $SU(d)$ is obviously transitive.

Again, it is easy to construct a Stinespring dilation $U_{d,\lambda} : \mathbb{C}^d \longrightarrow \mathbb{C}^d \otimes \mathbb{C}^{d^2}$ in block form:

$$U_{d,\lambda} = \begin{pmatrix} \sqrt{\lambda}\mathbf{1}_d \\ \sqrt{\frac{1-\lambda}{d^2-1}}\sigma_1 \\ \vdots \\ \sqrt{\frac{1-\lambda}{d^2-1}}\sigma_{d^2-1} \end{pmatrix}.$$

Recall the additivity of Holevo capacity of $T_{depol}^{d,\lambda}$ is shown in [30]. Due to results in the preceding section, for all states ρ_i supported on $U_{d_i,\lambda_i}\mathbb{C}^{d_i}$,

$$E_f(\rho_i) = S_{\min}(T_{depol}^{d_i,\lambda_i}), \quad E_f\left(\bigotimes_i \rho_i\right) = \sum_i E_f(\rho_i).$$

In particular, for any state supported on $U_{d,\lambda}\mathbb{C}^d$,

$$E_c(\rho) = E_f(\rho).$$

In [30], what is actually shown is

$$E_m(\rho) \geq \sum_i E_m\left(\rho|_{U_{d_i,\lambda_i}\mathbb{C}^{d_i}}\right),$$

for all ρ supported on $\bigotimes_i U_{d_i,\lambda_i}\mathbb{C}^{d_i}$. Hence, we have

$$E_f(\rho) \geq \sum_i E_f\left(\rho|_{U_{d_i,\lambda_i}\mathbb{C}^{d_i}}\right).$$

3.5.3 Example 3

This channel is contravariant with respect to the action of $SU(d)$,

$$T_{WH}^d(U\rho U^\dagger) = \overline{U}T_{WH}^d(\rho)\overline{U}^\dagger.$$

Due to contravariancy of the channel T_{WH}^d, we have

$$S(T_{WH}^d(|\psi\rangle\langle\psi|)) = S(UT_{WH}^d(|i\rangle_{aa}\langle i|)U^\dagger) = S\left(T_{WH}^d(|i\rangle_{aa}\langle i|)\right)$$
$$= S\left(\frac{1}{d-1}(\mathbf{1}_d - |i\rangle\langle i|)\right) = \log(d-1).$$

Therefore,

$$C(T_{WH}^d(|\psi\rangle\langle\psi|)) = \log d - \log(d-1) = \log\frac{d}{d-1},$$

and

$$E_f(\rho) = E_m(\rho) = \log(d-1),$$

with ρ is supported on Stinespring dilation $\mathcal{K}$, which is homogeneous to $\mathcal{K}_d$,

$$\mathcal{K}_d := \mathbb{C}_*^d = \mathrm{span}_{\mathbb{C}}\{|1\rangle_a, |2\rangle_a, \ldots, |d\rangle_a\} \subset \mathbb{C}^{d\otimes(d-1)},$$

where $|i_1\rangle_a := \frac{1}{\sqrt{(d-1)!}}\sum_{i_2,\ldots,i_d}\epsilon_{i_1i_2\ldots i_d}|i_2\rangle\otimes\cdots\otimes|i_d\rangle$, $1 \leq i_1, i_2, \ldots, i_d \leq d$, and ϵ is totally antisymmetric tensor. Here, let $\mathcal{H}_A \otimes \mathcal{H}_B = \mathbb{C}^{d\otimes d-1}$, $\mathcal{H}_A := \mathbb{C}^d$, $\mathcal{H}_B := \mathbb{C}^{d\otimes(d-2)}$, and consider the entanglement between A-B split.

Suppose $U \in SU(d)$ acts on $\mathbb{C}^d$ as $U|i\rangle = \sum_j U_j^i |j\rangle$, then on $\mathbb{C}_*^d$,

$$U|i_1\rangle_a = \frac{1}{\sqrt{(d-1)!}} \sum_{i_2,\cdots,i_d} U^{\otimes(d-1)} \epsilon_{i_1\ldots i_d} |i_2 \ldots i_d\rangle,$$

$$= \frac{1}{\sqrt{(d-1)!}} \sum_{j_1,\cdots,j_d} (U^\dagger)_{i_1}^{j_1} \epsilon_{j_1\ldots j_d} |j_2 \cdots j_d\rangle = \sum_{j_1} (U^\dagger)_{i_1}^{j_1} |j_1\rangle_a, \quad (26)$$

where we have used the fact that the totally antisymmetric tensor $\epsilon_{j_1\ldots j_d}$ is invariant under $U^{\otimes d}$. For example, for $d = 3$,

$$|1\rangle_a = \frac{1}{\sqrt{2}}(|2\rangle|3\rangle - |3\rangle|2\rangle), \quad |2\rangle_a = \frac{1}{\sqrt{2}}(|3\rangle|1\rangle - |1\rangle|3\rangle),$$

$$|3\rangle_a = \frac{1}{\sqrt{2}}(|1\rangle|2\rangle - |2\rangle|1\rangle),$$

which we use to identify $\mathcal{A}$ with $\mathbb{C}^3$. The Hermitian conjugate of U in the right-hand side suggests that $\mathbb{C}_*^d$ is the dual (contragredient) space of $\mathbb{C}^d$ [45]. The corresponding Young diagrams are

$$\mathbf{d} = \square, \quad \mathbf{d}_* = \left.\begin{array}{c}\square\\\vdots\\\square\end{array}\right\} d-1.$$

Due to the discussions in the preceding section, it suffices to show $E_m(\rho) \geq \sum_i E_m\left(\rho|_{U_{d_i},\lambda_i \mathbb{C}^{d_i}}\right)$, which implies all the rest of additivity conjectures. This inequality will be proved in the succeeding section.

4 Additivity for Special Cases

4.1 Additivity of WH Channel

The WH channel is proposed as a counterexample to general multiplicativity conjecture of the maximum output p-norm of a channel. Additivity of Holevo capacity thus had been a famous unsolved problem until our project finally settled the problem, preceding other similar results by almost a year.

Prior to this final result, we had been studied additivity question about EoF of a state supported on $\mathbb{C}_*^3$. First, *Shimono* [5, 6, 9] showed

$$E_f(\rho \otimes \rho) = 2E_f(\rho),$$

and later Yura obtained [7]

$$E_f(\rho^{\otimes n}) = nE_f(\rho).$$

Indeed, *Matsumoto* et al. [2] developed their main idea in these researches. Finally, in [8] it is found out that the basic idea of [7] generalized to a state supported on $\mathbb{C}_*^d$. In the same result, utilizing the discussions in [2], this result is related to additivity of Holevo capacity of WH channels. In [46], the multiplicativity of maximum p-norm of WH channels was proved for $1 \leq p \leq 2$.

Later on, our results are rediscovered by several authors, almost a year later. First, *Datta* et al. [47] proved the weaker assertion:

$$C(T_{WH}^{d\otimes 2}) = 2C(T_{WH}^d),$$

followed by the work in [46], which proved the additivity of WH channels in the same sense as ours.

Our final goal of the subsection is to show:

Lemma 4 *For any $\rho \in \mathcal{S}\left(\bigotimes_{i=1}^n \mathbb{C}_*^{d_i}\right)$,*

$$E_m(\rho) \geq \sum_{i=1}^n \log(d_i - 1) = \sum_{i=1}^n E_m(\rho|_{\mathbb{C}_*^{d_i}}).$$

As a result of Theorem 6, this implies following theorems:

Theorem 7 *For any $\rho_i \in \mathcal{S}(\mathbb{C}_*^{d_i})$,*

$$E_f(\otimes_{i=1}^n \rho_i) = \sum_{i=1}^n \log(d_i - 1) = \sum_{i=1}^n E_f(\rho_i).$$

Also, for $\rho \in \mathcal{S}\left(\bigotimes_{i=1}^n \mathbb{C}_*^{d_i}\right)$,

$$E_f(\rho) \geq \sum_{i=1}^n \log(d_i - 1) = \sum_{i=1}^n E_f(\rho|_{\mathbb{C}_*^{d_i}}).$$

Theorem 8

$$C\left(\bigotimes_i T_{WH}^{d_i}\right) = \sum_i C\left(T_{WH}^{d_i}\right).$$

To prove Lemma 4, we use the following lemma:

Lemma 5 (see also [7]) *Let X be a positive semidefinite operator such that $\mathrm{Tr}\, X = 1$. Then $\mathrm{Tr}[-X \log X] \geq -\log(\mathrm{Tr}\, X^2)$.*

Proof Let $f(x) := -\log x$ over $\mathbb{R}_+$. It follows from the convexity of the function f that $f(\sum_i p_i x_i) \leq \sum_i p_i f(x_i)$, where $\sum_i p_i = 1$, $p_i \geq 0$ and $x_i > 0$. By setting $x_i = p_i (\forall i)$, we have $-\sum_i x_i \log x_i \geq -\log\left(\sum_i x_i^2\right)$. This inequality holds even for some x_i equal to zero under the convention $0 \log 0 = 0$. □

In what follows, we denote the identity map from $\mathcal{S}(\mathcal{K})$ to $\mathcal{S}(\mathcal{K})$ by $\mathbf{I}_\mathcal{K}$, and $\sum |X_{ij}|^2$ by $\|X\|^2$.

Lemma 6 *For an arbitrary state ρ in $\mathcal{S}(\mathcal{K} \otimes \mathbb{C}_*^d)$, we have $\|\mathbf{I}_\mathcal{K} \otimes T_{WH}^d(\rho)\|^2 = \frac{1}{(d-1)^2}\left\{(d-2)\|\mathrm{Tr}_{\mathbb{C}_*^d}\rho\|^2 + \|\rho\|^2\right\}$. Here, the dimension of $\mathcal{K}$ is arbitrary.*

Proof Decompose $\rho \in \mathcal{S}(\mathcal{K} \otimes \mathbb{C}_*^d)$ into the sum $\sum_{i,j} |i\rangle_{aa}\langle j| \otimes \rho_{ij}$, where ρ_{ij} are operators in $\mathcal{K}$. Due to the definition of T_{WH}^d, we have

$$\|\left(\mathbf{I}_\mathcal{K} \otimes T_{WH}^d\right)(\rho)\|^2 = \left\|\frac{1}{d-1}\sum_i \sum_{j \neq i} |i\rangle\langle i| \otimes \rho_{jj} - \frac{1}{d-1}\sum_{i,j \neq i} |i\rangle\langle j| \otimes \rho_{ji}\right\|^2,$$

$$= \frac{1}{(d-1)^2}\left\{\sum_k \left\|\sum_{i \neq k} \rho_{ii}\right\|^2 + \sum_{i \neq j} \|\rho_{ij}\|^2\right\}.$$

The first term of the right side of the equation is rewritten as follows:

$$\sum_k \left\|\sum_{i \neq k} \rho_{ii}\right\|^2 = \sum_k \sum_{i \neq k, j \neq k} \mathrm{Tr}\rho_{ii}\rho_{jj},$$

$$= (d-1)\sum_i \|\rho_{ii}\|^2 + (d-2)\sum_{i \neq j} \mathrm{Tr}\rho_{ii}\rho_{jj},$$

$$= (d-2)\left\|\sum_i \rho_{ii}\right\|^2 + \sum_i \|\rho_{ii}\|^2.$$

Hence, after all we have,

$$\|\mathbf{I}_\mathcal{K} \otimes T_{WH}^d(\rho)\|^2 = \frac{1}{(d-1)^2}\left\{(d-2)\left\|\sum_i \rho_{ii}\right\|^2 + \sum_{i,j} \|\rho_{ij}\|^2\right\},$$

$$= \frac{1}{(d-1)^2}\left\{(d-2)\|\mathrm{Tr}_{\mathbb{C}_*^d}\rho\|^2 + \|\rho\|^2\right\},$$

and the lemma is proven. □

Lemma 7 *For any $\rho \in \mathcal{S}\left(\mathcal{K} \otimes \bigotimes_{i=1}^n \mathbb{C}_*^{d_i}\right)$,*

$$\left\|\left(\mathbf{I}_\mathcal{K} \otimes \bigotimes_{i=1}^n T_{WH}^{d_i}\right)(\rho)\right\|^2 \leq \prod_{i=1}^n \frac{1}{d_i - 1},$$

where the dimension of $\mathcal{K}$ is arbitrary.

Proof Induction is used for the proof. First, for $n = 1$, the assertion follows directly from Lemma 6, because $\|\sigma\| \leq 1$ holds for any density matrix σ. Second, let us assume the assertion is true for $n-1$. Then, Lemma 6 implies,

$$\left\| \left(\mathbf{I}_\mathcal{K} \otimes \bigotimes_{i=1}^n T_{WH}^{d_i} \right)(\rho) \right\|^2$$

$$= \frac{1}{(d_n - 1)^2} \left\{ (d_n - 2) \left\| \left(\mathbf{I}_\mathcal{K} \otimes \bigotimes_{i=1}^{n-1} T_{WH}^{d_i} \right)(\operatorname{Tr}_{\mathbb{C}_*^{d_n}} \rho) \right\|^2 \right.$$

$$\left. + \left\| \left(\mathbf{I}_{\mathcal{K} \otimes \mathbb{C}_*^{d_n}} \otimes \bigotimes_{i=1}^{n-1} \right) T_{WH}^{d_i}(\rho) \right\|^2 \right\}$$

$$\leq \frac{1}{(d_n - 1)^2} \left\{ (d_n - 2) \prod_{i=1}^{n-1} \frac{1}{d_i - 1} + \prod_{i=1}^{n-1} \frac{1}{d_i - 1} \right\} = \prod_{i=1}^{n} \frac{1}{d_i - 1},$$

where the inequality in the second line comes from the assumption of induction. Thus, the lemma is proven. □

This lemma leads to our final Lemma 4 as follows:

$$E_m(\rho) \geq -\min_\rho \log \operatorname{Tr} \left(\left(\bigotimes_{i=1}^n T_{WH}^{d_i} \right)(\rho) \right)^2,$$

$$\geq -\min_\rho \log \operatorname{Tr} \left(\left(\mathbf{I}_\mathcal{K} \otimes \bigotimes_{i=1}^n T_{WH}^{d_i} \right)(\rho) \right)^2,$$

$$\geq \sum_{i=1}^n \log(d_i - 1) = \sum_{i=1}^n E_m\left(\rho|_{\mathbb{C}^{d_i}}\right).$$

Note what is done here is essentially

$$\max_\rho \operatorname{Tr} \left(\left(\bigotimes_{i=1}^n T_{WH}^{d_i} \right)(\rho) \right)^2 \leq \prod_{i=1}^n \frac{1}{d_i - 1} = \prod_{i=1}^n \max_{\rho_i} \operatorname{Tr} \left(T_{WH}^{d_i}(\rho_i) \right)^2,$$

or

$$\nu_2 \left(\bigotimes_{i=1}^n T_{WH}^{d_i} \right) \leq \prod_{i=1}^n \nu_2 \left(T_{WH}^{d_i} \right).$$

For "$\geq$" is trivial, we have multiplicativity of the maximal output p-norm for $p = 2$.

5 Numerical Studies on Additivity Questions

5.1 A Qubit Channel That Requires Four Input States

In this subsection, we numerically study additivity question (2) for a qubit channel. It is a consequence of Carathéodory's theorem and the convex structure of the left hand side of (1) that the supremum in (1) can be replaced with the maximum over four pairs of $\{p_i, \pi_i\}$. It was demonstrated in [48] that there exist qubit channels requiring three input states to attain the maximum. However, it was left open whether or not there are one-qubit channels requiring four input states to achieve the maximum. Our project showed that such four-input channels do exist by presenting an example.

In addition to being of interest in their own right, four-state channels are good candidates for testing the additivity conjecture of the Holevo capacity for qubit channels. We present numerical evidence for additivity which, in view of special properties of the channels, gives extremely strong evidence for additivity of both capacity and minimal output entropy for qubit channels.

5.1.1 Some Useful Facts

Let us denote by $D(\rho||\sigma)$ a Umegaki relative entropy $\mathrm{Tr}\rho(\log\rho - \log\sigma)$ of two states. It was shown in [49] and [50] that

$$C(T) = \inf_\rho \sup_\omega D\left(T(\omega)||T(\rho)\right). \tag{27}$$

It is known that infimum is achieved when ρ is the optimal average input, and

$$C(T) = D\left(T(\pi_i)||T(\rho)\right) \tag{28}$$

for all i with $\{p_i, \pi_i\}$ being an optimal ensemble. Equation (27) implies

$$C(T) \leq \sup_\omega D\left(T(\omega)||T(\rho)\right), \forall \rho.$$

This can be used to check a numerical result. Let ρ be an approximate optimal average input state ρ obtained by a simulation. The result of numerical simulation is not reliable if the right-hand side considerably exceeds an approximate Holevo capacity obtained by the numerical simulation.

By virtue of nice properties of minimizing convex functions (e.g., Theorem 27.4 in [51]), ρ is an optimal average input state if and only if there is a Hermitian matrix Ξ such that for any σ,

$$\mathrm{Tr}\Xi(\sigma - \rho) + E_f(\rho) \leq E_f(\sigma) \tag{29}$$
$$\leq S(T(\sigma)) \leq \mathrm{Tr}\Xi(\sigma - \rho) + S(T(\rho)).$$

Here, E_f is considered in Stinespring dilation. This condition supplies another check of validity of a numerical result.

Equation (27) can also be used to check additivity without need to carry out the full variation in (1). In fact, applying (27) to the product channel $T \otimes T$ gives

$$2C(T) \leq C(T \otimes T) \leq \sup_{\omega} D\left((T \otimes T)(\omega) \, || \, T(\rho) \otimes T(\rho)\right). \tag{30}$$

If the supremum on the right equals $2C(T)$, then the channel is additive, and vice versa.

5.1.2 Setup

In the case of qubits, it is well-known that the set D of density matrices is isomorphic to the unit ball in $\mathbb{R}^3$ via the Bloch sphere representation. We will use the notation $\rho(\boldsymbol{x}) = \rho(x,y,z)$ to denote the density matrix $\frac{1}{2}[I + x\sigma_x + y\sigma_y + z\sigma_z]$. It was shown in [25, 32] that, up to specification of bases, a qubit channel can be written in the form

$$T[\rho(x,y,z))] = \rho(\lambda_1 x + t_1, \lambda_2 y + t_2, \lambda_3 z + t_3), \tag{31}$$

which gives an affine transformation on the Bloch sphere. In fact, it maps the Bloch sphere $\{\boldsymbol{x} = (x,y,z) \mid x^2 + y^2 + z^2 \leq 1\}$ to an ellipsoid with axes of lengths $\lambda_1, \lambda_2, \lambda_3$ and center $t_1, t_2 y, t_3$. Complete positivity poses additional constraints on the parameters $\{\lambda_k, t_k\}$ that are given in [32, 52]. The strict concavity of $S(\rho)$ implies that $S[T(\rho)]$ is also strictly concave for channels which are one-to-one. In the case of qubits, this will hold *unless* the channel maps the Bloch sphere into a one- or two-dimensional subset, which can only happen when one of the parameters $\lambda_k = 0$.

In the Bloch sphere representation, (29) reads as follows: $\rho(\mathbf{x})$ is a optimal input if and only if there is a $\xi \in \mathbb{R}^3$ such that for any $\boldsymbol{x}' \in \mathbb{R}^3$,

$$\xi^{\mathrm{T}}(\boldsymbol{x}' - \boldsymbol{x}) + E_f(\rho(\boldsymbol{x})) \leq E_f(\rho(\boldsymbol{x}'))$$
$$\leq S(T(\rho(\boldsymbol{x}'))) \leq \xi^{\mathrm{T}}(\boldsymbol{x}' - \boldsymbol{x}) + S(T(\rho(\boldsymbol{x}))).$$

5.1.3 Heuristic Construction of a Four-State Channel

The existence of four state channels of the type found above can be understood as emerging from small deformations of 3-state channels with a high level of symmetry. As noted above, a channel of the form (31) maps the Bloch sphere to an ellipsoid with axes of lengths $\lambda_1, \lambda_2, \lambda_3$ and center t_1, t_2, t_3. When $t_1 = t_2 = t_3$, the ellipsoid is centered at the original and the capacity is achieved with a pair of orthogonal inputs which map to the endpoints of the longest axis of the ellipsoid. However, when some t_k are nonzero, this no longer holds.

One of the 3-state channels in [48] is

$$T_3(\rho(x,y,z)) := \rho(0.6x, 0.6y, 0.5z + 0.5), \qquad (32)$$

which has rotational symmetry about the z-axis of the Bloch sphere.

To find a true four-state channel, the symmetry must be lowered so that the full three-dimensional geometry of the Bloch sphere is required. Keeping the CP condition [17, 25, 52] in mind, we deform the channel (32), and obtain

$$T_4[\rho(x,y,z))] := \rho(0.6x + 0.21, 0.601y, 0.5z + 0.495). \qquad (33)$$

Observe that replacing *all* inputs $\rho_i(x,y,z)$ by $\rho_i(x,-y,z)$ leaves the capacity unchanged. Therefore, *either* all optimal inputs lie in the x–z plane *or* the set of optimal inputs contains pairs of the form $\rho(x, \pm y, z)$ with the same probability. (This follows easily from a small modification of the convexity argument in [48].)

In view of the discussion above, it is reasonable to expect that one can find a family of four-state channels which have the form $T(\rho(x,y,z)) = \rho(\lambda_1 x + \epsilon_1, (\lambda_1 + \epsilon_2)y, \lambda_z + t_3)$ with ϵ_k suitable small constants, $\lambda_3 + t_3 = 1 - \epsilon_3$, and $\lambda_1 > \lambda_3$ chosen so that $T(\rho(x,y,z)) = \rho(\lambda_1 x, \lambda_2 y, \lambda_3 z + t_3)$ is close to a three-state channel.

In the class of channels above, one always has $t_2 = 0$, which raises the question of whether or not there exist four-state channels with all t_k all nonzero. Therefore, maps of the form $T(\rho(x,y,z)) = \rho(0.6x + 0.021, 0.601y + t_2, 0.5z + 0.495)$ were considered with $t_2 \neq 0$. With $t_2 < 0.48$ such maps are completely positive, and we have showed that the channel with $t_2 = 0.00005$ requires four inputs to achieve capacity.

5.1.4 Approximation Algorithm to Compute the Holevo Capacity

To study the issue numerically, one has to have an algorithm to compute Holevo capacity. The first algorithm is a quantum version of well-known Arimoto–Blahut algorithm developed in [53]. Later on, use of interior-point methods is suggested by our project [54]. A method is presented in [55] for computing the capacity by combining linear programming techniques, including column generation, with convex optimization.

In our study, we used the following approximation algorithm, which is almost sufficient to compute the Holevo capacity of a one-qubit channel in practice.

In (1), let $\{\rho_i\} = D$, with i being a continuous variable. This infinite set may be regarded as fixed, leaving only p_i as variables. The objective function is concave with respect to p_i.

Owing to the concavity of the von Neumann entropy, $\boldsymbol{x}$ can be restricted to a pure state. In case of a one-qubit channel, this corresponds to $x^2 + y^2 +$

$z^2 = 1$ in terms of the Bloch sphere. The sphere is two-dimensional, and is approximated by a square mesh of $k^2 - k + 2$ points, with $k = 100$. Then, a close lower bound to the real maximum is given by considering this concave maximization problem with respect to $\{p_i\}(1 \leq i \leq 100^2 - 100 + 2)$.

Interior-point methods can be applied to this high-dimensional concave maximization programming problem (e.g., [56]). This was done utilizing a mathematical programming package NUOPT [57] (Mathematical Systems, Inc.). These results, accurate to at most 7–8 significant figures, were further refined by using them as starting points in a program to find a critical point of the capacity by applying Newton's method to the gradient.

5.1.5 Numerical Verification of Four-State Channel

We first check whether the channel T_4 requires four input states. The results are shown in Table 1.

Table 1. Data for four-state channel T_4. ϕ, θ denote the angular coordinates

capacity $= 0.321\,485\,158\,9$
$S(T(\rho_i(\boldsymbol{x}))) - \xi^{\mathrm{T}})\boldsymbol{x} = 0.978\,505\,562\,1 \quad \forall i$
$D(T(\rho_i)||T(\rho)) = 0.321\,485\,158\,9 \quad \forall i$

Probability	Optimal input (x, y, z)			ϕ	θ
0.232 282 570 5	(0.253 075 986 2,	−0.000 000 000 0,	0.967 446 404 3)	0.127 929	0
0.213 322 081 9	(0.978 395 099 9,	0.000 000 000 0,	0.206 743 871 8)	0.681 275	0.0
0.277 197 673 8	(−0.473 408 753 3,	0.864 646 138 9,	−0.168 140 437 6)	0.869 870	2.071 131
0.277 197 673 8	(−0.473 408 753 3,	−0.864 646 138 9,	−0.168 140 437 6)	0.869 870	−2.071 131
Average	(0.005 042 809 9,	0.000 000 000 0,	0.175 607 694 4)		

To verify that these results give a true four-state optimum, the function $S(T_4(\rho(\boldsymbol{x}))) - \xi^{\mathrm{T}}\boldsymbol{x}$ was computed and plotted, where

$$\xi = (-0.039\,662\,202\,2, 0, -0.962\,107\,144\,0).$$

These results are shown in Fig. 1 and confirm the condition that the hyperplane $(\xi, -1) \cdot (x, y, z, w) = -0.978\,505\,562\,1$ passes through the four points $((x_i, y_i, z_i, S(T_4[\rho(x_i, y_i, z_i)])))$ and the condition that the hyperplane lies below the surface $(x, y, z, S(T_4[\rho(x, y, z)]))$ in $\mathbb{R}^4$. (The components ξ_x, ξ_y, ξ_z of ξ are obtained by solving the four simultaneous equations $\xi^{\mathrm{T}}\boldsymbol{x} + \xi_0 = S(T_4[\rho(x_i, y_i, z_i)])$ ($k = 1, 2, 3, 4$) for the variables $(\xi_x, \xi_y, \xi_z, \xi_0)$.)

In addition, the optimal three-state capacity was also computed and shown to be < 0.321461, which is strictly less than the four-state capacity of 0.321485. As an optimization problem, the capacity has other local maxima in addition to the three-state and four-state results discussed above. For example, there are several two-state optima, but these have lower capacity and are not relevant to the work presented here.

It was also checked that the four-state optimal ensemble satisfies (28), and $D\left(T_4(\pi_i^4)\|T_4(\rho^4)\right) = 0.321485159$ for all i. The three-state optimal ensemble also satisfies the same condition with $D[T_4(\pi_i^3), T_4(\rho^3)] = 0.321460988 \; \forall \; i$. However,

$$\sup_\omega D\left(T_4(\omega)\|T_4(\rho^3)\right) > 0.3215 > D\left(T_4(\pi_i)\|T_4(\rho^3)\right),$$

showing that the three-state ensemble is *not* optimal.

5.1.6 Numerical Check of Additivity

As mentioned earlier, four-state channels might be good candidates for examining the additivity of channel capacity. Those considered here have the property $\lambda_2 > \max_{i=1,3} |\lambda_i|$, $t_2 = 0$ and $t_1, t_3 \neq 0$. Channels of this type do not belong to one of the classes of qubit maps for which multiplicativity of the maximal p-norm has been proved and its geometry seems resistant to simple analysis. (See [58] for a summary and further references.)

We will use (30). The function $g(\rho) = D\left(T_4(\rho)\|T_4(\rho^4)\right)$ has ten critical points (four maxima, four saddle points, and two (relative) minima), as shown in Fig. 2. This implies that $G(\omega) := D\left(T_4^{\otimes 2}(\omega)\|T_4^{\otimes 2}(\rho^{4\otimes 2})\right)$ has at least 100 critical points, 16 maxima, 4 (relative) minima, and 80 saddle-like critical points when one restricts ω to a product state. The complexity of this landscape seems greater than that of any other class of channels studied. If the capacity of any qubit channel is nonadditive, it seems likely that it would be a channel of this type. Therefore, a thorough numerical analysis is called for. Unfortunately, the large number of critical points also make a full optimization very challenging.

It suffices to optimize over pure states $\omega = |\Psi\rangle\langle\Psi|$, whose Schmidt form writes

$$|\Psi\rangle = \sqrt{p} \begin{pmatrix} \cos\theta_u \\ e^{i\phi_u}\sin\theta_u \end{pmatrix} \otimes \begin{pmatrix} \cos\theta_v \\ e^{i\phi_v}\sin\theta_v \end{pmatrix}$$
$$+ e^{i\nu}\sqrt{1-p} \begin{pmatrix} e^{-i\phi_u}\sin\theta_u \\ -\cos\theta_u \end{pmatrix} \otimes \begin{pmatrix} e^{-i\phi_v}\sin\theta_v \\ -\cos\theta_v \end{pmatrix}, \quad (34)$$

and $p \in [0,1]$, $\theta_u, \theta_v, \nu \in [0, 2\pi]$, $\phi_u, \phi_v \in [0, \frac{\pi}{2}]$.

Because of the difficulty of optimizing over all six parameters, plots of $G(\omega)$ were made as a function of only p, ν with u, v fixed and as a function of p with the remaining five parameters fixed. A typical example is shown in Fig. 3 and appears to be a convex function in p for several choices of nu. Many other examples were considered with u, v both corresponding to optimal inputs, u, v chosen randomly, u, v chosen to be highly nonoptimal, and various combinations of these. The shape of the curve seems to be extremely resilient for all inputs in Schmidt form (34) and suggests convexity in p with a deep minimum. Although the minimum lies above that for the corresponding

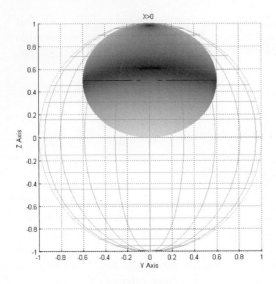

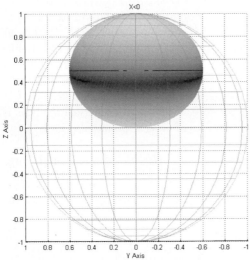

(a) output states $T(\rho(x,y,z))$ on the image ellipsoid. *top:* $x > 0$; *bottom:* $x < 0$

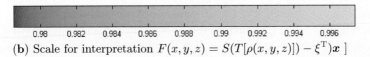

(b) Scale for interpretation $F(x,y,z) = S(T[\rho(x,y,z)]) - \xi^{\mathrm{T}})x$]

Fig. 1. Depiction of $F(x,y,z) = S(T_4[\rho(x,y,z)]) - \xi^{\mathrm{T}})x$ with respect to optimal average output in terms of gray scale on the boundary of the Bloch sphere and its image

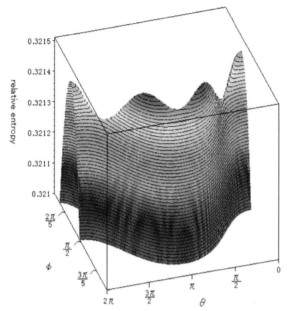

Plot of $D(T_4(\omega(\cos\theta\sin\phi, \sin\theta\sin\phi, \cos\phi)||T_4(\rho^4))$. Depicted only near the area $\phi = \frac{\pi}{2}$ showing 3 distinct maxima and saddle points.

Fig. 2. Plots of relative entropy of output states with respect to the optimal average output as a function of a pair of angles defining pure input states on the surface of the Bloch sphere

mixed state with $X = 0$, it is well below both endpoints. Changes as ν ranges from 0 to 2π are small.

States of the form $\frac{1}{\sqrt{2}}\left(|u_i\rangle \otimes |u_j\rangle + e^{i\nu}|u_k\rangle \otimes |u_\ell\rangle\right)$ with u_i corresponding to the four optimal inputs were also considered. Although the relative entropy has a slightly different shape as a function of p and ν, it still lies below the plane $2C(T_4)$ and has a deep minimum.

Thus, there seems to be little room for obtaining a counterexample by varying the channel parameters. This may give the strongest numerical evidence for additivity yet, at least in the case of qubit channels.

5.2 Strong Superadditivity of EoF of Pure States

One of central difficulty of numerical verification of (9) is that EoF of a state is in general hard to compute. However, we can partly sidestep this problem if the given four-partite state is pure: the left-hand side of (9) equals entropy of entanglement. This restriction is made without spoiling generality: (9) for all the pure states is sufficient to prove the relation for all the states.

In addition, we restrict ourselves to four-qubit states. Then, the right-hand side of (9) is sum of EoF of bipartite qubit states, which are easily

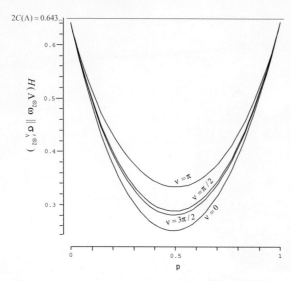

Fig. 3. Typical plot of $G(\omega) = D(T_4^{\otimes 2}(\omega)||T_4(\rho) \otimes T_4(\rho))$ as of function of p for $\nu = 0, \frac{\pi}{2}, \pi, \frac{3\pi}{2}$ using pure states of the form (34) and u, v fixed and $e^{i\nu} = 1, i, -i, -1$. Endpoints correspond to product states and $p = 0.5$ maximally entangled

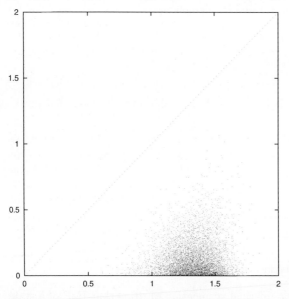

Fig. 4. Eight thousand states are randomly chosen from the whole space of four-qubit states. For each state, the corresponding point is plotted at $(x, y) = (lhs, rhs)$ of (9)

computed via Wootter's concurrence [37]. Therefore, (9) is easily verifiable for each four-qubit pure state.

In Fig. 4 we depict the result of our test of (9) for 8 000 points randomly chosen from whole four-qubit pure states. In all cases (1 000 000 points), violation of the inequality has not been observed.

Observe that most of the points are far below, $y = x$ line. To sample more points near the $y = x$ line, we tested (9) for 1 000 000 points chosen near the tensor product states, and again violation of the inequality was not observed.

Acknowledgements

The author thanks Dr. Toshiyuki Shimono for providing the figures.

References

[1] K. G. H. Vollbrecht, R. F. Werner: Entanglement measures under symmetry, Phys. Rev. A **64**, 062307 (2001)
[2] K. Matsumoto, T. Shimono, A. Winter: Remarks on additivity of the Holevo channel capacity and of the entanglement of formation, Commun. Math. Phys. **246**, 427–442 (2004)
[3] K. Matsumoto: Yet another additivity conjecture, Phys. Lett. A **350**, 179–181 (2006)
[4] P. W. Shor: Equivalence of additivity questions in quantum information theory, Commun. Math. Phys. **246**, 453–472 (2004)
[5] T. Shimono: Lower bound for entanglement cost of antisymmetric states (2002) URL http://lanl.arxiv.org/abs/quant-ph/0203039
[6] T. Shimono: Towards additivity of entanglement of formation, in C. Calude, M. J. Dinneen, F. Peper (Eds.): *Proc. Third International Conference on Unconventional Models of Computation*, LNCS **2509** (Springer 2002) pp. 252–263
[7] F. Yura: Entanglement cost of three-level antisymmetric states, J. Phys. A: Math. Gen. **36**, L237–L242 (2003)
[8] K. Matsumoto, F. Yura: Entanglement cost of antisymmetric states and additivity of capacity of some quantum channels, J. Phys. A **37**, L167 (2004) URL http://lanl.arxiv.org/abs/quant-ph/0306009
[9] T. Shimono: Additivity of entanglement of formation of two three-level-antisymmetric states, Int. J. Quant. Inform. **1**, 259–267 (2003)
[10] H. Fan: Additivity of entanglement of formation for some special cases (2002) URL http://lanl.arxiv.org/abs/quant-ph/0210169
[11] T. Hiroshima: Additivity and multiplicativity properties of some Gaussian channels for Gaussian inputs, Phys. Rev. A **73** (2006)
[12] M. Hayashi, H. Imai, K. Matsumoto, M. B. Ruskai, T. Shimono: Qubit channels which require four inputs to achieve capacity: Implications for additivity conjectures, Quantum Inf. Comput. **5**, 13–31 (2005)
[13] T. Shimono, H. Fan: Numerical test of superadditivity of entanglement of formation for 4 qubit states, in *Third Erato Conference on Quantum Information Science poster session* (2003) p. 119

[14] A. S. Holevo: The capacity of the quantum channel with general signal states, IEEE Trans. Inf. Theory **44**, 269–273 (1998)
[15] B. Schumacher, M. D. Westmoreland: Sending classical information via noisy quantum channels, Phys. Rev. A **56**, 131–138 (1997)
[16] E. B. Davies: Information and quantum measurements, IEEE Trans. Inf. Theory **24**, 596–599 (1978)
[17] A. Fujiwara, H. Nagaoka: Operational capacity and pseudoclassicality of a quantum channel, IEEE Trans. Inf. Theory **44**, 1071–1086 (1988)
[18] A. S. Holevo: Problems in the mathematical theory of quantum communication channels, Rep. Math. Phys. **12**, 273–278 (1979)
[19] C. H. Bennett, C. A. Fuchs, J. A. Smolin: Entanglement-enhanced classical communication on a noisy quantum channel, in O. Hirota, A. S. Holevo, C. M. Caves (Eds.): *Quantum Communication, Computing and Measurement* (Plenum, New York 1997) pp. 79–88
[20] S. Osawa, H. Nagaoka: Theoretical basis and applications of the quantum Arimoto–Blahut algorithms, in *Proc. the 2nd Workshop on Quantum Information Technologies* (1999) pp. 107–112
[21] S. Osawa, H. Nagaoka: Numerical experiments on the capacity of quantum channel with entangled input states, IEICE Trans. Fundamentals **E84-A**, 2583–2590 (2001)
[22] G. Amosov, A. S. Holevo, R. F. Werner: On the additivity hypothesis in quantum information theory, Problemy Peredachi Informatsii **36**, 25–34 (2000) in Russian. English translation in Probl. Inf. Transm. **4**, pp. 305–313 (2000)
[23] G. G. Amosov, A. S. Holevo: On the multiplicativity conjecture for quantum channels, Theor. Probab. Appl. **47**, 143–146 (2002)
[24] D. Bruss, L. Faoro, C. Macchiavello, M. Palma: Quantum entanglement and classical communication through a depolarizing channel, J. Mod. Optics **47**, 325–331 (2000)
[25] C. King, M. B. Ruskai: Minimal entropy of states emerging from noisy quantum channels, IEEE Trans. Inf. Theory **47**, 192–209 (2001)
[26] A. Fujiwara, T. Hashizume: Additivity of the capacity of depolarizing channels, Phys. Lett. A **299**, 469–475 (2002)
[27] C. King: Maximization of capacity and ℓ_p norms for some product channels, J. Math. Phys. **43**, 1247–1260 (2002)
[28] C. King: Additivity for unital qubit channels, J. Math. Phys. **43**, 4641–4653 (2002)
[29] P. W. Shor: Additivity of the classical capacity of entanglement-breaking quantum channels, J. Math. Phys. **43**, 4334–4340 (2002)
[30] C. King: The capacity of the quantum depolarizing channel (2002) URL http://lanl.arxiv.org/abs/quant-ph/0204172
[31] R. F. Werner, A. S. Holevo: Counterexample to an additivity conjecture for output purity of quantum channels, J. Math. Phys. **43**, 4353–4357 (2002)
[32] A. Fujiwara, P. Algoet: One-to-one parameterization of quantum channels, Phys. Rev. A **59**, 3290–3294 (1999)
[33] R. F. Werner: All teleportation and dense coding schemes, J. Phys. A **34**, 7081–7094 (2001)
[34] C. H. Bennett, D. P. DiVincenzo, J. A. Smolin, W. K. Wootters: Mixed-state entanglement and quantum error correction, Phys. Rev. A **54**, 3824–3851 (1996)

[35] G. Vidal: Entanglement cost of mixed states, Phys. Rev. Lett. **89**, 027901 (2002)
[36] P. M. Hayden, M. Horodecki, B. M. Terhal: The asymptotic entanglement cost of preparing a quantum state, J. Phys. A: Math. Gen. **34**, 6891–6898 (2001)
[37] W. K. Wootters: Entanglement of formation of an arbitrary state of two qubits, Phys. Rev. Lett. **80**, 2245–2248 (1998)
[38] B. M. Terhal, K. G. H. Vollbrecht: The entanglement of formation for isotropic states, Phys. Rev. Lett. **85**, 2625–2628 (2000)
[39] F. Benatti, H. Narnhofer: On the additivity of the entanglement of formation, Phys. Rev. A **63**, 042306 (2001)
[40] W. F. Stinespring: Positive functions on c^*-algebras, Proc. Amer. Math. Soc. **6**, 211–216 (1955)
[41] M. B. Ruskai: Inequalities for quantum entropy: A review with conditions for equality, J. Math. Phys. **43**, 4358–4375 (2002)
[42] K. Kraus: *States, Effect and Operations: Fundamental Notions of Quantum Theory* (Springer, Berlin, Heidelberg 1983)
[43] A. S. Holevo: On quantum communication channels with constrained inputs (1997)
URL http://lanl.arxiv.org/abs/quant-ph/9705054
[44] A. Winter: Coding theorem and strong converse for quantum channels, IEEE Trans. Inf. Theory **45**, 2481–2485 (1999)
[45] R. Goodman, N. R. Wallach: *Representations and Invariants of the Classical Groups* (Cambridge Univ. Press, Cambridge 1998)
[46] R. Alicki, M. Fannes: Note on multiple additivity of minimal output entropy output of extreme $su(d)$-covariant channels, Open Systems and Information Dynamics **11**, 339–342 (2004)
[47] N. Datta, A. Holevo, Y. Suhov: A quantum channel with additive minimum output entropy (2004)
URL http://lanl.arxiv.org/abs/quant-ph/0403072
[48] C. King, M. Nathanson, M. B. Ruskai: Qubit channels can require more than two inputs to achieve capacity, Phys. Rev. Lett. **88**, 057901 (2002)
[49] M. Ohya, D. Petz, N. Watanabe: On capacities of quantum channels, Prob. Math. Stats. **17**, 170–196 (1997)
[50] B. Schumacher, M. D. Westmoreland: Optimal signal ensembles, Phys. Rev. A **63**, 022308 (2001)
[51] R. T. Rockafellar: *Convex Analysis* (Princeton Univ. Press, Princeton 1970)
[52] M. B. Ruskai, S. Szarek, E. Werner: An analysis of completely-positive trace-preserving maps on $\mathcal{M}_2$, Linear Algebra and its Applications **347**, 159–187 (2002)
[53] H. Nagaoka: Algorithms of Arimoto–Blahut type for computing quantum channel capacity, in *Proc. IEEE International Symposium on Information Theory* (1998) p. 354
[54] H. Imai, M. Hachimori, M. Hamada, H. Kobayashi, K. Matsumoto: Optimization in quantum computation and information, in *Proc. of the 2nd Japanese-Hungarian Symposium on Discrete Mathematics and Its Applications* (2001) pp. 0–69
[55] P. W. Shor: Capacities of quantum channels and how to find them, Mathematical Programming **97**, 311–335 (2003)

[56] F. Potra, Y. Ye: A quadratically convergent polynomial algorithm for solving entropy optimization problems, SIAM J. Optim. **3**, 843–860 (1993)
[57] M. S. I. NUOPT.: Technical report
URL http://www.msi.co.jp/en/home.html
[58] C. King, M. B. Ruskai: Comments on multiplicativity of p-norms for $p = 2$, in O. Hirota (Ed.): *Quantum Information, Statistics and Probability* (Rinton Press 2004) pp. 102–114
URL http://lanl.arxiv.org/abs/quant-ph/0401026

Index

channel state, 139
convex roof, 137

entanglement cost, 137
entanglement of formation (EoF), 137
entropy of entanglement, 137

Holevo capacity, 134

linear cost constraint, 140

minimal output entropy, 136
minimum output entropy, 135

optimal average input, 153

separable, 134
Strong superadditivity, 137

Part III

Quantum Security

Quantum Computational Cryptography

Akinori Kawachi[1] and Takeshi Koshiba[2]

[1] Graduate School of Information Science and Engineering, Tokyo Institute of Technology, Ookayama 2-12-1, Meguro-ku, Tokyo 152-8552, Japan
kawachi@is.titech.ac.jp
[2] Department of Information and Computer Sciences, Saitama University, 255 Shimo-Ohkubo, Sakura-ku, Saitama 338-8570, Japan
koshiba@tcs.ics.saitama-u.ac.jp

Abstract. As computational approaches to classical cryptography have succeeded in the establishment of the foundation of the network security, computational approaches even to quantum cryptography are promising, since quantum computational cryptography could offer richer applications than the quantum key distribution. Our project focused especially on the quantum one-wayness and quantum public-key cryptosystems. The one-wayness of functions (or permutations) is one of the most important notions in computational cryptography. First, we give an algorithmic characterization of quantum one-way permutations. In other words, we show a necessary and sufficient condition for quantum one-way permutations in terms of reflection operators. Second, we introduce a problem of distinguishing between two quantum states as a new underlying problem that is harder to solve than the graph automorphism problem. The new problem is a natural generalization of the distinguishability problem between two probability distributions, which are commonly used in computational cryptography. We show that the problem has several cryptographic properties and they enable us to construct a quantum public-key cryptosystem, which is likely to withstand any attack of a quantum adversary.

1 Introduction

Cryptographic technology plays an important role in guaranteeing the network security. Current cryptographic systems are partitioned into symmetric-key systems and asymmetric-key systems. The former has information theoretical flavor, and the latter has computational flavor. While both types of systems are heterogeneous, the quantum mechanism affects them in their security.

Since *Diffie* and *Hellman* [1] first used a computationally intractable problem to build a key exchange protocol, computational cryptography has been extensively investigated. In particular, a number of practical cryptographic systems (e.g., public-key cryptosystems (PKCs), bit commitment schemes (BCSs), pseudorandom generators, and digital signature schemes) have been constructed under reasonable computational assumptions, such as the hardness of the integer factorization problem (IFP) and the discrete logarithm problem (DLP), where we have not found any efficient classical (deterministic or probabilistic) algorithm. Nevertheless, if an adversary runs a *quantum computer* (we call such an adversary a *quantum adversary*), he can efficiently solve various problems, including IFP [2], DLP [2, 3, 4], and the principal ideal problem [5]. Therefore, the quantum adversary can easily break any cryptosystem whose security relies on the hardness of these problems.

A new area of cryptography, so-called *quantum cryptography*, has emerged to deal with quantum adversaries and has been dramatically developed over the past two decades. In 1984, *Bennett* and *Brassard* [6] proposed a *quantum key distribution* scheme, which is a key distribution protocol using quantum communication. Later, *Mayers* [7] proved its unconditional security. Nevertheless, *Mayers* [8] and *Lo* and *Chau* [9] independently demonstrated that quantum mechanics cannot necessarily make all cryptographic schemes information-theoretically secure. In particular, they proved that no quantum BCS can be both concealing and binding unconditionally. Therefore, it is still important to take "computational" approaches to quantum cryptography. In the literature, there are a number of quantum cryptographic properties discussed from the complexity-theoretic point of view [10, 11, 12, 13, 14, 15].

Our project focused especially on the quantum one-wayness and quantum public-key cryptosystems. In what follows, we review our results on quantum one-wayness [16, 17] and quantum public-key cryptosystems [18], including related results. In [16, 17], we gave an algorithmic characterization of quantum one-way permutations. In other words, we showed a necessary and sufficient condition for quantum one-way permutations in terms of reflection operators, which are successfully used in the Grover algorithm [19] and the quantum amplitude amplification technique [20]. In [18], we introduced a problem of distinguishing between two quantum states as a new underlying problem to build a computational cryptographic scheme that is "secure" against quantum adversaries. Our problem is a natural generalization of the distinguishability problem between two probability distributions, which are commonly used in computational cryptography. Our problem has several cryptographic properties. It should be especially mentioned that our problem is at least as hard in the worst case as the graph automorphism problem. The cryptographic properties of our problem enable us to construct a public-key cryptosystem, which is likely to withstand any attack of a quantum adversary.

2 Quantum One-Wayness of Permutations

One-way functions are functions f such that, for each x, $f(x)$ is efficiently computable but $f^{-1}(y)$ is computationally tractable only for a negligible fraction of all y's. While modern cryptography depends heavily on one-way functions, the existence of one-way functions is one of the most important open problems in theoretical computer science. On the other hand, *Shor* [2] showed that famous candidates of one-way functions such as the RSA function or the discrete logarithm function are no longer one-way in the quantum computation model. Nonetheless, some cryptographic applications based on quantum one-way functions have been considered (see, e.g., [10, 14]).

As a cryptographic primitive other than one-way functions, pseudorandom generators have been studied well. *Blum* and *Micali* [21] proposed how to construct pseudorandom generators from one-way permutations and introduced the next-bit test for pseudorandom generators. (They actually constructed a pseudorandom generator assuming the hardness of the discrete logarithm problem.) Since *Yao* [22] proved that the next-bit test is a universal test for pseudorandom generators, the Blum–Micali construction paradigm of pseudorandom generators from one-way permutations was proved to work properly. In the case of pseudorandom generators based on one-way permutations, the next-bit unpredictability can be proved by using hard-core predicates for one-way permutations. After that, *Goldreich* and *Levin* [23] showed that there exists a hard-core predicate for any one-way function (and also permutation) and *Håstad* et al. [24] showed that the existence of pseudorandom generators is equivalent to that of one-way functions.

Yao's result on the universality of the next-bit test assumes that all bits appearing among the pseudorandom bits are computationally unbiased. *Schrift* and *Shamir* [25] extended Yao's result to the biased case and proposed universal tests for nonuniform distributions. On the other hand, no universal test for the one-wayness of a function (or a permutation) is known, although pseudorandom generators and one-way functions (or permutations) are closely related.

In the quantum computation model, *Kashefi* et al. [26] gave a necessary and sufficient condition for the existence of *worst-case* quantum one-way permutations. They also considered the *cryptographic* (i.e., *average-case*) quantum one-way permutations and gave a sufficient condition of (cryptographic) quantum one-way permutations, and posed a conjecture that the condition would be necessary. Their conditions are based on the efficient implementability of reflection operators about some class of quantum states. Note that the reflection operators are successfully used in the Grover algorithm [19] and the quantum amplitude amplification technique [20]. To obtain a sufficient condition of cryptographic quantum one-way permutations, a notion of "pseudo identity" operators was introduced [26]. Since the worst-case hardness of reflection operators is concerned with the worst-case hardness of the inversion of the permutation f, we need some technical tool with which the inversion process of f becomes tolerant of some computational errors in order to obtain a sufficient condition of cryptographic quantum one-way permutations. Actually, pseudo identity operators permit *exponentially* small errors during the inversion process [26].

In this section, we complete a necessary and sufficient condition of cryptographic quantum one-way permutations conjectured in [26]. We incorporate their basic ideas with a probabilistic argument in order to obtain a technical tool to permit *polynomially* small errors during the inversion process. Roughly speaking, pseudo identity operators are close to the identity operator in a sense. The similarity is defined by an intermediate notion between the statistical distance and the computational distance. In [26], it is "by upper-bounding the similarity" that the sufficient condition of cryptographic quantum one-way permutations was obtained. By using a probabilistic argument, we can estimate the expectation of the similarity and then handle polynomially small errors during the inversion of the permutation f.

2.1 Notations and Basic Operators

Since our study is an extension of the results by *Kashefi* et al. [26], we use the same notions, definitions and notations.

We say that a unitary operator U (on n qubits) is *easy* if there exists a quantum circuit implementing U of size polynomial in n. Similarly, a set $\mathcal{F}$ of unitary operators is *easy* if every $U \in \mathcal{F}$ is easy. Throughout this section, we assume that $f : \{0,1\}^* \to \{0,1\}^*$ is a length-preserving permutation unless otherwise stated. Namely, for any $x \in \{0,1\}^n$, $f(x)$ is an n-bit string and the set $\{f(x) : x \in \{0,1\}^n\}$ is of cardinality 2^n for every n. First, we mention some useful operators in describing the previous and our results. The *tagging* operators O_j are defined as follows:

$$O_j |x\rangle |y\rangle = \begin{cases} -|x\rangle |y\rangle, & \text{if } f(y)_{(2j+1,2j+2)} = x_{(2j+1,2j+2)}, \\ |x\rangle |y\rangle, & \text{if } f(y)_{(2j+1,2j+2)} \neq x_{(2j+1,2j+2)}, \end{cases}$$

where $y_{(i,j)}$ denotes the substring from the ith bit to the jth bit of the bit string y if $i \leq j$, and the null string otherwise. Note that these unitary operators O_j are easy if f is efficiently computable. Next, we consider the *reflection* operators $Q_j(f)$ as follows:

$$Q_j(f) = \sum_{x \in \{0,1\}^n} |x\rangle \langle x| \otimes (2 |\psi_{j,x}\rangle \langle \psi_{j,x}| - I),$$

where

$$|\psi_{j,x}\rangle = \frac{1}{\sqrt{2^{n-2j}}} \sum_{y : f(y)_{(1,2j)} = x_{(1,2j)}} |y\rangle.$$

We sometimes use the notation Q_j instead of $Q_j(f)$.

2.2 Worst-Case Characterization

Informally speaking, a function f is said to be worst-case quantum one-way if f can be computed by an efficient quantum machine and f^{-1} cannot be computed by any efficient quantum machine. One of the results in [26] is the following characterization of worst-case quantum one-way permutations:

Theorem 1. (*Kashefi* et al. [26]) *Let* $f : \{0,1\}^n \to \{0,1\}^n$ *be a permutation. Then* f *is worst-case quantum one-way if and only if the set* $\mathcal{F}_n = \{Q_j(f)\}_{j=0,1,\ldots,n/2-1}$ *of unitary operators is not easy.*

As a part of the proof of Theorem 1, *Kashefi* et al. [26] give a quantum algorithm, which we call Algorithm INV in what follows, that computes $f^{-1}(x)$ by using unitary operators O_j and Q_j. The initial input state to INV is assumed to be

$$\frac{1}{\sqrt{2^n}} |x\rangle \sum_{y \in \{0,1\}^n} |y\rangle \quad \left(= |x\rangle |\psi_{0,x}\rangle \right).$$

Then INV performs the following steps:

> **foreach** $j = 0$ to $n/2 - 1$
> (step W.j.1) Apply O_j to the first and the second registers;
> (step W.j.2) Apply Q_j to the first and the second registers.

After each step, we have the following:

$$\begin{pmatrix} \text{the state after} \\ \text{step W.j.1} \end{pmatrix} = \frac{2^j}{\sqrt{2^n}} |x\rangle \left(\sqrt{2^{n-2j}} |\psi_{j,x}\rangle - 2 \sum_{y: f(y)_{(1,2j+2)} = x_{(1,2j+2)}} |y\rangle \right).$$

$$\begin{pmatrix} \text{the state after} \\ \text{step W.j.2} \end{pmatrix} = \frac{2^{j+1}}{\sqrt{2^n}} |x\rangle \sum_{y: f(y)_{(1,2j+2)} = x_{(1,2j+2)}} |y\rangle.$$

The above properties are with respect to "worst-case" (i.e., noncryptographic) quantum one-way permutations, but they also play essential roles in the case of "average-case" (i.e., cryptographic) quantum one-way permutations.

2.3 Average-Case Characterization

First, we define two types of cryptographic "one-wayness" in the quantum computational setting.

Definition 1. *A permutation f is weakly quantum one-way if the following conditions are satisfied:*

1. f can be computed by a polynomial-size classical circuit.
2. There exists a polynomial $p(\,\cdot\,)$ such that for every polynomial-size quantum circuit A and all sufficiently large n's,

   $$\Pr[\, A(f(U_n)) \neq U_n \,] > 1/p(n),$$

   where U_n is the uniform distribution over $\{0,1\}^n$.

Definition 2. A permutation f is *strongly quantum one-way* if the following conditions are satisfied:

1. f can be computed by a polynomial-size classical circuit.
2. For every polynomial-size quantum circuit A and every polynomial $p(\cdot)$ and all sufficiently large n's,

$$\Pr[\, A(f(U_n)) = U_n\,] < 1/p(n).$$

As in the classical one-way permutations, we can show that the existence of weakly quantum one-way permutations is equivalent to that of strongly quantum one-way permutations. Thus, we consider the weakly quantum one-way permutations only. While Theorem 1 is a necessary and sufficient condition of *worst-case* quantum one-way permutations, Kashefi et al. [26] also gave a sufficient condition of *cryptographic* quantum one-way permutations by using the following notion.

Definition 3. Let $d(n) \geq n$ be a polynomial in n and J_n be a $d(n)$-qubit unitary operator. J_n is called $(a(n), b(n))$-*pseudo identity* if there exists a set $X_n \subseteq \{0,1\}^n$ such that $|X_n|/2^n \leq b(n)$ and for every $z \in \{0,1\}^n \setminus X_n$

$$|1 - (\langle z|_1 \langle 0|_2) J_n (|z\rangle_1 |0\rangle_2)| \leq a(n),$$

where $|z\rangle_1$ is the n-qubit basis state for each z and $|0\rangle_2$ corresponds to the ancillae of $d(n) - n$ qubits.

The closeness between a pseudo identity operator and the identity operator is measured by a pair of parameters $a(n)$ and $b(n)$. The first parameter $a(n)$ is a measure of a statistical property, and the second one $b(n)$ is the ratio of "ill-behaved" elements. Note that we do not care where each $z \in X_n$ is mapped by the pseudo identity operator J_n. While we will give a necessary and sufficient condition of quantum one-way permutations by using the notion of pseudo identity, we next introduce a new notion, which may be helpful to understand intuitions of our and previous conditions.

Definition 4. Let $d'(n) \geq n$ be a polynomial in n and P_n be a $d'(n)$-qubit unitary operator. P_n is called $(a(n), b(n))$-*pseudo reflection* (with respect to $|\psi(z)\rangle$) if there exists a set $X_n \subseteq \{0,1\}^n$ such that $|X_n|/2^n \leq b(n)$ and for every $z \in \{0,1\}^n \setminus X_n$ and every n-dimensional vector w

$$\left| 1 - \left(\langle z|_1 \otimes \langle w|_2 \left(\sum_{y \in \{0,1\}^n} |y\rangle \langle y|_1 \otimes (2 |\psi(y)\rangle \langle \psi(y)| - I)_2 \right) \right. \right.$$

$$\left. \left. \otimes \langle 0|_3 \right) P_n (|z\rangle_1 |w\rangle_2 |0\rangle_3) \right| \leq a(n). \quad (1)$$

Let J_n be a $d(n)$-qubit $(a(n), b(n))$-pseudo identity operator. Then $(I_n \otimes J_n)^\dagger (Q_j \otimes I_{d(n)-n})(I_n \otimes J_n)$ is a $(d(n)+n)$-qubit $(a'(n), b'(n))$-pseudo reflection operator with respect to $|\psi_{j,x}\rangle$, where $a'(n) \leq 2a(n)$ and $b'(n) \leq 2b(n)$. These estimations of $a'(n)$ and $b'(n)$ are too rough to obtain a necessary and sufficient condition. Rigorous estimation of these parameters is a main technical issue.

Now, we are ready to mention results with respect to "average-case" quantum one-way permutations shown in [26].

Theorem 2. (Kashefi et al. [26]) *Let f be a permutation that can be computed by a polynomial-size quantum circuit. If f is not (weakly) quantum one-way, then for all polynomials p's and infinitely many n's, there exist a polynomial $r_p(n)$ and an $r_p(n)$-qubit $(1/2^{p(n)}, 1/p(n))$-pseudo identity operator J_n such that the family of pseudo reflection operators*

$$\mathcal{F}_{p,n}(f) = \{(I_n \otimes J_n)^\dagger (Q_j(f) \otimes I_{r_p(n)-n})(I_n \otimes J_n)\}_{j=0,1,\ldots,n/2-1}$$

is easy.

Note that the second parameter $1/p(n)$ of the pseudo identity operator stated in Theorem 2 comes from the error bound of inverting algorithms for weakly one-way quantum permutations. Kashefi et al. [26] conjectured that the converse of Theorem 2 should still hold and proved a weaker version (with respect to the error bound of pseudo identity operators) of the converse as follows:

Theorem 3. (Kashefi et al. [26]) *Let f be a permutation that can be computed by a polynomial-size quantum circuit. If for all polynomials p's and infinitely many n's there exist a polynomial $r_p(n)$ and an $r_p(n)$-qubit $(1/2^{p(n)}, p(n)/2^n)$-pseudo identity operator family $\{J_{j,n}\}_{j=0,1,\ldots,n/2-1}$ such that the family of pseudo reflection operators*

$$\mathcal{F}_{p,n}(f) = \{(I_n \otimes J_{j,n})^\dagger (Q_j(f) \otimes I_{r_p(n)-n})(I_n \otimes J_{j,n})\}_{j=0,1,\ldots,n/2-1}$$

is easy, then f is not (weakly) quantum one-way.

Remark 1. In the corresponding statement in [26], "single" pseudo identity operator rather than pseudo identity operator "family" is used. On the other hand, their actual proof in [26] is for "family", which is as a strong statement as Theorem 3.

Note that pseudo identity operators stated in Theorem 3 permit "exponentially" small errors while pseudo identity operators that will appear in our statement permit "polynomially" small errors. We mention why it is difficult to show the converse of Theorem 2 (or, equivalently, the resulting statement by replacing "$p(n)/2^n$" of Theorem 3 with "$1/p(n)$"). To prove it by contradiction, all we can assume is the existence of a pseudo identity operator. This means that we cannot know how the pseudo identity operator is close to the identity operator. To overcome this difficulty, we introduce a probabilistic technique and estimate the expected behavior of the pseudo identity operator. Eventually, we give a necessary and sufficient condition of the existence of cryptographic quantum one-way permutations in terms of reflection operators. This affirmatively settles their conjecture. We stress that results with respect to cryptographic functions are obtained by generalizing ones with respect to noncryptographic functions, since there are few connections between cryptographic and noncryptographic functions in the classical computation model.

Theorem 4. *Let f be a permutation that can be computed by a polynomial-size quantum circuit. If for all polynomials p's and infinitely many n's there exist a polynomial $r_p(n)$ and an $r_p(n)$-qubit $(1/2^{p(n)}, 1/p(n))$-pseudo identity operator family $\{J_{j,n}\}_{j=0,1,\ldots,n/2-1}$ such that the family of pseudo reflection operators*

$$\mathcal{F}_{p,n}(f) = \{\tilde{Q}_j(f)\}$$
$$= \{(I_n \otimes J_{j,n})^\dagger (Q_j(f) \otimes I_{r_p(n)-n})(I_n \otimes J_{j,n})\}_{j=0,1,\ldots,n/2-1}$$

is easy, then f is not (weakly) quantum one-way.

Assume that f is a weakly quantum one-way permutation. By a probabilistic argument, we can show that a contradiction follows from this assumption. Actually, we constructed an efficient inverter av-INV for f using $\mathcal{F}_{p,n}$ and then, if we choose a polynomial $p(n)$ appropriately, this efficient inverter can compute x from $f(x)$ for a large fraction of inputs, which violates the assumption that f is a weakly quantum one-way permutation.

Algorithm av-INV is similar to Algorithm INV except the following change: the operator Q_j is now replaced with $\tilde{Q}_j$. The initial input state to av-INV is also assumed to be

$$\frac{1}{\sqrt{2^n}} |x\rangle_1 \sum_{y \in \{0,1\}^n} |y\rangle_2 |0\rangle_3,$$

where $|z\rangle_1$ (resp., $|z\rangle_2$ and $|z\rangle_3$) denotes the first n-qubit (resp., the second n-qubit and the last $(r_p(n) - n)$-qubit) register.

Algorithm av-INV performs the following steps:

foreach $j = 0$ to $n/2 - 1$
    (step j.1) Apply O_j to the first and the second registers;
    (step j.2) Apply $\tilde{Q}_j$ to all the registers.

We gave a proof of Theorem 4 by showing the following two lemmas:

Lemma 1. *Suppose that f is a weakly quantum one-way permutation, i.e., there exists a polynomial $r(n) \geq 1$ such that for every polynomial-size quantum circuit A and all sufficiently large n's, $\Pr[A(f(U_n)) \neq U_n] > 1/r(n)$.*

Then, for every polynomial $q(n) > r^{1/2}(n)$, there are at least $2^n(1/r(n) - 1/q^2(n))/(1 - 1/q^2(n))$ x's such that A cannot compute x from $f(x)$ better than with probability $1 - 1/q^2(n)$.

Lemma 2. *Let $q(n) = p^{1/4}(n)/\sqrt{2n}$. There are at most $2^n/q(n)$ x's such that Algorithm* av-INV *cannot compute x from $f(x)$ with probability at least $1 - 1/q^2(n)$.*

2.4 Universal Tests

The necessary and sufficient condition of quantum one-way permutations can be regarded as a universal test for the quantum one-wayness of permutations. First, we explain what universal tests mean. Pseudorandom bits w's, which are drawn according to some probability distribution, can be defined as ones that pass "all" polynomial-time computable statistical tests. Since w passes "all" polynomial-time computable statistical tests if w passes the *next-bit test*, the next-bit test is said to be *universal* for (unbiased) pseudorandom generators. On the other hand, "passing through the next-bit test" means that the next-bit is computationally unpredictable from the previous bits read so far and the *unpredictability* is defined for "all" polynomial-time algorithms. In this sense, "passing through the next-bit test" is just a necessary and sufficient condition for pseudorandom generators. Furthermore, it is worthwhile to mention that the next-bit test is a family of subtests which are uniformly defined. Namely, the next-bit test means a family that consists of the second-bit test, the third-bit test, and so on. After all, the advantage of the next-bit test for pseudorandom generators is not only its universality but also the fact that it is defined in terms of more primitive uniform components.

We now move to universal tests for quantum one-way permutations. To test the quantum one-wayness for given a permutation f, we have to consider all the polynomial-time quantum algorithms. Theorem 1 provides a universal test for worst-case quantum one-way permutations. Namely, f has an efficient implementation of all reflection operators Q_j's with respect to f if and only if f is not one-way. The efficient implementability of all Q_j's also means the next quantum state computability. Thus, we call the universal test *next quantum state computability test*. Note that the next quantum state computability test for worst-case quantum one-way permutations is also defined in terms of more primitive uniform components, as the next-bit test for pseudorandom generators is.

Our average-case characterization gives a universal test for "cryptographic" quantum one-way permutations, because it is a generalization of the next quantum state computability test for worst-case quantum one-way permutations. Since, in our universal test we do not have to compute exactly the next quantum state, we may call our test *next quantum state approximability test*. Note that the next quantum state approximability test for average-case quantum one-way permutations is also defined in terms of more primitive uniform components.

3 Quantum Public-Key Cryptosystem

A quantum computer is capable of breaking many computational assumptions on which the security of existing cryptographic protocols such as public-key cryptosystems (PKCs) rely. To build a secure PKC against any attack of a polynomial-time quantum adversary, it is important to discover computationally-hard problems that can be used as a building block of the cryptosystem. For example, the subset sum (knapsack) problem and the shortest vector problem are used as a basis of knapsack-based cryptosystems [15, 27] and lattice-based cryptosystems [28, 29, 30]. Although quantum adversaries are currently ineffective in the attacks on these cryptosystems, it is unknown whether they can essentially withstand quantum adversaries. We therefore continue searching for better underlying problems to build quantum cryptosystems which can withstand any attack of polynomial-time quantum adversaries.

We propose a *new* problem, called quantum state computational distinguishability with fully flipped permutations (QSCD$_{ff}$), which satisfies useful cryptographic properties to build a quantum cryptosystem. Our problem QSCD$_{ff}$ generalizes the distinguishability problems between two probability distributions used in [21, 22, 31].

Definition 5. The *advantage* of a polynomial-time quantum algorithm $\mathcal{A}$ that distinguishes between two l-qubit states ρ_0 and ρ_1 is the function $\delta(l)$ defined as:

$$\delta(l) = \left| \Pr_{\mathcal{A}}[\mathcal{A}(\rho_0) = 1] - \Pr_{\mathcal{A}}[\mathcal{A}(\rho_1) = 1] \right|,$$

where the subscript $\mathcal{A}$ means that outputs of $\mathcal{A}$ are determined randomly by measuring the final state of $\mathcal{A}$ in the computational basis. The distinguishability problem between ρ_0 and ρ_1 is said to be *solvable by $\mathcal{A}$ with advantage $\delta(l)$* if the above equation holds for any number l.

The problem QSCD$_{ff}$ is defined as the distinguishability problem between two sequences of random coset states ρ_π^+ and ρ_π^- with a hidden permutation π. Let S_n be the symmetric group of degree n and let

$$\mathcal{K}_n = \{\pi \in S_n : \pi^2 = id \text{ and } \forall i \in \{1, ..., n\}[\pi(i) \neq i]\},$$

where n is described as $2(2n' + 1)$ for some $n' \in \mathbb{N}$.

Definition 6. k-QSCD$_{ff}$ is the distinguishability problem between $\rho_\pi^{+\otimes k}$ and $\rho_\pi^{-\otimes k}$, where $k = k(n)$ is a polynomial in n and the quantum states ρ_π^+ and ρ_π^- are defined as:

$$\rho_\pi^+ = \frac{1}{2n!} \sum_{\sigma \in S_n} (|\sigma\rangle + |\sigma\pi\rangle)(\langle\sigma| + \langle\sigma\pi|), \quad \text{and}$$

$$\rho_\pi^- = \frac{1}{2n!} \sum_{\sigma \in S_n} (|\sigma\rangle - |\sigma\pi\rangle)(\langle\sigma| - \langle\sigma\pi|),$$

for $\pi \in \mathcal{K}_n$. We often call the problem QSCD$_{ff}$ simply if there is no confusion.

The parameter n of the above definition is used to measure the computational complexity of our problem and is called the *security parameter* in the cryptographic context. From a technical reason, this security parameter must be of the form $2(2n' + 1)$ for a certain $n' \in \mathbb{N}$ as stated above. Moreover, we assume that any permutation σ can be represented in binary using $O(n \log n)$ bits.

3.1 Cryptographic Properties of QSCD$_{ff}$

We show three cryptographic properties of QSCD$_{ff}$ and its application to quantum cryptography. These properties are summarized as follows:

1. QSCD$_{ff}$ has the trapdoor property; namely, given a hidden permutation π, we can efficiently distinguish between ρ_π^+ and ρ_π^-;
2. The average-case hardness of QSCD$_{ff}$ over randomly chosen permutations $\pi \in \mathcal{K}_n$ coincides with its worst-case hardness.
3. The hardness of QSCD$_{ff}$ is lower-bounded by the worst-case hardness of the graph automorphism problem, defined as
   GRAPH AUTOMORPHISM PROBLEM: (GA)
   input: an undirected graph $G = (V, E)$;
   output: YES if G has a nontrivial automorphism, and NO otherwise.
   Since GA is not known to be solved efficiently, QSCD$_{ff}$ seems hard to solve. Moreover, we show that QSCD$_{ff}$ cannot be efficiently solved by any quantum algorithm that naturally extends Shor's factorization algorithm.

Technically speaking, the cryptographic properties of QSCD$_{ff}$ follow mainly from the definition of the set $\mathcal{K}_n$ of the hidden permutations. Although the definition seems somewhat artificial, the following properties of $\mathcal{K}_n$ lead to cryptographic and complexity-theoretic properties of QSCD$_{ff}$:

- $\pi \in \mathcal{K}_n$ is of order 2, which provides the trapdoor property of QSCD$_{ff}$.
- For any $\pi \in \mathcal{K}_n$, the conjugacy class of π is equal to $\mathcal{K}_n$, which enables us to prove the equivalence between the worst-case/average-case hardness of QSCD$_{ff}$.
- GA is (polynomial-time Turing) equivalent to its subproblem with the promise that a given graph has a unique nontrivial automorphism in $\mathcal{K}_n$ or none at all. This equivalence is exploited to give a complexity-theoretic lower bound of QSCD$_{ff}$, that is, the worst-case hardness of GA.

For these proofs, we introduce new techniques: a new version of the so-called *coset sampling method*, which is broadly used in extensions of Shor's algorithm (see, e.g., [32]), and a quantum version of the hybrid argument, which is a strong tool for security reduction in modern cryptography.

From now on, we show the above cryptographic properties more precisely. For simplicity, let ι denote the maximally mixed state, i.e.,

$$\iota = \frac{1}{n!} \sum_{\sigma \in S_n} |\sigma\rangle\langle\sigma|,$$

which appears later as a technical tool.

3.2 Trapdoor Property

We prove that QSCD$_{ff}$ has the *trapdoor property*, which plays a key role in various cryptosystems. Actually, the following is an efficient distinction algorithm between ρ_π^+ and ρ_π^- with a hidden permutation π in $\mathcal{K}_n$ with certainty.

[Distinction Algorithm]
Input: unknown state χ which is either ρ_π^+ or ρ_π^-.
Procedure:

Step 1. Prepare two quantum registers: The first register holds a control bit and the second one holds χ. Apply the Hadamard transformation H to the first register. The state of the system now becomes

$$H|0\rangle\langle 0|H \otimes \chi.$$

Step 2. Apply the Controlled-π operator C_π to the two registers, where $C_\pi |0\rangle |\sigma\rangle = |0\rangle |\sigma\rangle$ and $C_\pi |1\rangle |\sigma\rangle = |1\rangle |\sigma\pi\rangle$ for any $\sigma \in S_n$.
Step 3. Apply the Hadamard transformation to the first register.
Step 4. Measure the first register in the computational basis. If the result is 0, output YES; otherwise, output NO.

3.3 Reduction From the Worst Case to the Average Case

We reduce the worst-case hardness of QSCD$_{ff}$ to its average-case hardness. Such a reduction implies that QSCD$_{ff}$ with a random π is at least as hard as QSCD$_{ff}$ with the most difficult π.

Theorem 5. Let $k = k(n)$ be any polynomial in n. Assume that there exists a polynomial-time quantum algorithm $\mathcal{A}$ that solves k-QSCD$_{ff}$ with non-negligible advantage for a uniformly random $\pi \in \mathcal{K}_n$; namely, there exists a polynomial p such that, for any n,

$$\left| \Pr_{\pi,\mathcal{A}}[\mathcal{A}(\rho_\pi^{+\otimes k}) = 1] - \Pr_{\pi,\mathcal{A}}[\mathcal{A}(\rho_\pi^{-\otimes k}) = 1] \right| > 1/p(n),$$

where π is chosen uniformly at random from $\mathcal{K}_n$. Then, there exists a polynomial-time quantum algorithm $\mathcal{B}$ that solves k-QSCD$_{ff}$ with non-negligible advantage in the worst case.

3.4 Hardness of QSCD$_{ff}$

We show that the computational complexity of QSCD$_{ff}$ is lower-bounded by that of GA by constructing an efficient reduction from GA to QSCD$_{ff}$. Our reduction constitutes two parts: a reduction from GA to a variant of GA, called UniqueGA$_{ff}$, and a reduction from UniqueGA$_{ff}$ to QSCD$_{ff}$. We also discuss a relationship between QSCD$_{ff}$ and the symmetric hidden subgroup problem (SHSP), which suggests that QSCD$_{ff}$ may be hard for polynomial-time quantum algorithms to solve. Next, we discuss the so-called *coset sampling method*, which has been largely used in many extensions of Shor's algorithm.

Lemma 3. There exists a polynomial-time quantum algorithm that, given an instance G of UniqueGA$_{ff}$, generates a quantum state ρ_π^+ if G is an "YES" instance with its unique nontrivial automorphism π, or $\iota = \frac{1}{n!}\sum_{\sigma \in S_n}|\sigma\rangle\langle\sigma|$ if G is a "NO" instance.

Now, we introduce a new version of the coset sampling method as a technical tool for our reduction. Note that this algorithm essentially requires the fact that the hidden π is an odd permutation, which is one of the special properties of $\mathcal{K}_n$.

Lemma 4. There exists a polynomial-time quantum algorithm that, given an instance G of UniqueGA$_{ff}$, generates a quantum state ρ_π^- if G is an "YES" instance with the unique nontrivial automorphism π, or ι if G is a "NO" instance.

We are now ready to present a reduction from GA to QSCD$_{ff}$, which implies that QSCD$_{ff}$ is computationally at least as hard as GA.

Theorem 6. If there exist a polynomial $k = k(n)$ and a polynomial-time quantum algorithm that solves k-QSCD$_{ff}$ with non-negligible advantage, there exists a polynomial-time quantum algorithm that solves any instance of GA in the worst case with non-negligible probability.

The distinguishability problem QSCD$_{ff}$ is rooted in SHSP. It is shown that a natural extension of Shor's algorithm cannot solve the distinguishability problem between ρ_π^+ and ι in [33, 34, 35]. Here, we give a theorem on a relationship between QSCD$_{ff}$ and the distinguishability problem between ρ_π^+ and ι.

Most recently, *Moore* and *Russell* [36] and *Hallgren* et al. [37] proved the impossibility of distinguishing between two certain random coset states over the symmetric group with multiple copies. They showed that there exists no quantum algorithm distinguishing between $\rho_\pi^{+\otimes k}$ and $\iota^{\otimes k}$ with non-negligible advantage if $k = o(n \log n)$ in our context. In fact, we can obtain a similar result on QSCD$_{ff}$. The following theorem implies that QSCD$_{ff}$ can be reduced to their distinguishability problem, which supports that QSCD$_{ff}$ cannot be efficiently solved by any algorithm that naturally extends Shor's factoring algorithm. To prove the theorem, we need a quantum version of the so-called *hybrid argument*.

Theorem 7. Let $k = k(n)$ be any polynomial in n. If there exists a polynomial-time quantum algorithm that solves k-QSCD$_{f\!f}$ with non-negligible advantage, then there exists a polynomial-time quantum algorithm that solves the distinguishability problem between $\rho_\pi^{+\otimes k}$ and $\iota^{\otimes k}$ with non-negligible advantage.

3.5 Construction

We have shown useful cryptographic properties of QSCD$_{f\!f}$. As an application of QSCD$_{f\!f}$, we build a quantum PKC whose security relies on the hardness of QSCD$_{f\!f}$. First, we give two quantum algorithms for the construction: One is a quantum algorithm that generates ρ_π^+ from π with certainty and the other is a quantum algorithm that converts ρ_π^+ to ρ_π^- without π.

> [Public-Key Generation Algorithm]
> Input: $\pi \in \mathcal{K}_n$
> Procedure:
> **Step 1.** Choose a permutation σ from S_n uniformly at random and store it in the second register. Then, the entire system is in the state $|0\rangle |\sigma\rangle$.
> **Step 2.** Apply the Hadamard transformation to the first register.
> **Step 3.** Apply the Controlled-π to the both registers.
> **Step 4.** Apply the Hadamard transformation to the first register again.
> **Step 5.** Measure the first register in the computational basis. If 0 is observed, then the quantum state in the second register is ρ_π^+. Otherwise, the state of the second register is ρ_π^-. Now, apply the conversion algorithm to ρ_π^-.

[Conversion Algorithm]
The following transformation inverts, given ρ_π^+, its phase according to the sign of the permutation with certainty.

$$|\sigma\rangle + |\sigma\pi\rangle \longmapsto (-1)^{\mathrm{sgn}(\sigma)} |\sigma\rangle + (-1)^{\mathrm{sgn}(\sigma\pi)} |\sigma\pi\rangle.$$

Since π is odd, the above algorithm converts ρ_π^+ into ρ_π^-.

Next, we describe our quantum PKC and give its security proof. For the security proof, we need to specify the model of attacks. Of all attack models in [38], we pay our attention to a quantum analogue of *the indistinguishability against the chosen plaintext attack (IND-CPA)*. In particular, we adopt the weakest scenario in quantum counterparts of IND-CPA as follows.

Alice (sender) wants to send securely a classical message to Bob (receiver) via a quantum channel. Assume that Alice and Bob are polynomial-time quantum Turing machines. Bob first generates certain quantum states for encryption keys. Alice then requests Bob for his encryption keys. Note that anyone can request him for the encryption keys. Now, we assume that Eve (adversary) can pick up the encrypted messages from the quantum channel, and tries to extract the original message using her quantum computer, i.e., a polynomial-time quantum Turing machine. Since Eve can also obtain Bob's encryption keys as well as Alice does, she can exploit polynomially many encryption keys to distinguish the encrypted message. Thus, we assume that Eve attacks the protocol during the message transmission phase to reveal the content of the encrypted message.

The protocol to transmit a message using our PKC consists of two phases: the key transmission phase and the message transmission phase. We will give a reduction from the worst-case hardness of GA to the case of Eve's attack.

We first describe the protocol of our quantum PKC as follows:

[Key Transmission Phase]
1. Bob chooses a decryption key π uniformly at random from $\mathcal{K}_n$.
2. Bob generates sufficiently many copies of the encryption key ρ_π^+ by using the public-key generation algorithm.
3. Alice obtains encryption keys from Bob.

[Message Transmission Phase]
1. Alice encrypts 0 or 1 into ρ_π^+ or ρ_π^-, respectively, by using the conversion algorithm, and sends it to Bob.
2. Bob decrypts Alice's message using the distinction algorithm.

Step 1 in Key Transmission Phase can be easily implemented by uniformly choosing transpositions one by one in such a way that all transpositions are different and by forming the product of these transpositions.

The security of our PKC is shown by reducing GA to Eve's attack during Message Transmission Phase. Our reduction is a modification of the reduction given in Theorem 6.

Proposition 1. Assume that there exists a polynomial-time quantum adversary $\mathcal{A}$ in the message transmission phase that, for any n, satisfies the following inequality:

$$\left| \Pr_{\pi,\mathcal{A}}[\mathcal{A}(\rho_\pi^+, \rho_\pi^{+\otimes l(n)}) = 1] - \Pr_{\pi,\mathcal{A}}[\mathcal{A}(\rho_\pi^-, \rho_\pi^{+\otimes l(n)}) = 1] \right| > 1/p(n)$$

for a certain polynomial $l(n)$ indicating the number of the encryption keys in use by $\mathcal{A}$ and another polynomial $p(n)$. Then, there exists a polynomial-time quantum algorithm that solves any instance of GA in the worst case with non-negligible probability.

3.6 Remarks

The computational distinguishability problem $QSCD_{\textit{ff}}$ has shown useful properties to build a computational PKC whose security is based on the computational hardness of GA. Although GA is reducible to $QSCD_{\textit{ff}}$, the gap between the hardness of GA and that of $QSCD_{\textit{ff}}$ seems large because a combinatorial structure of its underlying graphs which GA enjoys is completely lost in $QSCD_{\textit{ff}}$. It is therefore important to discover a classical problem, such as the problems of finding a centralizer or finding a normalizer [39], which captures the true hardness of $QSCD_{\textit{ff}}$. Discovering an efficient quantum algorithm for $QSCD_{\textit{ff}}$ is likely to require a new tool and a new technique, which also bring a breakthrough in quantum computation. It is important to discover useful quantum states whose computational distinguishability is used for constructing a more secure cryptosystem.

References

[1] W. Diffie, M. E. Hellman: New directions in cryptography, IEEE Trans. Inf. Theory **22**, 644–654 (1976)
[2] L. K. Grover: Polynomial-time algorithms for prime factorization and discrete logarithms on a quantum computer, SIAM J. Comput. **26**, 1484–1509 (1997)
[3] D. Boneh, R. J. Lipton: Quantum cryptanalysis of hidden linear functions, in *LNCS* **963** (1995) pp. 424–437
[4] A. Kitaev: Quantum measurements and the abelian stabilizer problem, LANL Archive quant-ph/9511026 (1995)
[5] S. Hallgren: Polynomial-time quantum algorithms for Pell's equation and the principal ideal problem, *Proc. 34th ACM Symp. Theory of Computing*, pp. 653–658 (2002)
[6] C. H. Bennett, G. Brassard: Quantum cryptography: Public key distribution and coin tossing, *Proc. IEEE International Conf. Computers, Systems and Signal Processing*, pp. 175–179 (1984)
[7] D. Mayers: Unconditional security in quantum cryptography, J. Assoc. Comput. Mach. **48**, 351–406 (2001)
[8] D. Mayers: Unconditionally secure quantum bit commitment is impossible, Phys. Rev. Lett. **78**, 3414–3417 (1997)
[9] H. K. Lo, H. F. Chau: Is quantum bit commitment really possible?, Phys. Rev. Lett. **78**, 3410–3413 (1997)
[10] M. Adcock, R. Cleve: A quantum Goldreich–Levin theorem with cryptographic applications, in *LNCS* **2285** (2002) pp. 323–334
[11] C. Crépeau, P. Dumais, D. Mayers, L. Salvail: Computational collapse of quantum state with application to oblivious transfer, in *LNCS* **2951** (2004) pp. 374–393
[12] C. Crépeau, F. Légaré, L. Salvail: How to convert the flavor of a quantum bit commitment, in *LNCS* **2045** (2001) pp. 60–77
[13] I. Damgård, S. Fehr, L. Salvail: Zero-knowledge proofs and string commitments withstanding quantum attacks, in *LNCS* **3152** (2004) pp. 254–272

[14] P. Dumais, D. Mayers, L. Salvail: Perfectly concealing quantum bit commitment from any quantum one-way permutation, in *LNCS* **1807** (2000) pp. 300–315
[15] T. Okamoto, K. Tanaka, S. Uchiyama: Quantum public-key cryptosystems, in *LNCS* **1880** (2000) pp. 147–165
[16] A. Kawachi, H. Kobayashi, T. Koshiba, R. H. Putra: Universal test for quantum one-way permutations, in *LNCS* **3153** (2004) pp. 839–850
[17] A. Kawachi, H. Kobayashi, T. Koshiba, R. H. Putra: Universal test for quantum one-way permutations, Theor. Comput. Sci. **345**, 370–385 (2005)
[18] A. Kawachi, T. Koshiba, H. Nishimura, T. Yamakami: Computational indistinguishability between quantum states and its cryptographic application, in *LNCS* **3494** (2005) pp. 268–284
[19] L. K. Grover: A fast quantum mechanical algorithm for database search, *Proc. 28th ACM Symp. Theory of Computing*, pp. 212–219 (1996)
[20] G. Brassard, P. Høyer, M. Mosca, A. Tapp: Quantum amplitude amplification and estimation, in S. J. Lomonaco, Jr., H. E. Brandt (Eds.): *Quantum Computation and Information*, AMS Contemporary Mathematics **305** (2002) pp. 53–74
[21] M. Blum, S. Micali: How to generate cryptographically strong sequences of pseudo-random bits, SIAM J. Comput. **13**, 850–864 (1984)
[22] A. C. C. Yao: Theory and applications of trapdoor functions, *Proc. 23rd IEEE Symp. Foundations of Computer Science*, pp. 80–91 (1982)
[23] O. Goldreich, L. A. Levin: A hard-core predicate for all one way functions, *Proc. 21st ACM Symp. Theory of Computing*, pp. 25–32 (1989)
[24] J. Håstad, R. Impagliazzo, L. A. Levin, M. Luby: A pseudorandom generator from any one-way function, SIAM J. Comput. **28**, 1364–1396 (1999)
[25] A. W. Schrift, A. Shamir: Universal tests for nonuniform distributions, J. Cryptol. **6**, 119–133 (1993)
[26] E. Kashefi, H. Nishimura, V. Vedral: On quantum one-way permutations, Quantum Inf. Comput. **2**, 379–398 (2002)
[27] R. Impagliazzo, M. Naor: Efficient cryptographic schemes provably as secure as subset sum, J. Cryptol. **9**, 199–216 (1996)
[28] M. Ajtai, C. Dwork: A public-key cryptosystem with worst-case/average-case equivalence, *Proc. 29th ACM Symp. Theory of Computing*, pp. 284–293 (1997)
[29] O. Regev: New lattice-based cryptographic constructions, J. Assoc. Comput. Mach. **51**, 899–942 (2004)
[30] O. Regev: On lattices, learning with errors, random linear codes and cryptography, *Proc. 37th ACM Symp. Theory of Computing*, pp. 84–93 (2005)
[31] S. Goldwasser, S. Micali: Probabilistic encryption, J. Comput. Syst. Sci. **28**, 270–299 (1984)
[32] O. Regev: Quantum computation and lattice problems, SIAM J. Comput. **33**, 738–760 (2004)
[33] S. Hallgren, A. Russell, A. Ta-Shma: The hidden subgroup problem and quantum computation using group representations, SIAM J. Comput. **32**, 916–934 (2003)
[34] M. Grigni, L. J. Schulman, M. Vazirani, U. Vazirani: Quantum mechanical algorithms for the nonabelian hidden subgroup problem, Combinatorica **24**, 137–154 (2004)

[35] J. Kempe, A. Shalev: The hidden subgroup problem and permutation group theory, *Proc. 16th ACM-SIAM Symp. Discrete Algorithms*, pp. 1118–1125 (2005)
[36] C. Moore, A. Russell: Tight results on multiregister fourier sampling: Quantum measurements for graph isomorphism require entanglement, LANL Archive quant-ph/0511149 (2005)
[37] S. Hallgren, M. Rötteler, P. Sen: Limitations of quantum coset states for graph isomorphism, LANL Archive quant-ph/0511148 (2005)
[38] M. Bellare, A. Desai, D. Pointcheval, P. Rogaway: Relations among notions of security for public-key encryption schemes, in *LNCS* **1462** (1998) pp. 26–45
[39] E. M. Luks: Permutation groups and polynomial-time computation, in L. Finkelstein, W. M. Kantor (Eds.): *Groups and Computation*, DIMACS Series in Discrete Mathematics and Theoretical Computer Science **11** (1993) pp. 139–175

Index

coset sampling method, 177, 179

graph automorphism problem, 177

hard-core predicate, 169
hidden subgroup problem, 179
hybrid argument, 180

one-way function, 168
one-way permutation, 169–171

pseudo identity operator, 172
pseudo reflection operator, 172
pseudorandom generator, 169

public-key cryptosystem, 176

quantum state computational distinguishability, 176

strongly quantum one-way permutation, 172

trapdoor property, 178

universal test, 169, 175

weakly quantum one-way permutation, 171

Quantum Key Distribution: Security, Feasibility and Robustness

Xiang-Bin Wang

Quantum Computation and Information Project, ERATO, JST Daini Hongo
White Bldg. 201, 5-28-3, Hongo, Bunkyo-ku, Tokyo 113-0033, Japan
wang@qci.jst.go.jp

Abstract. Unlike from the existing classical protocols, a quantum key distribution (QKD) protocol can help two remote parties do secure private communication with proven security. We study various QKD protocols with special emphasis on security, feasibility and robustness. The security proof of the standard four-state protocol can be done based on the virtual entanglement purification by classical hashing. QKD can be done with unconditional security even when an imperfect source such as the coherent light from a traditional laser device is used. In particular, we show that one can do QKD effectively with a traditional laser device through the decoy-state method, i.e., changing the intensity of each pulse randomly among three values. One can improve the key rate and noise threshold if the channel noise is asymmetric. One can also do QKD more robustly with a two-qubit quantum code given that the channel noise is collective or independent.

1 Introduction

Our goal here is to find a way for two remote parties, Alice and Bob, to do secure private communication. There can be an eavesdropper (Eve) in the middle who can do whatever does not violate natural laws to attack the protocol.

So far, there is no classical protocol that offers the *proven* security for such types of private communication. Intuitively, Alice and Bob may do private communication through a secret channel. However, they have no way to verify that the data transmitted through their assumed secret channel have not been read by Eve in the middle because Eve does not have to cause any noise to the classical data when she reads them. This is to say, in principle, Eve may have access to the assumed secret channel with her presence being entirely hidden. In the 1970s, the so-called public key system was invented. Instead of using secret channels, one may use a public channel for secure communication if there exists any one-way function. Say, Alice announces a public key k and keeps the secret key x. With the public key k, any remote party Bob can encode his message M by function $E_k(M)$. Without the secure key, it is too complex for anybody to decode $E_k(M)$, i.e., to compute M given $E_k(M)$. However, given the secret key X, Alice may easily decode it, i.e., to compute M by $M = D_x[E_k(M)]$. If function $E_k(M)$ is indeed a one-way function, then the protocol is secure. However, so far no *proven* one-way function has been found although some candidates are widely *believed* to be so. However, widely believed security is not proven security, therefore the private communication with a public key system could actually be insecure. For example, the most widely used public key system, RSA, is insecure if Eve has a quantum computer. Given a quantum computer, Eve can factorize any larger number N in $O(\log_2 N)$ steps [1, 2]. So far it is unclear whether the RSA system is secure if Eve only uses a classical computer. However, many other classical cryptographic systems have been broken by classical algorithms [3, 4, 5]. In short, the security status of currently existing protocols in classical private communication is this: Although many of them are widely believed to be secure, none of them has been *proven* to be secure even if Eve is restricted to use classical algorithms. The most widely used system, the RSA system, has been proven to be *insecure* with a quantum algorithm. Some other systems have been proven to be *insecure* even with a classical computer.

A question naturally arising is whether we have another choice which offers *proven* security for private communication. The answer is yes, if we use a quantum key distribution (QKD) protocol [6, 7] to set up the private key first and then use this private key as a one-time pad for encryption and decryption in transmitting a message. Instead of the computational complexity, here the security is guaranteed by principles of quantum mechanics. Let us recall the first QKD protocol, the BB84 which was proposed by *Bennett* and *Brassard* [7] in 1984. Bit values 0, 1 are carried by the photon polarization: The horizontal or $\pi/4$ polarizations represent bit value 0, and the vertical or $3\pi/4$ polarizations represent bit value 1. Each time, Alice randomly chooses one polarization and sends the photon to Bob. Bob measures it in either the horizontal–vertical basis or the $\pi/4 - 3\pi/4$ basis. Later, Bob announces his measurement basis for each photon, and they discard those bits with basis mismatch. They then randomly choose some remaining bits and publicly announce them for error test, and they discard these test bits. If the error rate is too large, they abort the protocol; otherwise they continue to distill the final key with appropriate error correction and privacy amplification for the remaining bits.

Roughly speaking, if Eve wants to know any bit value, she must watch (measure) the polarization of a certain photon in transmission. The polarization is a *quantum state*. According to quantum mechanics, a measurement in principle disturbs a quantum state unless the state is measured in its eigenbasis. Here the polarization state of any photon has two *possible* eigen-bases. Eve has a probability of 50% to measure a photon in a wrong basis, which means Eve will inevitably cause disturbance to those qubits if she measures them. This is the crucial difference between QKD and any classical private communication protocol.

However, a strict security proof for QKD in practice is not so simple. For example, in practice the channel is always noisy. Although the channel for Alice and Bob is noisy, in principle Eve can have a noiseless channel. In such a case, Eve may watch the transmitted photons and pretend her action is channel noise by sending those photons to Bob through a noiseless channel. Besides the noisy channel, there are some other restrictions in practice. For another example, the source can be imperfect and the channel can be lossy. We shall show the security of QKD in practice, including the case of noisy and lossy channels and an imperfect source. Although a QKD protocol offers unconditional security, it does not always run successfully with larger than zero key rate. If the channel noise is larger than a certain threshold value, no final key can be produced. The error rate threshold value for the BB84 protocol is 11%. If the channel noise is asymmetric, there are ways to raise the key rate and noise threshold. However, by using two-way classical communication and/or using encoded BB84 states, this threshold value can be raised significantly, therefore the quantum key distribution can be done more robustly.

2 Security Proof of BB84 QKD With Perfect Single-Photon Source

An unconditionally secure QKD protocol requires Eve's information about the final key to be exponentially close to 0, i.e.,

$$I_{AE} \leq \epsilon_1. \tag{1}$$

Alternatively, we can define the security by the probability that Eve can obtain a nonnegligible amount of information about the final key, i.e.,

$$P(I_{EA} \geq \epsilon_2) \leq \epsilon_3. \tag{2}$$

Here $\epsilon_1, \epsilon_2, \epsilon_3$ are exponentially close to 0, say, e.g., 2^{-50}. The strict proof of unconditional security is strongly nontrivial because we must assume that Eve can do *whatever* in principle exists. In particular, Eve can do coherent attack to many qubits transmitted from Alice, and Eve can delay her measurement until Alice and Bob complete the final key distillation and then directly attack their final key by a certain optimized measurement. The security proof should not be only based on a special attacking scheme.

The first security proof of quantum key distribution was given by *Mayers* [8,9]. Later, it was simplified by *Shor* and *Preskill* [10] based on the simple idea of first doing entanglement purification with a Calderbank Shor–Steane (CSS) code [10, 11] and then reduce it to a prepare-and-measure protocol of distilling the measured data by a *classical* (CSS) code. An excellent tutorial of Shor and Preskill's proof has been presented by *Gottesman* and *Preskill* [12]. Suppose the channel-flipping rate for those qubits of both bases is t. By Shor and Preskill's argument, we can use code C_1 to do error correction and use code C_2 to do privacy amplification, if we can find such C_1, C_2 satisfying

$$C_2^\perp \subset C_1, \tag{3}$$

and C_1, C_2 must be able to correct tN errors if there are N raw bits for distillation. Due to this constraint, so far it is not known how to construct a large classical CSS code efficiently. The construction task is even more complicated when we have another constraint: C_1 must be efficiently decodable. Actually, before talking about constructing a CSS code, even the existence of a CSS code needs a nontrivial proof [13] in the most general case, say, the channel-flipping rates for qubits of different bases are different.

Here we modify Shor and Preskill's proof, and we present an alternative approach which is based on the idea of (modified) BDSW entanglement purification protocol [14] given by *Bennett, Divincenzo, Smolin* and *Wooters*, which we shall call BDSW protocol hereafter. By using our modified proof, the problems such as how to construct a CSS code or the existence of good CSS code is avoided [15]. (Mayers' proof does not have these problems, but it is rather complex.)

If Alice and Bob share a number of perfectly entangled pairs (i.e., states of $|\phi^+\rangle = \frac{1}{\sqrt{2}}(|00\rangle + |11\rangle)$), they can measure them at each side in Z basis ($\{|0\rangle, |1\rangle\}$) and use the measurement outcome as the shared secure key, say, outcome $|0\rangle$ for bit value 0 and $|1\rangle$ for bit value 1. This key will be perfect and no third party can have any information about it. If the pairs are almost perfect (exponentially close to perfect ones), they can also use the measurement outcome as the secure key since any third party or Eve's information about the key is exponentially close to 0. If the preshared pairs are noisy but not too noisy, they can first purify them into a small number of almost perfect pairs which are exponentially close to perfect pairs and then obtain the final key. This is the entanglement-purification-based protocol. We shall use the modified BDSW protocol [14] for entanglement purification and then show how to reduce it to a prepare-and-measure one [10, 12]. The main idea of the BDSW protocol is to detect the positions of wrong pairs by hashing. In what follows, we shall first consider the classical hashing, and then we can state the entanglement purification by hashing if we know the bit-flip rate and phase-flip rate of raw pairs. Finally, we "classicalize" the purification protocol so that it is equivalent to the BB84 QKD protocol, which only needs one to prepare and measure a qubit.

2.1 Hashing and Error-Correction in Classical Communication

Suppose Alice sends a binary string s_A which contains N classical bits to Bob through a noisy private channel. Some bits in the string are flipped due to the channel noise; therefore Bob has actually received an N-bit string s_B. They can now do error correction with classical communication through a noiseless public channel. Naively, Bob (or) Alice may just announce the value of each bit, and they can then correct all errors in s_B. However, they do not want to reveal too many bit values to the public, and they can actually do it more sophisticatedly by *hashing*. Say, if the bit-flip rate is bounded by $t_b < 50\%$, asymptotically, there are only

$$\omega = 2^{N \cdot H(t_b)} \tag{4}$$

likely candidate strings for s_B because those candidates with flip-rate larger than t_b are impossible. Through this paper, we use notation $H(x)$ for the function of entropy, and

$$H(x) = -x \log_2 x - (1-x) \log_2 (1-x). \tag{5}$$

Moreover, we can use the concept of *bit-flip string* to represent the information of positions of flips at Bob's string,

$$s_f = s_A \oplus s_B, \tag{6}$$

and notation $\oplus$ is bit-wise summation, i.e., $1 \oplus 1 = 0 \oplus 0 = 0$ and $0 \oplus 1 = 1$. Explicitly, suppose $s_A = a_1 a_2 \cdots a_N$ and $s_B = b_1 b_2 \cdots b_N$, we have

$$s_f = c_1 c_2 \cdots c_N = (a_1 \oplus b_1)(a_2 \oplus b_2) \cdots (a_N \oplus b_N). \tag{7}$$

Remark: The bit-flip rate t_b can be known by error test. For example, Alice can mix m more bits with her string s_A before sending. After Bob receives $N + m$ bits, they announce the bit values of those m test bits and find the upper bound value of error rate of the remaining N bits by classical statistics. They discard those m test bits.

They can then share a shorter identical string after error correction through hashing. If Bob knows string s_f explicitly, Bob has actually known the positions of wrong bits in his string s_B, and he can obtain s_A by flipping those wrong bits. Similarly, if Bob knows explicitly a substring of s_f, he can then correct a substring of s_B and share a shorter identical string with Alice. He can know explicitly a substring of s_f by hashing. Suppose every time the bit positions and the parity of a random substring from s_f are revealed to Bob and one bit from the subset is deleted. As it has been shown in [14], after $k = \log_2 \omega + \delta N$ times, Bob can compute the remaining bits of s_f almost exactly: After any step i, the remaining string s_{fi} contains $N - i$ bits. After step k, there are k parity values of different random subsets have been revealed. Remember that there are only ω likely candidates for the initial N-bit string s_f. Suppose Bob finds a specific N-bit string s_{f0} that can produce all those k parity values revealed. From s_{f0}, we can deduce an explicit substring s_{fk} containing $N - k$ bits, and this outcome can be used for error correction of a substring in s_B. One may argue that the true substring of s_f after step k could be different from the deduced one, s_{f0}, because there could be many strings satisfying the same conditions of parity values of a number of substrings. It can be easily shown that this probability is almost 0: Suppose among all those ω likely candidates, there is another string $s'_{f0} \neq s_{f0}$ that can also produce all those revealed parity values. Suppose at certain step, the remaining string from s_{f0} is s_i and the remaining string from s'_{f0} is s'_i. The joint probability that they produce the same parity value for a random subset and $s_i \neq s'_i$ is less than a half. After k steps, the joint probability that two different initial strings satisfying all conditions as stated above is less than $2^{-(\log_2 \omega + \delta N)}$. There are ω likely candidates for the initial string. Therefore, after k steps, the probability that the remaining string is actually different from the deduced one is bounded by $p \leq 2^{-\delta N}$. Define set $E = \{e_1, e_2, \ldots, e_u\}$ to be the positions of bits chosen for a random substring. Strings $s_A(E)$, $s_B(E)$, $s_f(E)$ contains those bits in positions given by E from string s_A, s_B, s_f, respectively. Explicitly, we have $s_A(E) = a_{e_1} a_{e_2} \cdots a_{e_u}$, $s_B(E) = b_{e_1} b_{e_2} \cdots b_{e_u}$ $(u \leq N)$ and

$$s_f(E) = s_A(E) \oplus s_B(E). \tag{8}$$

Therefore, every time to reveal the parity value of a random substring given by positions E from the bit-flip string, Bob only needs Alice to announce $s_A(E)$. The following protocol of classical hashing can be used in classical error correction: (1) Alice transmits $N + m$ bits to Bob through a noisy private channel. (2) Alice announces m bit values, and they discard these m test bits. Now they know that the bit-flip rate for the remaining N bits is bounded by t_b. (Suppose t_b is not too large.) (3) Alice randomly chooses a subset of the remaining bits. She announces their positions and parity value. Bob and Alice randomly delete one bit from the subset. (4) They repeat step (3) for $k = \log_2 \omega + \delta N$ times, and Bob can compute the positions of wrong bits in his remaining shorter string explicitly.

2.2 The Main Idea of Entanglement Purification

To have a clear picture, we state the main idea of (modified) BDSW [14] entanglement purification here. In our notation, $|0\rangle, |1\rangle$ are eigenstates of operator Z with eigenvalues $0, 1$; and $|\bar{0}\rangle, |\bar{1}\rangle$ are eigenstates of operator X with eigenvalues $0, 1$. When we say bit value of any qubit in Z or X basis, we mean the value we shall obtain if we measure that qubit in Z or X basis.

We shall use notation $Z_{A \oplus B}$ for the collective measurement basis of parity value in Z basis. In particular, given any state $\alpha |a\rangle_A |b\rangle_B + \beta |a \oplus 1\rangle_A |b \oplus 1\rangle_B$, the measurement outcome in $Z_{A \oplus B}$ basis will be $z_{A \oplus B} = a \oplus b$ and the state is unchanged after the measurement. If we do individual measurement at each side in Z basis, there would be two outcomes, z_A and z_B at each side, and we have

$$z_A \oplus z_B = z_{A \oplus B}. \tag{9}$$

However, the individual measurements at each side will in general destroy the original state. Similarly, we can also denote the parity measurement in X basis by $X_{A \oplus B}$ and its measurement outcome is

$$x_{A \oplus B} = x_A \oplus x_B. \tag{10}$$

Note that $Z_{A\oplus B}$ and $X_{A\oplus B}$ commute. Given any pair, we can measure it in both $Z_{A\oplus B}$ and $X_{A\oplus B}$. The pair carries a bit-flip error if $z_{A\oplus B} = 1$, a phase-flip error if $x_{A\oplus B} = 1$.

Consider the 4 states in Bell basis:

$$|\phi^{\pm}\rangle = \frac{1}{\sqrt{2}}(|00\rangle \pm |11\rangle); \quad |\psi^{\pm}\rangle = \frac{1}{\sqrt{2}}(|01\rangle \pm |10\rangle). \tag{11}$$

A pair state must be $|\phi^+\rangle$ if we are sure that both $z_{A\oplus B} = 0$ and $x_{A\oplus B} = 0$ for this state. Moreover, if the value of both $z_{A\oplus B}$ and $x_{A\oplus B}$ are known, the state of that pair is also known and we can change it into $|\phi^+\rangle$ by a unitary to one qubit of tha pair. In particular, $|\psi^+\rangle$ carries a net bit-flip, $|\phi^-\rangle$ carries a net phase-flip and $|\psi^-\rangle$ carries both a bit-flip and a phase-flip.

The idea of hashing in classical error correction can be used for entanglement purification where instead of starting from classical strings in the beginning, Alice and Bob share a number of raw pairs and their goal is to distill a smaller number of almost perfect entangled pairs. Given N raw pairs, we can use two N-bit classical binary strings, the *bit-flip string* s_b and the *phase-flip string* s_p to represent the position of bit-flips and phase-flips. For the ith pair, if it has a bit-flip error, the ith element in string s_b is 1, otherwise it is 0; if it has a phase-flip, the ith element in string s_p is 1, otherwise it is 0. For example, given 5 pairs $|\phi^+\rangle|\phi^+\rangle|\psi^+\rangle|\phi^-\rangle|\psi^-\rangle$, the two classical strings are

$$s_b = 00101; \quad s_p = 00011. \tag{12}$$

Our goal is to obtain some pairs which are all in state $|\phi^+\rangle$. We shall then by revealing parity values of some substring from s_b, s_p determine the bit-flip string and phase-flip string explicitly for the remaining pairs therefore obtain a smaller number of perfect pairs after flipping the wrong pairs appropriately.

Remark: The purification through hashing method here does not require the state of each pair be independent. It works even though there are complex entanglement among different raw pairs initially. We don't have to know the state of all pairs initially. We only need to know the number of likely candidate bit-flip string and phase-flip string in the beginning and then we can determine the explicit bit-flip string and phase-flip string for the remaining n pairs in the end with a probability exponentially close to 1, therefore obtain some perfect pairs. In other words, the state of the remaining n pairs satisfies the following fidelity condition

$$\langle \Phi_n^+ | \rho | \Phi_n^+ \rangle \geq 1 - \epsilon, \tag{13}$$

and $|\Phi_n^+\rangle = |\phi^+\rangle^{\otimes n}$, ϵ is exponentially close to 0.

To start hashing, we must first know the number of likely strings in the beginning. We do so by error test, and we now show why classical statistics works for the error test, even for quantum states here [12].

2.3 Error Test

Given $N + m$ raw pairs, we can randomly choose m pairs as test pairs. We want to know the error rates (including bit-flip rate and phase-flip rate) of the remaining N pairs by examining these m test pairs. This is called *error test*. The initial $N+m$ shared raw pairs can be in any state, e.g., there could be complex entanglement among them. But here our task is not to know the state of these pairs. We only want to know the error rate, say, if we measure each pair in both $Z_{A \oplus B}$ and $X_{A \oplus B}$ basis, what is the fraction of outcome $z_{A \oplus B} = 1$ and $x_{A \oplus B} = 1$ we would obtain.

Since $X_{A \oplus B}$ and $Z_{A \oplus B}$ commute, we can imagine that we had first measured each pair in both bases and then estimate the error rates of N pairs by examining the measurement outcome of m test pairs. Classical statistics work perfectly here because it is equivalent to the case of estimating the error rates of classical data [12]. Further, we can divide the m test pairs into two subsets and each subset contains $m/2$ pairs. We can measure each of $m/2$ pairs in the first subset of test pairs in basis $Z_{A \oplus B}$ to see the bit-flip rate of those N pairs and measure each pair in the second subset of test pairs in $X_{A \oplus B}$ basis to see the phase-flip rate for those N pairs. After these, we know the upper bound of both bit-flip rate and phase-flip rate of the remaining N pairs if each of them were measured in both $X_{A \oplus B}$ and $Z_{A \oplus B}$. Also, since all pairs for error test will be discarded, we can actually use *whatever* measurement to those test pairs provided that the outcome tells us $z_{A \oplus B}$ (or $x_{A \oplus B}$). Therefore we can replace $Z_{A \oplus B}$ for the first subset by individual measurements Z at each side, and $X_{A \oplus B}$ for the second subset by individual measurements X at each side, for, the outcome of individual measurements determines the outcome of collective measurements by (9) and (10). Further, it makes no difference if Alice measures her halves of those test pairs before entanglement distribution. This means, instead of sharing m pairs with Bob, she can simply sends Bob m single qubits with each state being randomly chosen from $|0\rangle, |1\rangle$ and $|\bar{0}\rangle, |\bar{1}\rangle$ (i.e., BB84 states) for future error test.

Remark: Classical statistics works perfectly here given whatever state of raw pairs, even there are entanglement among raw pairs. The purpose of error test is only to find the error rate of the remaining pairs rather than find the state of the raw pairs.

2.4 Entanglement Purification by Hashing

Initially, the number of likely bit-flip string and phase-flip string for N raw pairs are bounded by

$$\omega_b = 2^{N \cdot H(t_b)}, \quad \omega_p = 2^{N \cdot H(t_{ph})}, \qquad (14)$$

respectively and $H(x) = -x \log_2 x - (1-x) \log_2 (1-x)$. (After the error test, they know the upper bound of bit-flip rate and phase-flip rate to be t_b, t_{ph} respectively, for the remaining N pairs.)

2.4.1 Bit-Flip Error Correction

To use the classical result of hashing here, we only need to know how to detect the parity value of any subset of the bit-flip string s_b, i.e., the parity of a shorter bit-flip string for any subset of N raw pairs. This can be done by biCNOT operations and local measurement in Z basis.

A CNOT gate in Z basis is a gate that takes the following transformation:

$$|z_1\rangle|z_2\rangle \longrightarrow |z_1\rangle|z_1 \oplus z_2\rangle. \tag{15}$$

Here the first state is the control state, the second state is the target state, and z_1, z_2 can be any value from $\{0, 1\}$. A biCNOT gate is to do CNOT gates at each side. Explicitly, we first denote $|\phi^+\rangle, |\phi^-\rangle, |\psi^+\rangle, |\psi^-\rangle$ by $|\chi_{00}\rangle, |\chi_{01}\rangle, |\chi_{10}\rangle, |\chi_{11}\rangle$, respectively. Given two pair state $|\chi_{a,b}\rangle|\chi_{a',b'}\rangle$, if we do biCNOT in Z basis on these two pairs with the second pair being the target, we have

$$\text{biCNOT}_Z(|\chi_{a,b}\rangle_1|\chi_{a',b'}\rangle_2) = |\chi_{a,b\oplus b'}\rangle_1|\chi_{a'\oplus a, b'}\rangle_2. \tag{16}$$

Suppose $E = \{e_1, e_2, \ldots, e_u\}$ is a subset that contains pair $e_1, e_2, \ldots, e_u$. The bit-flip string for this subset is $s_b(E) = c_{e_1}c_{e_2}\cdots c_{e_u}$, and the bit-flip string of N pairs are $s_b = c_1 c_2 \cdots c_N$ and $N \geq u$. We want to know parity of string $s_b(E)$, i.e., $\sum_{l \in E} c_l = c_{e_1} \oplus c_{e_2} \oplus \cdots \oplus c_{e_u}$. We have the following fact

$$\sum_{l \in E} c_l = \sum_{l \in E} z_{Al \oplus Bl} = \sum_{l \in E}(z_{Al} \oplus z_{Bl}) = \left(\sum_{l \in E} z_{Al}\right) \oplus \left(\sum_{l \in E} z_{Bl}\right). \tag{17}$$

Here z_{Al} is the bit value of Alice's qubit in pair l, if it were measured in Z basis; z_{Bl} is the bit value of Bob's qubit in pair l, if it were measured in Z; $z_{Al \oplus Bl}$ is the parity value for pair l, if it were measured in $Z_{A \oplus B}$. Given (17), one can know the parity of string $s_b(E)$ if $\sum_{l \in E} z_{Al}$ and $\sum_{l \in E} z_{Bl}$ are known. To know these, it is not necessary to know the value of each z_{Al} and each z_{Bl}. Instead, they can gather $\sum_{l \in E} z_{Al}$ and $\sum_{l \in E} z_{Bl}$ into one qubit at each side by biCNOT gates and then measure the target pair. For example, suppose $E = \{1, 4, 5\}$. To know the parity of bit-flip substring $s_b(E) = c_1 c_4 c_5$, they can first do biCNOT operations in Z basis as shown in Fig. 1 and then measure pair 5 in basis $Z_{A \oplus B}$.

Since the target pair will be discarded, they can actually measure the target pair in Z basis at each side and obtain $\sum_{l \in E} z_{Al}$ and $\sum_{l \in E} z_{Bl}$ separately. The bit-flip error correction is now reduced to the following: At any step i, Alice announces Ei, which is a random subset from the remaining pairs and pair $d_i \in Ei$. They do biCNOT operations in Z basis to collect the parity value $\sum_{l \in Ei}(z_{Al} \oplus z_{Bl})$ at pair d_j. They measure target pair in Z basis at each side. Alice announces her measurement outcome, i.e., $\sum_{l \in Ei} z_{Al}$. Bob calculates $(\sum_{l \in Ei} z_{Al}) \oplus (\sum_{l \in Ei} z_{Bl})$, and this is just the parity value of string $s_b(Ei)$. They discard pair d_i. They repeat this for $q = N[H(t_b) + \delta_1]$ times and Bob can compute the bit-flip string for the remaining $N - q$ pairs, i.e., he knows all positions where the pair bears a bit-flip error. Bob flips those

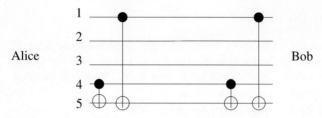

Fig. 1. Alice and Bob first do biCNOT operations in Z basis and then measure the target pair (pair 5) to obtain the parity value for the bit-flip string of a subset E which contains pair $\{1,4,5\}$

of his qubits from bit-flipped pairs in Z basis. The bit-flip error correction is completed here. They only need local operations and classical communication in doing error correction.

However, the phase-flip string for the remaining $N - q$ pairs is not determined yet. In doing the hashing for bit-flip string, the phase-flip string for the remaining pairs changes. However, the number of likely phase-flip string for the remaining pairs cannot be larger than ω_p, the number of likely phase-flip string for the initial N raw pairs. Because the initial phase-flip string of N pairs determines the phase-flip string of the remaining pairs after bit-flip hashing. Equation (16) shows that given the initial phase-flip string, the later phase-flip string for the remaining pairs is determined exactly. Keeping this fact in the mind, they can then correct phase-flip errors.

2.4.2 Phase-Flip Error Correction

They can also do hashing in X basis. In particular, the CNOT operation in X basis is defined by

$$\mathrm{CNOT}_X(|\bar{a}\rangle_1|\bar{b}\rangle_2) = |\bar{a}\rangle_1|\overline{a \oplus b}\rangle_2 . \tag{18}$$

The state subscribed by 1 is control qubit, and the one subscribed by 2 is the target qubit, and the bar upon a, b is only an indication that these states are eigenstates of X basis. With this they can do hashing in X basis. Everything here is the same as that in the bit-flip error correction except that they do biCNOT operations and measurement in X basis. Explicitly, at any step $j > q$, Alice announces Ej, which is a random subset from all remaining pairs and pair $d_j \in Ej$. They do biCNOT operations in X basis to collect the parity value $\sum_{l \in Ej}(x_{Al} \oplus x_{Bl})$ at pair d_i. They measure target pair in X basis at each side. Alice announces her measurement outcome, i.e., $\sum_{l \in Ej} x_{Al}$. Bob calculates $(\sum_{l \in Ej} x_{Al}) \oplus (\sum_{l \in Ej} x_{Bl})$, and this is the parity of string $s_p(Ej)$, the phase-flip string of pairs in subset Ej. They discard pair d_j. They repeat this for $p = N[H(t_{ph}) + \delta_2]$ times and Bob can compute the phase-flip string for the remaining $N - p - q$ pairs, i.e., he knows all positions where the pair bears a phase-flip error. Bob corrects those phase-flipped pairs by a taking a phase shift operation to his own qubits. The purification is ended here.

2.5 Classicalization

The classicalization is to reduce the protocol to the prepare-and-measure type. Since their only purpose is to obtain a secure final key, they need not really correct the phase-flip errors. Therefore Bob does not need to compute the positions of the phase errors. Consequently, Alice does not need to announce her measurement outcome at each step of the phase-flip error correction, for Eve cannot do it better with less information. This is to say, at each step of hashing for phase-flip string, they *only* need to do some biCNOT operations in X basis and directly discard pair d_j. Similar to (16), we have the following fact for biCNOT operation in X basis:

$$\text{biCNOT}_x(l \to d_j) = \text{biCNOT}_z(d_j \to l). \tag{19}$$

The left side of the equation means a biCNOT in X basis with pair d_j being the target and pair l being the control, the right side is a biCNOT in Z basis with pair l being the target and pair d_j being the control. This formula shows that, instead of doing biCNOT operation in X basis, we can actually do biCNOT operation in Z with the target pair and the control pair being exchanged. All operations can now be done in Z basis. Therefore, Alice can choose to measure all her halves of N pairs in Z basis before sending anything to Bob, i.e., she can directly send Bob N random qubits in Z basis and Bob can measure each of them before key distillation. The final key distillation becomes the distillation of the data of the measurement outcome in Z basis at Bob's side.

We have the following simplified bit-flip **Error correction**: At step i, Alice announces the parity of a random subset of her qubits and a randomly selected bit d_i in that subset. They discard that bit. Alice repeats this for q times and Bob can then compute Alice's key for the remaining $N - q$ bits. Bob uses this as his key. We shall simply use *error correction* for the term *bit-flip error correction* because the phase-flip error correction is now reduced to the **Privacy amplification**: At any step $j > q$, Alice's bits and Bob's bits are identical. Alice announces a random subset Ej and bit $d_j \in Ej$. For any $l \in Ej$, they replace z_l by $z_l \oplus z_{d_j}$ and they discard bit d_j. Here z_l is the bit value of lth bit. They need to repeat so by p steps and obtain the final key.

If they do not use a quantum memory, Bob must measure each qubit once he receives it. Therefore Alice has no way to tell Bob the right basis for each one. Bob can randomly choose basis Z or X. Therefore they must discard those outcomes that form a wrong basis. We have the following prepare-and-measure protocol:

(0) Alice and Bob have agreed that states $|0\rangle, |\bar{0}\rangle$ represent for bit value 0 and states $|1\rangle, |\bar{1}\rangle$ represent for bit value 1. (1) Alice prepares $2(N+m)$ qubits. The preparation basis of each qubit is random, with a prior probability of $P_z = \frac{2N+m}{2(N+m)}$ for Z basis, and $P_x = \frac{m}{2(N+m)}$ for X basis. The bit value of

each individual qubit is randomly chose from 0 and 1. (2) Alice sends these qubits to Bob and Bob measures each of them in a basis randomly chosen from $\{X, Z\}$. Bob announces his measurement basis for each qubits and they discard those outcome from a measurement basis that is different from Alice's preparation basis. Approximately, there should be $N+m$ classical bits remaining among which about $m/2$ are X bits (outcome of measurement in X basis) and $N+m/2$ are Z bits (outcome of measurement in Z basis). (3) Bob announces the bit values of all X bits and the same number of Z bits for error test. They discard all the announced bits, and there are about N bits remaining. After this error test, they know that the bit-flip rate and phase-flip rate for the remaining N bits are bounded by t_b, t_ph, respectively. (4) They do bit-flip error correction and privacy amplification to the remaining N classical data as we have stated previously and obtain the final key. The final key rate is

$$f = 1 - H(t_\mathrm{b}) - H(t_\mathrm{ph}) - \delta_1 - \delta_2 \,. \tag{20}$$

3 Secure Key Distillation With a Known Fraction of Tagged Bits

Although many standard QKD protocols such as BB84 [7] have been proven to be unconditionally secure [8,9,10,16,17], this does not guarantee the security of QKD in practice, due to various types of imperfections in a practical setup. In practice, the source is often imperfect. Say, it may produce multiphoton pulses with a small probability. Normally weak coherent states are used in a real setup for QKD, see, e.g., [18,19,20,21,22,23,24,25]. The probability of multiphoton pulses is around 10% among all nonvacuum pulses. Here we shall show how to make secure final key even with an imperfect source, i.e., with a small probability that the source sends multiphoton pulses. The bits caused by those multiphoton pulses from the source are regarded as *tagged bits* of which Eve can, in principle, have full information without causing any disturbance.

3.1 Final Key Distillation With a Fraction of Tagged Bits

It was shown by ILM-GLLP [26, 27] that we can also distill the secure final key by a CSS code even with an imperfect source, if we know the bit-flip rate, phase-flip rate and upper bound value of the fraction of tagged bits, Δ. Here we give a simple proof. For simplicity, let us recall the protocol reduced from the entanglement purification protocol in Sect. 2. Alice now sends Bob BB84 states (with most of them being prepared in Z basis) from a perfect single-photon source. But, before sending, she randomly chooses a small fraction of the qubits and tells Eve the right bases (and bit values) of them. This small fraction of qubits is called *tagged qubits*. In general, the channel can be lossy. Suppose after Bob has measured each qubits he received and they discarded those outcomes with basis mismatch, there are $N+m$ classical bits remaining. About m of them will be used for error test, among which half are Z bits and half are X bits. The remaining N bits are all Z bits and they will be used to distill the final key. We shall call them *untested bits*. At this stage, any bit that is caused by a tagged qubit is defined as a *tagged bit*. We assume Alice and Bob carry out the protocol as if they did not know which ones are tagged bits but they know the fraction of tagged bits Δ. They now distill the final key with those N untested bits. They need to know the number of likely bit-flip strings and the number of phase-flip strings for those N untested bits. The error correction part is of no difference from the ideal protocol where there are no tagged bits, i.e., after they do the error test, they know that the upper bound of the bit-flip rate of those N untested bits is t_b. We shall use notations $t_{b,tag}$, $t_{b,untag}$ for the upper bounds of bit-flip rates of the tagged bits and untagged bits from those N untested bits, respectively. We have

$$\Delta t_{b,\,tag} + (1-\Delta) t_{b,\,untag} = t_b. \tag{21}$$

And the number of likely bit-flip strings is bounded by

$$\omega_b = 2^{\Delta N H(t_{b,\,tag})} \cdot 2^{(1-\Delta) N H(t_{b,\,untag})} \leq 2^{N H(t_b)}. \tag{22}$$

Since they do not know the refined information of error rates for tagged bits and untagged bits separately, they can only use the value $2^{NH(t_b)}$ as the upper bound of the number of likely bit-flip string for error correction. What is a bit tricky is the number of the likely phase-flip string. Also, there are two groups of bits in the N untested bits, tagged bits and untagged bits. If the number of phase-flip string for tagged bits is bounded by $\omega_{ph,\,tag}$ and the number of phase-flip string for those untagged bits is bounded $\omega_{ph,\,untag}$, then the number of phase-flip string for all N bits is bounded by

$$\omega_{ph} = \omega_{ph,\,untag} \cdot \omega_{ph,\,tag}. \tag{23}$$

Give any n-bit binary string, in whatever case the number of likely strings is bounded by 2^n. Therefore we have

$$\omega_{ph,\,tag} \leq 2^{\Delta N}. \tag{24}$$

To bound the value of $\omega_{ph,\,untag}$, we must bound the phase-flip rate of those *untagged* untested bits. Asymptotically, this value is given by the error rate of those *untagged* test bits in X basis. But the observed value t_x is the *averaged* error rate of all X bits, among which only $(1-\Delta)N$ are untagged bits. Therefore, the separate error rate for the untagged X bits could actually be larger than the averaged one, t_x, but it must be bounded by $t_x/(1-\Delta)$, which corresponds to the worst case that all errors are carried by the untagged X bits. Therefore the phase-flip rate of those untagged untested bits is bounded by

$$t_{ph} \leq \frac{t_x}{1-\Delta} + \delta_0. \tag{25}$$

Therefore the number of likely phase-flip string of all untested bits is bounded by

$$\omega_{\rm ph} \leq 2^{\Delta N} \cdot 2^{(1-\Delta)NH(\frac{t_x}{1-\Delta}+\delta_0)}. \tag{26}$$

If $N(1-\Delta)$ is very large, δ_0 is very small and t_b is very close to t_z, the observed error rate of those test bits in Z basis. Asymptotically, the final key rate is

$$f = 1 - H(t_z) - \Delta - (1-\Delta)H\left(\frac{t_x}{1-\Delta}\right), \tag{27}$$

which confirms the result of ILM-GLLP [26, 27].

Remark: The bit-flip and phase-flip for a tagged bit are rather different. Consider the case with real entanglement. Once Alice tells Eve in advance the bit value in Z basis of certain pair, Alice must have already measured it in Z basis. The phase-flip is immediate very large even Eve does not touch it in the future. However, the bit-flip can still be 0 and give a noiseless channel. This is the reason we have to treat the number of two types of likely strings differently.

The above model applies to the important situation that Alice uses an imperfect source. Say, the source may sometimes produce multiphotons, i.e., some copies of single-photon states. Obviously, to Alice and Bob, the situation here cannot be worse than the situation of *tagged-bit* model where Alice announces some of the basis (and bit values), therefore it must be secure here if they use the model of *tagged bits* to treat the imperfect source, provided that they know the value Δ for the raw bits. In the case of using coherent light as the source, if a bit is created at Bob's side when Alice sends out a multiphoton pulse, that bit is regarded as a tagged bit. If the channel is lossy, it is not a trivial task to know a tight upper bound of the Δ value. Since we need to assume the channel to be Eve's channel for security, the channel transmittance can be *different* for single-photon pulses (untagged qubits) and multiphoton pulses (tagged qubits). Therefore the fraction of multiphoton pulses of the source can not be used for Δ value, the fraction of tagged bits in Bob's raw key. In particular, since they do not know which pulses are single-photon pulses, it is possible that the channel transmittance to single-photon pulses is 0 and *all* bits in Bob's raw key are tagged bits. This means, without a good method to upper-bound Δ tightly, the maximum secure distance for QKD with weak coherent light is about only 20 km [26, 27], given the existing detection technology.

3.2 PNS Attack

In practice, the channel can be very lossy. For example, if we want to do QKD over a distance longer than 100 km using an optical fiber and light pulses of wavelength 1.55 μm, the overall transmittance can be on the order of 10^{-3} or even 10^{-4}. (Suppose the detection efficiency is 15%.) This opens a door for Eve by the so called photon-number-splitting (PNS) attack [28, 29], as shown in Fig. 2. Suppose sometimes Alice sends a multiphoton pulse. Every photon inside the same pulse is in the same state. Eve can keep one photon from the pulse and sends other photons of the pulse to Bob through a transparent channel. This action will not cause any error, but Eve may have full information about Bob's bit: After the measurement basis is announced by Alice or Bob, Eve will be always able to measure the photon she has kept in the correct basis. If any bit is caused by a multiphoton pulse, in principle Eve can fully know it without causing any error. Therefore, here all those bits caused by multiphoton pulses are *tagged bits*. According to the model given by ILM-GLLP [26, 27] as we have studied, if the fraction of tagged bits in Bob's raw key is not too large, we can still obtain a secure final key by (27). In using this result, we must first know the value Δ. Alice does not know which pulse contains multiphotons; she only knows a distribution over different photon numbers for all pulses. As we have mentioned, the fraction of multiphoton pulses from the source can be very *different* from the fraction of tagged bits in Bob's raw key, for Eve's channel transmittance can be dependent on the photon number of the pulse from Alice. Naively, one can assume the worst case to estimate Δ: The channel transmittance for multiphoton pulses is 1 and we check how many raw bits are generated. If the number is larger than the number of multiphoton pulses, there must be some untagged bits. However, in such a way, the intensity of source must be rather weak in order to have some untagged bits. Suppose the channel transmittance is η and the light pulse intensity is μ. After phase randomization, the source is a mixed state of photon numbers:

$$\rho_\mu = e^{-\mu} \sum \frac{\mu^n}{n!} |n\rangle\langle n|. \tag{28}$$

Consider the normal case in which there is no Eve: The channel transmittance is η to every photon, therefore if Alice sends N_0 pulses, Bob will find $N_0(1 - e^{-\eta\mu})$ counts at his side. However, for security, they have to assume this to be Eve's channel. Using the naive worst-case estimation, we require

$$1 - e^{-\eta\mu} > 1 - e^{-\mu} - \mu e^{-\mu} \approx \mu^2/2, \tag{29}$$

i.e.,

$$\mu < 2\eta. \tag{30}$$

This shows that to guarantee that not all raw bits are tagged, the efficiency must be bounded by η^2. Given that $\eta \leq 10^{-3}$, the key rate is almost 0. Moreover, the dark count of photon detectors can lead to a very high error rate and the final key rate by (27) must be 0, given existing technology of single-photon detectors. To obtain the secure final key with a meaningful key

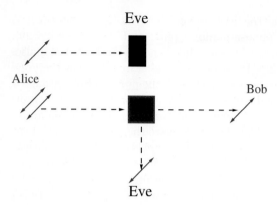

Fig. 2. A schematic diagram for photon-number-splitting attack. Eve is in the middle and controls the channel. Whenever Alice sends a single-photon pulse (untagged qubit), Eve may absorb it. Whenever Alice sends a multiphoton pulse, Eve may split it and keeps one photon. Of courses, Eve has many other choices

rate, we must have a better way to verify the Δ value, i.e., the fraction of tagged bits.

Remark: Bob cannot verify the tagged bits at his side by measuring the photon number in each coming pulses. Suppose he finds a certain pulse contains only one photon. The bit caused by that pulse could be still a tagged bit because the pulse could have contained two photons when Alice sent it.

In short, given the ILM-GLLP [26, 27] result, the remaining task is to verify the fraction of tagged bits in Bob's raw key faithfully. A reliable tight verification is nontrivial. The central task for the decoy-state method is to make a tight verification of Δ, upper bound of the fraction of tagged bits or equivalently, Δ_1, lower bound of the fraction of untagged bits in Bob's raw key.

Remark: In the above, in showing that a certain old protocol is *insecure*, we have used some specific attacking schemes. Of course, Eve can have many choices in the attack, e.g., methods in [30]. Definitely, in showing that certain protocol is *secure*, one should not assume any specific attacking scheme. We are now going to present the decoy-state method which does not assume any specific attacking scheme. The security of this method is only based on principles of quantum mechanics, and classical statistics therefore is *unconditional*.

4 The Decoy-State Method

The first idea of the decoy-state method is presented by *Hwang* [31]. Hwang proposed to do the nontrivial verification by randomly switching the intensity of each pulse between $\mu = 0.3$ (for signal pulses) and $\mu' = 1$ (for decoy pulses). By watching the counting rate of decoy pulses, one can deduce the upper bound of the fraction of tagged bits among all those bits caused by signal pulses.

The decoy-state method was not immediately useful in practice until *Wang* [32] presented the tight estimation of Δ by treating the counting rates of different intensities jointly with nontrivial inequalities and density operator convex technique. Quantitative results of a tighten verification including both analytical formulas and numerical results are also given there [32]. Earlier, a review on Hwang's result with some rough ideas was presented [33]. However, there is no quantitative result, though it proposed to use vacuum states to test dark count and *very* weak coherent states to test single-photon counting rates. Also, this idea seems to be inefficient in practice due to the possible fluctuation of dark count [34, 35]. Naturally, one can expect a higher key rate if one uses more intensities. The key rate in the limit of using infinite intensities is studied in [36].

Here we are most interested in a protocol that is *practically* efficient. Obviously, there are several criteria for a practically efficient protocol. First, the protocol must be clearly stated. For example, there should be *quantitative* description about the intensities used and *quantitative* results about the verification because we need the *explicit* information of intensities in the implementation and the *explicit* value of Δ for key distillation. Second, the result of verified value Δ should be tight in the normal case when there is no Eve. This criterion is to guarantee a good final key rate. Third, it should only use a few different intensities. In practice, it is impossible to switch the intensity among infinite numbers of different values. Finally, it should be robust to possible statistical fluctuations. Note that the counting rates are very small parameters. The effects of possible statistical fluctuations can be very important because the repetition rate of any real system is limited and therefore we cannot assume too large a number of pulses. Concerning the above criterion, a 3-intensity protocol is then proposed [32]. The protocol uses 3 intensities: vacuum, μ and μ' for the verification.

For conceptually clarity and mathematical simplicity, here we give up Hwang's original statement and derivation and we start from the classical statistical principle. And we shall directly use the technique of density matrix convex [32] for which only a few parameters are involved [32].

4.1 The Main Ideas and Results

Principle 1. Given a large number of identical and independent pulses, the averaged value per pulse of any physical quantity for some randomly sampled pulses must be (almost) equal to that of the remaining pulses, if both the number of sampled pulses and the number of the remaining pulses are sufficiently large.

In standard QKD protocols with perfect single-photon sources, this principle is used for the error test: They check the error rate of a random subset, and use this as the error rate of the remaining bits. Also, this principle can be used for estimation of the averaged value of any other physical quantities, such as the *counting rate*. In the protocol, Alice sends pulses to Bob. Given a lossy channel, whenever a pulse is sent out from Alice, Bob's detector may click during a certain time window. If his detector clicks, a raw bit is generated. The counting rate is the ratio of the number of Bob's clicks and the number of pulses sent out from Alice. More specifically, if source x sends out N_x pulses and Bob's detector clicks n_x times in appropriate time windows, the counting rate for pulses from source x is

$$S_x = \frac{n_x}{N_x}. \tag{31}$$

Obviously, this quantity for any real source can be directly observed in the protocol itself. In the QKD protocol, Alice controls the source. We shall use the concept of a *mixed source*. Source X and source Y together make a *mixed source* if the following conditions are satisfied: (1) Each individual pulse has a probability p_X to be produced from source X and probability p_Y to be produced from Y. (And $p_X + p_Y = 1$.) (2) Each individual pulse is independent. (3) Except for the states in photon-number space, pulses from source X and pulses from source Y are indistinguishable by any other physical quantities, e.g., the wavelength, the polarization, the transmission path, and so on.

In particular, if X and Y make a *mixed source* and states of pulses from each source are identical in photon-number space, then the counting rate of source X must be equal to that of source Y, since in such a case all pulses are identical and pulses from X can be regarded as sampled pulses and pulses from Y can be regarded as the remaining pulses in using *principle 1*.

Principle 2. Given light pulses from a mixed source that contains source X and source Y, if X and Y produce the same states in photon-number space, the counting rate for pulses from source X must be equal to that of source Y, provided that the number of pulses from each source is sufficiently large.

In our protocol, Alice randomly switches the intensities between μ, μ' and 0 for each pulse. We shall regard them as three independent sources, A_μ, $A_{\mu'}$ and A_0, respectively. The state for source A_μ can be rewritten in the following equivalent convex form:

$$\rho_\mu = e^{-\mu}|0\rangle\langle 0| + \mu e^{-\mu}|1\rangle\langle 1| + c\rho_c, \tag{32}$$

and $c = 1 - e^{-\mu} - \mu e^{-\mu} > 0$,

$$\rho_c = \frac{1}{c}\sum_{n=2}^{\infty} P_n(\mu)|n\rangle\langle n|, \tag{33}$$

and $P_n(\mu) = \frac{e^{-\mu}\mu^n}{n!}$. This convex form shows that source A_μ sends out three types of pulses: sometimes sends out vacuum, sometimes sends out

$|1\rangle\langle 1|$, sometimes sends pulses of state ρ_c. Therefore, source A_μ can be equivalently regarded as three sources, $A_{\mu 0}$ producing those vacuum pulses, $A_{\mu 1}$ producing those single-photon pulses and $A_{\mu c}$ producing those pulses in state ρ_c. Nobody outside Alice's lab can tell whether Alice has actually used these three sources or A_μ, the real source. Bob's counts caused by ρ_c from Alice are regarded as *tagged* bits. Since we know explicitly the probability of pulses ρ_c for our source, we shall know the fraction of tagged bits if we know s_c, the counting rate of state ρ_c. The *counting rate* of any state ρ is the probability that Bob's detector counts whenever Alice sends ρ.

For convenience, we shall always assume

$$\mu' > \mu; \mu' e^{-\mu'} > \mu e^{-\mu} \tag{34}$$

in this paper.

$$\rho_{\mu'} = e^{-\mu'}|0\rangle\langle 0| + \mu' e^{-\mu'}|1\rangle\langle 1| + c\frac{\mu'^2 e^{-\mu'}}{\mu^2 e^{-\mu}}\rho_c + d\rho_d. \tag{35}$$

Here $d \geq 0$ and ρ_d is a density operator. We do not need the explicit formula for ρ_d; we shall only need the fact that state $\rho_{\mu'}$ can be written in the above convex form. Source $A_{\mu'}$ can be equivalently regarded as four virtual sources: $A_{\mu' 0}$, which contains all vacuum pulses from $A_{\mu'}$, $A_{\mu' 1}$, which contains all single-photon pulses from $A_{\mu'}$, $A_{\mu' c}$, which contains all ρ_c pulses of source $A_{\mu'}$, and $A_{\mu' d}$, which contains all ρ_d pulses of source $A_{\mu'}$. Also, source A_μ can be equivalently regarded as three sources, as we have mentioned before. Of course, $A_{\mu' c}$ and source $A_{\mu c}$ make a mixed source and they produce identical states in photon number space. This is to say, the counting rate of state ρ_c from source A_μ is equal to the counting rate of ρ_c from source $A_{\mu'}$, i.e.,

$$s_c(\mu) = s_c(\mu') = s_c. \tag{36}$$

By the same reason we have a more general formula for counting rate:

$$s_\alpha(\mu) = s_\alpha(\mu') = s_\alpha. \tag{37}$$

Here the subscript α represents for a state. In our article, we shall use $\alpha = 0, 1, c$ for vacuum state, single-photon state and state ρ_c, respectively.

Remark: Equation (36) *does not* mean that pulses of intensity μ and pulses of intensity μ' are indistinguishable to Eve. We only mean that, given the subset of those pulses whose states are ρ_c, Eve cannot treat each pulse in this subset differently according to which source (A_μ or $A_{\mu'}$) they come from. Consider a similar classical story: there are many people in set C. Each of them in the set is either a professional basketball player or a professional football player. The inspector does not know the profession of each one, but she knows the fact that each one there is either a basketball player or a football player. She has a task to do her best to label each one in set C differently according to his profession, say, if she believes somebody is a basketball player, she labels him by B, and she labels him by F if she believes that he is a football player. In general, the inspector can label them differently to a certain extent because of the fact of height difference between people from those two professions: In average, a basketball player is taller than a football player, though this is not always correct to each individual case. Equation (36) does not claim anything for this general situation. What it has claimed is a special condition: Among set C, there exists a subgroup $C_{1.88}$ which contains all those persons who are 1.88 m tall. In subset $C_{1.88}$, some of them are basketball players and some of them are football players. To each person inside this subset, the inspector has no way to label them differently according to their professions. Say, inside the subset $C_{1.88}$, if $x\%$ of basketball players are labeled by B, then there must be $x\%$ of football players also labeled by B, if the number of persons of each profession in subset $C_{1.88}$ is large enough. In short, the inspector here cannot treat persons of the same height differently according to their professions. In (36), we have claimed that Eve cannot treat pulses of the same state differently according to which sources they are from.

Remark: Here Alice has only used two real sources, A_μ, $A_{\mu'}$. The counting rate of a real source can be observed directly in the protocol. Therefore, s_0, S_μ and $S_{\mu'}$, the counting rates of vacuum, a pulse from A_μ and a pulse from $A_{\mu'}$ are regarded as known parameters hereafter. But counting rate of a virtual source, e.g., $A_{\mu c}$ cannot be directly observed directly. We must deduce it mathematically based on the observed results of the protocol. We shall use notations of s_1, s_c, s_d for counting rate of a single-photon pulse, state ρ_c and state ρ_d. In the asymptotic case, we have the following equations:

$$S_\mu = e^{-\mu} s_0 + \mu e^{-\mu} s_1 + c s_c, \tag{38}$$

$$S_{\mu'} = e^{-\mu'} s_0 + \mu' e^{-\mu'} s_1 + c \frac{\mu'^2 e^{-\mu'}}{\mu^2 e^{-\mu}} s_c + d s_d. \tag{39}$$

In the above we have used the same notations S_0, s_1, s_c in both equations. This is because we have assumed that the counting rates of the same state from different sources are equal. S_0 is known, and s_1 and s_d are unknown, but they are never less than 0. Therefore setting s_d to be zero, we can obtain the following crude result by using (39) alone:

$$c s_c \leq \frac{\mu^2 e^{-\mu}}{\mu'^2 e^{-\mu'}} \left(S_{\mu'} - e^{-\mu'} s_0 - \mu' e^{-\mu'} s_1 \right). \tag{40}$$

Since $s_1 \geq 0$, we obtain a crude result for the upper bound of s_c:

$$c s_c \leq \frac{\mu^2 e^{-\mu}}{\mu'^2 e^{-\mu'}} \left(S_{\mu'} - e^{-\mu'} s_0 \right). \tag{41}$$

However, we can tighten the verification by using (38). Having obtained the crude result above, we now show that the verification can be done more sophisticatedly and one can further tighten the bound significantly. In obtaining inequality (41), we have dropped terms s_1 and s_d, since we only have trivial knowledge about s_1 and s_d there, i.e., $s_1 \geq 0$ and $s_d \geq 0$. Therefore, inequality (41) has no advantage to Hwang's result at that moment. However, after we have obtained the crude upper bound of s_c, we can have a larger-than-0 lower bound for s_1 by (38), provided that our crude upper bound for s_c given by (41) is not too large. Combining the crude upper bound for s_c given by (41) and (38), we have the nontrivial lower bound for s_1:

$$s_1 \geq S_\mu - e^{-\mu}s_0 - cs_c > 0. \qquad (42)$$

With this new lower bound of s_1, we can further tighten the upper bound of s_c by (40) and obtain a more tightened s_c. With the new s_c, we can again improve the lower bound of s_1 by (38). The final bound values are determined by infinite iteration. Therefore tight values for s_c and s_1 can be obtained by solving the simultaneous constraints of (38) and (40). We have the following final bound after solving them:

$$\Delta \leq \frac{\mu}{\mu' - \mu}\left(\frac{\mu e^{-\mu}S_{\mu'}}{\mu' e^{-\mu'}S_\mu} - 1\right) + \frac{\mu e^{-\mu}s_0}{\mu' S_\mu}. \qquad (43)$$

Here we have used $\Delta = cs_c/S_\mu$. In the case of $s_0 \ll \eta$, if there is no Eve., $S'_\mu/S_\mu = \mu'/\mu$. Alice and Bob must be able to verify

$$\Delta = \left.\frac{\mu\left(e^{\mu'-\mu} - 1\right)}{\mu' - \mu}\right|_{\mu'-\mu\to 0} = \mu \qquad (44)$$

in the protocol. This is close to the real value of fraction of multiphoton counts: $1 - e^{-\mu} \geq \mu - \mu^2/2$, given that $\eta \ll 1$. In this three-intensity protocol for the verification, both μ and μ' can be set in a reasonable range, therefore both of them can be used for final key distillation. Given Δ, one can immediately calculate the value Δ_1, the lower bound of the fraction of untagged bits by [32]:

$$\Delta_1 = 1 - \Delta - e^{-\mu}S_0/S_\mu. \qquad (45)$$

Of course, if we also want to use source $A_{\mu'}$ for key distillation, we need the value Δ', the upper bound of the fraction of tagged bits for source $A_{\mu'}$. Given s_c, we can calculate the lower bound of s_1 through (42). Given s_1, we can also calculate the upper bound of Δ', the fraction of multiphoton count among all counts caused by pulses from source $A_{\mu'}$. Explicitly,

$$\Delta' \leq 1 - (1 - \Delta - \frac{e^{-\mu}s_0}{S_\mu})e^{\mu-\mu'} - \frac{e^{-\mu'}s_0}{S_{\mu'}}. \qquad (46)$$

The values of μ, μ' should be chosen in a reasonable range, e.g., from 0.2 to 0.5.

4.2 The Issue of Unconditional Security

The security of the decoy-state method is a direct consequence of the separate prior art result in [26, 27], which we shall call ILM-GLLP, the initials of authors of the result. Therefore a separate security proof is not necessary. ILM-GLLP [26, 27] offers methods to distill the unconditionally secure final key from the raw key if the upper bound of fraction of tagged bits is known, given whatever imperfect source and channel. The decoy-state method verifies such an upper bound for a coherent-state source. We can consider an analog using the model of pure water distillation: Our task is to distill pure water by heating from raw water that may contain certain poison constituents. Suppose it is known that the poison constituents will be evaporated in the heating. We want to know how long the heating is needed to obtain the pure water for certain. If we blindly heat the raw water for too long, all the raw water will be evaporated and we obtain nothing. If we heat the raw water for a too short period, the water could be still poisonous. ILM-GLLP finds an explicit formula for the heating time which is a function of the upper bound of the fraction of poison constituents. They have proven that one can always obtain pure water if we use that formula for the heating time. However, the formula itself does not tell how to examine the fraction of poison constituents. The decoy-state method is a method to verify a tight upper bound of the fraction of the poison constitute. It is guaranteed by the classical statistical principle that the verified upper bound by the decoy-state method is (always) larger than the true value. Using this analog, the next question is how to obtain a tighter upper bound: if the verified value overestimates too much, it is secure but it is inefficient.

Hwang's result is a large step toward the efficient and secure QKD with the existing setup. However, there is still a gap for immediate application. The estimated Δ value is still too large. Given such a value, one cannot obtain a meaningful key rate in practice for long-distance QKD with existing setups. We want a faithful and *tighter* estimation. We want a way to obtain a value that is only a bit larger than the true value in the normal case that there is no Eve (for efficiency), *and* it is *always* larger than the true value in whatever case (for security). This can be achieved by the improved decoy-state method.

4.3 Robustness Analysis

The results above are only for the asymptotic case. In practice, there are both operational errors and statistical fluctuations. All these types of errors can be formulated in one unified picture. Consider those pulses of intensity μ'. At each individual time i, Alice could actually have produced a coherent pulse of intensity μ'_i when she *wants* to produce $\rho_{\mu'}$. Therefore, the probability distribution in photon number space for source $A_{\mu'}$ is actually a bit different from the assumed distribution. On the other hand, even if there is no operational error, there could be statistical fluctuation for the distribution of pulses from each source in photon number space, i.e., the state for source A_x is a bit different from the assumed state ρ_x, since the number of pulses in any real protocol is always finite. Moreover, there is another type of statistical fluctuation: The counting rate of the same state for pulses from different source could be a bit different, i.e., Eve has nonnegligibly small probability to treat the pulses from different sources a little bit differently, even though the pulses have the same state.

We shall show that these three types of errors can be described in one unified picture, and we shall demonstrate numerically that the decoy-state method is rather robust if the operational error is not too large.

We have assumed a distribution

$$P_\mu(n) = e^{-\mu}\frac{\mu^n}{n!} \quad (47)$$

in photon number space for coherent states with intensity μ. However, at any time Alice decides to set the intensity of an individual pulse to be μ or μ', the actual intensity could be μ_i, which can be a bit different from μ or μ'. Therefore we should replace $P_\mu(n)$ and $P_{\mu'}(n)$ by slightly different distributions of $\frac{1}{N_\mu}\sum_{i=0}^{N_\mu} P_{\mu_i}(n)$ and $\tilde{P}_{\mu'}(n)\frac{1}{N_{\mu'}}\sum_{i=0}^{N_{\mu'}} P_{\mu'_i}(n) = P_{\mu'}(n)(1+c'_n)$. Also, since the number of pulses of each intensity is finite, there are statistical fluctuations to the probability of each state. Considering these two kinds of errors, each time we have assumed to produce a state ρ_μ, the actual state is

$$\tilde{\rho}_\mu = \sum \tilde{P}_\mu(n)|n\rangle\langle n|, \quad (48)$$

and $\tilde{P}_\mu(n) = P_\mu(n)(1+\epsilon_n)$. Moreover, since ϵ_n is small, we have a convex form with only a few variables:

$$\tilde{\rho}_\mu = \tilde{P}_0|0\rangle\langle 0| + \tilde{P}_1|1\rangle\langle 1| + \tilde{c}\tilde{\rho}_c, \quad (49)$$

and $\tilde{P}_{0,1} = \tilde{P}_\mu(0,1)$; $\tilde{c} = 1 - \tilde{P}_0 - \tilde{P}_1$. Say, e.g., given N_μ pulses of intensity μ, the number of vacuum, single-photon state and multiphoton state ρ_c could be a bit different from the assumed values of $N_\mu P_0(\mu), N_\mu P_1(\mu), N_\mu P_c(\mu)$, respectively. The similar deviations also apply to the pulses of intensity μ'. A numerical study has been performed in [37] based on infinite parameters. Here we only need a few parameters. Besides this, there are operational errors. In practice, the main error to the distribution could be from the operational error in each switching of the intensity.

Because of the small operational error, the intensity of light pulses in source A_0 could be slightly larger than 0. This does not matter because a little bit overestimation on the vacuum count only decreases the efficiency a little bit, but does not undermine the security at all. Therefore we do not care about the operational error of this part. Say, given n_0 counts for all the pulses from source A_0, we then simply assume the tested vacuum counting rate is $s_0 = n_0/N_0$, though we know that the actual value of vacuum counting rate is less than this.

Since the new distributions are only a bit different from the old ones, there must exist a positive number $\tilde{d}$ and a density operator $\tilde{\rho}_d$ so that the state from source $A_{\mu'}$ is convexed by

$$\tilde{\rho}_{\mu'} = \tilde{P}'_0|0\rangle\langle 0| + \tilde{P}'_1|1\rangle\langle 1| + \tilde{c}'\tilde{\rho}_c + \tilde{d}\tilde{\rho}_d. \tag{50}$$

This, together with the possible statistical fluctuation to counting rates gives the following simultaneous constraint:

$$\begin{cases} S_\mu = P_\mu(0)(1+\epsilon_0)s_0 + P_\mu(1)(1+\epsilon_1)s_1 + c(1+\epsilon_c)s_c, \\ S_{\mu'} \geq P_{\mu'}(0)(1+\epsilon'_0)s'_0 + P_{\mu'}(1)(1+\epsilon'_1)s'_1 + \frac{\mu'^2 e^{-\mu'}}{\mu^2 e^{-\mu}} c(1+\epsilon'_c)s'_c, \end{cases} \tag{51}$$

and $c = 1 - P_\mu(0) - P_\mu(1); c' = \frac{\mu'^2 e^{-\mu'}}{\mu^2 e^{-\mu}} c$. Suppose we know the upper bounds of all values of $|\epsilon_x|, |\epsilon'_x|, x = 0, 1, c$, we can calculate the lower bound of s_1 and the upper bound of s_c.

Mathematically, the counting rates of the same state for pulses from different sources can be a bit different, say

$$s_x(\mu') = (1 + r_x)s_\rho(\mu), \tag{52}$$

and the real number r_x is the relative statistical fluctuation for counting rate of different states. In particular, $x = 0, 1, c$ indicates that of state of vacuum, single-photon pulses and ρ_c, respectively.

Our task remaining is to verify a tight upper bound of Δ and the probability that the real value of Δ breaks the verified upper bound is exponentially close to 0. This is to solve the following constraints

$$\begin{cases} S_\mu = P_\mu(0)(1+\epsilon_0)s_0 + P_\mu(1)(1+\epsilon_1)s_1 + c(1+\epsilon_c)s_c, \\ S_{\mu'} \geq P_{\mu'}(0)(1+\epsilon'_0)(1+r_0)s'_0 + P_{\mu'}(1)(1+\epsilon'_1)(1+r_1)s'_1 \\ \qquad + \frac{\mu'^2 e^{-\mu'}}{\mu^2 e^{-\mu}} c(1+\epsilon'_c)(1+r_c)s'_c. \end{cases} \tag{53}$$

Given $N_1 + N_2$ copies of state ρ, suppose the counting rate for N_1 randomly chosen states is s_ρ and the counting rate for the remaining states is s'_ρ, the probability that $|s_\rho - s'_\rho| > \delta_\rho$ is less than $\exp(-O\delta_\rho^2 N_0/s_\rho)$ and $N_0 = \text{Min}(N_1, N_2)$. To make a faithful estimation with exponential certainty, we require $\delta_\rho^2 N_0/s_\rho = 100$. This causes a relative fluctuation

$$|r_\rho| \leq \frac{\delta_\rho}{s_\rho} \leq 10\sqrt{\frac{1}{s_\rho N_0}}. \tag{54}$$

The probability of violation is less than $e^{-O(100)}$. To formulate the relative fluctuation r_1, r_c by s_c and s_1, we only need to check the number of pulses in

ρ_c, $|1\rangle\langle 1|$ in each sources in the protocol. That is, using (54), we can replace r_1, r_c in (53) by $10e^{\mu/2}\sqrt{\frac{1}{\mu s_1 N}}$, $10\sqrt{\frac{1}{cs_c N}}$, respectively, and N is the number of pulses in source A_μ. Since we assume the case where the vacuum-counting rate is much less than the counting rate of state ρ_μ, we omit the effect of fluctuation in vacuum counting, i.e., we set $r_0 = 0$.

Also, we can replace $(1+\epsilon'_x)(1+r'_x)$ by $(1+r'_x)$ and raise the value of r'_x a little bit. Similarly, we can formulate the parameters $\epsilon_x \le 10\sqrt{\frac{1}{P_x N_x}}$ and $\epsilon'_x \le 10\sqrt{\frac{1}{P'_x N'_x}}$. Given that the channel transmittance is very small, r_x is much larger than ϵ_x. Here we simply take all $\epsilon_x, \epsilon'_x = 0$.

With these inputs, (53) can now be solved numerically for the largest value of s_c in the likely range of stistical fluctuation, i.e., the fluctuation beyond the assumed range is in the magnitude order of $e^{-O(100)}$. Good numerical results for a tighter estimation of Δ value have been obtained [32] in many parameter settings based on existing technology [18, 19, 20, 21, 22, 23, 24, 25]. For example, given $\mu = 0.3$, $\mu' = 0.43$, and $\eta = 10^{-3}$, we obtain $\Delta = 34.4\%$, which is greatly less than Hwang's asymptotic result 60.4%, though it is still a bit larger than the true value, 25.9%. The fraction of multiphoton counts for pulses of intensity μ' have also been tightly verified [32]. The above estimation about the robustness with respect to small operational errors and errors of source state does not produce the optimized key rate, but it has shown the robustness of decoy-state method already.

4.4 Final Key Rate and Further Studies

Given the methods to verify Δ, the upper bound of fraction of tagged bits or Δ_1, the lower bound of fraction of untagged bits, we can calculate the final key rate by (27). However, there are more efficient formulas. As is pointed out in [36], one need only correct the phase-flip errors of single-photon counts and remove tagged bits for privacy amplification. In particular, the following formula is recommended [36] for key rate of any intensity μ_x:

$$Q_1(\mu_x) = H(E(\mu_x)) - \Delta_1(\mu_x) H(e_1). \tag{55}$$

Here $\Delta_1(\mu_x)$ is the fraction of single-photon counts for the intensity μ_x, which can be either μ or μ', $E(\mu_x)$ is the observed error rate of bits caused by source μ_x, and e_1 is the error-rate of counts caused by single-photon pulses. This formula gives a higher key rate than that of (27). Given a Δ value, $\Delta_1(\mu_x)$ can be calculated straightly by [32] from with (45). Also, e_1 can be estimated efficiently. Therefore, we only need to derive the e_1 value. We can do so by using the weaker source, μ. Obviously, if the observed total error rate is $E(\mu)$ for source μ, then e_1 can be calculated by

$$e_1 = \frac{E(\mu) - \frac{e^{-\mu} S_0}{S_\mu}}{\Delta_1(\mu)}. \tag{56}$$

After the major works presented in [31, 32, 36], the decoy-state method has been further studied. *Harrington* et al. [37] numerically studied the effect of fluctuation of the state itself. *Wang* [35] proposed a four-state protocol: using three of them to make optimized verification, and using the other one μ_s as the main signal pulses. This is because, if we want to optimize the verification of the Δ value, μ, μ' cannot be chosen freely. Therefore we use another intensity μ_s to optimize the final key rate. It is shown numerically how to choose the intensity for the main signal pulses (μ_s), and good key rates are obtained in a number of specific conditions. The results of final key rates [32, 35] show that good key rates can be obtained even if the channel transmittance is around 10^{-4}. This corresponds to a distance of 120 km to 150 km for practical QKD with coherent states.

4.5 Summary

Given the result of ILM-GLLP [26, 27], one knows how to distill the secure final key if he knows the fraction of tagged bits. The purpose of the decoy-state method is to do a tight verification of the the fraction of tagged bits. The main idea of the decoy-state method is to use different intensities of source light, and one can verify the fraction of tagged bits of certain intensity by watching the the counting rates of pulses of different intensities [31]. With the technique of density operator convex and jointly treating the counting rates of different intensities with nontrivial inequalities [32], the upper bound of the fraction of tagged bits or the lower bound of the fraction of untagged bits can be verified tightly [32]. Since the counting rates are small quantities, and in any real setup the number of intensities and pulses are limited, the effect of statistical fluctuation is very important. It has been shown that the three-state decoy-state method in practice can work even with the fluctuations and other errors [32]. If one uses an infinite number of different intensities and each intensity consists of infinite pulses, one can actually verify it perfectly [36]. The decoy-state method has promised a distance of 120 km to 150 km for practical QKD [32, 35]. To further increase the distance, we need to improve the technologies, which includes decreasing the dark counts, and raising the detection efficiency and the system repetition rate.

Remark: Although the decoy-state method is promising for the practical quantum key distribution, it is not the only choice. Other promising methods for practical QKD include the method using strong reference light [38, 39], the mixed B92 protocol [40] (i.e., SARG04 protocol), and so on.

5 QKD With Asymmetric Channel Noise

Most of the known prepare-and-measure protocols assume the symmetric channel to estimate the noise threshold for the protocol. Also, most of the protocols use the symmetrization method: Alice randomly chooses bases from a certain set to prepare her initial state. All bases in the set are chosen with equal probability. In such a way, even the noise of the channel (Eve's channel) is not symmetric; the symmetrization make the error rate be *always* symmetric. In this work we show that actually we can let all key bits be prepared in a single basis and we can have advantages in key rate or noise threshold provided that the channel noise is asymmetric. We shall propose protocols with higher key rate and larger channel error rate threshold, given an asymmetric physical channel and an invisible Eve.

5.1 Channel Error, Tested Error and Key-Bits Error

Normally, Alice will transmit qubits in different basis (e.g., Z basis and X basis) to Bob, and Bob will also measure them in different basis. The Hadamard transformation $H = \frac{1}{\sqrt{2}}\begin{pmatrix} 1 & 1 \\ 1 & -1 \end{pmatrix}$ interchanges the Z-basis $\{|0\rangle, |1\rangle\}$ and X-basis $\{|\pm\rangle = \frac{1}{\sqrt{2}}(|0\rangle \pm |1\rangle)\}$. We shall use the term key bits for those raw bits which are used to distill the final key and the term check bits for those bits whose values are announced publicly to test the error rates. Our purpose is to know the bit-flip rate and phase-flip rate to key-bits. We do it in this way: first test flipping rates of the check bits, then deduce the channel flip rate, and then determine the error rates of key-bits. The flipping rates of qubits in different bases are in general different, due to the basis transformation. Here we give a more detailed study on this issue. We first consider the four-state protocol with CSS code, where only two basis, Z-basis and X-basis are involved in operation. For such a case of four-state protocol, we define asymmetric channel as the channel with its bit-flip error rate being different from its phase-flip error rate. The check bits will be discarded after the error test. We use the term Z-bits for those qubits which are prepared and measured in Z basis, the term X-bits for those bits which are prepared and measured in X basis. For clarity we shall regard Alice's action of preparing a state in X basis as the joint actions of state preparation in Z basis followed by a Hadamard transform. We shall also regard Bob's measurement in X basis as the joint action of first taking a Hadamard transform and then taking the measurement in Z basis. Therefore we shall also call those Z-bits as I-qubits and those X-bits as H-bits. To let the CSS code work properly, we need to know the value of average bit-flip rate and average phase-flip rate over all key bits. We define three Pauli matrices:

$$\sigma_x = \begin{pmatrix} 0 & 1 \\ 1 & 0 \end{pmatrix}, \quad \sigma_y = \begin{pmatrix} 0 & -i \\ i & 0 \end{pmatrix}, \quad \sigma_z = \begin{pmatrix} 1 & 0 \\ 0 & -1 \end{pmatrix}.$$

The matrix σ_x applies a bit-flip error but no phase-flip error to a qubit, σ_z applies a phase-flip error but no bit-flip error, and σ_y applies both errors. We assume the $\sigma_x, \sigma_y, \sigma_z$ rates of the physical channel are q_{x0}, q_{y0}, q_{z0}, respectively. Note that the phase-flip rate or bit-flip rate of the channel is the summation of q_{z0}, q_{y0} or q_{x0}, q_{y0}, respectively. Explicitly we have

$$p_{x0} = q_{x0} + q_{y0},$$
$$p_{z0} = q_{z0} + q_{y0}, \qquad (57)$$
$$p_{y0} = q_{x0} + q_{z0}, \qquad (58)$$

and q_{y0} is defined as the channel flipping rate to the qubits prepared in Y-basis ($\frac{1}{\sqrt{2}}(|0\rangle \pm i|1\rangle)$). After the error test on check bits, Alice and Bob know the value of p_{z0}, p_{x0} of the channel. Given the channel flipping rates q_{x0}, q_{y0}, q_{z0}, one can calculate the bit-flip rate and phase-flip rate for the remaining qubits. For those I-bits, the bit-flip rate and phase-flip rate are just $q_{x0}+q_{y0}$ and $q_{z0}+q_{y0}$, respectively, which are just the channel bit flip rate and phase flip rate. Therefore the channel bit-flip rate p_{x0} is identical to the tested flip rate of I-bits. The channel phase-flip rate p_{z0} can be determined by testing the flip rate of those H-bits. An H-qubit is a qubit treated in the following order: *prepared in Z basis, Hadamard transform, transmitted over the noisy channel, Hadamard transform, measurement in Z basis.* If the channel noise offers a σ_y error, the net effect is

$$H\sigma_y H \begin{pmatrix} c|0\rangle \\ |1\rangle \end{pmatrix} = \sigma_y \begin{pmatrix} c|0\rangle \\ |1\rangle \end{pmatrix}. \qquad (59)$$

This shows the channel σ_y error will also cause a σ_y error to an H-qubit. Similarly, due to the fact of

$$H\sigma_z H = \sigma_x, \quad H\sigma_x H = \sigma_z, \qquad (60)$$

a channel σ_x flip or a channel σ_z flip will cause a net σ_z error or σ_x error to an H-bit. Consequently, a channel phase flip causes a bit flip error to H-bit, a channel bit-flip causes a phase flip error to H-bit. This is to say, *the measured error of H-bits is just the channel phase flipping rate.* Therefore the average bit-flip error rate and phase-flip error rate to each types of key bits will be

$$p_z^I = p_x^H = p_{z0}, \quad p_z^H = p_x^I = p_{x0}. \qquad (61)$$

Here $p_x^H, p_x^I (p_z^H, p_z^I)$ are for the bit flip (phase flip) error of H-bits and I-bits from those key-bits, respectively. Suppose the key bits consist of η I-bits and $1 - \eta$ H-bits, the average flip error of the key bits is

$$p_x = \eta p_x^I + (1-\eta) p_z^H = \eta p_{x0} + (1-\eta) p_{z0},$$
$$p_z = \eta p_z^I + (1-\eta) p_x^H = \eta p_{z0} + (1-\eta) p_{x0}. \qquad (62)$$

5.2 QKD With One-Way Classical Communication

Here we consider the almost trivial application of our analysis of an asymmetric channel above for the case of key distillation with one-way classical communication. The result here itself is not new, but it demonstrates how one can improve the key rate with an asymetric channel noise In the standard BB84 protocol, since the preparation basis of key bits are symmetrized, the average bit-flip error and phase-flip error to those key bits are always equal no matter whether the channel noise itself is symmetric or not. That is to say, when half of the key-bits are X-bits and half of them are Z-bits, the average flip rates over all key bits are always

$$p_x = p_z = (p_{x0} + p_{z0})/2. \tag{63}$$

Therefore the key rate for the standard BB84 protocol (Shor–Preskill protocol) [10] with whatever asymmetric channel is $1 - 2H(\frac{p_{x0}+p_{z0}}{2})$ [10], where $H(t) = -(t \log_2 t + (1-t) \log_2(1-t))$. (Note that in the four-state protocol, an asymmetric channel is simply defined by $p_{x0} \neq p_{z0}$.) However, if all key bits had been prepared and measured in Z-basis, then the bit-flip and phase-flip rates to key bits would be equal to those flipping values of the channel itself. In such a case the key rate is

$$R = 1 - H(p_{x0}) - H(p_{z0}), \tag{64}$$

Obviously, this is, except for the special case of $p_{x0} = p_{z0}$, always larger than the key rate in standard BB84 protocol with CSS code (Shor–Preskill protocol), where the key-bits are prepared in Z-basis and X-basis with equal probability. For a higher key rate, one should *always* use the above modified BB84 protocol with all key bits prepared and measured in a single basis, Z-basis.

In the standard protocol [41], the key bits are equally distributed over all three different bases. In distilling the final key, each type of flipping error used is the averaged value over three bases, i.e.,

$$\bar{q}_x = \frac{q_{x0} + 2q_{z0}}{3}, \quad \bar{q}_z = \frac{q_{x0} + q_{z0} + q_{y0}}{3}, \quad \bar{q}_y = \frac{q_{x0} + 2q_{y0}}{3}. \tag{65}$$

The key rate is

$$r = 1 - H(\bar{q})$$
$$H(\bar{q}) = -\bar{q}_x \log_2 \bar{q}_x - \bar{q}_y \log_2 \bar{q}_y - \bar{q}_z \log_2 \bar{q}_z - q_{I0} \log_2 q_{I0}. \tag{66}$$

This is the key rate for standard six-state protocol where we mix all qubits in diffent bases together. However, such a mixing is unnecessary. We can choose to simply distill three batches of final keys from Z-bits, X-bits and Y-bits separately. If we do it in such a way, the key rate will be increased to

$$r' = 1 - q_{x0} \log_2 q_{x0} - q_{y0} \log_2 q_{y0} - q_{z0} \log_2 q_{z0} - q_{I0} \log_2 q_{I0}. \tag{67}$$

Obviously, r' is *never* less than r since the mixing operation never decreases the entropy. The only case where $r' = r$ is $q_{x0} = q_{y0} = q_{z0}$. Therefore for a higher key rate, we propose to *always* distill 3 batches of final keys separately. The advantage in such a case is unconditional, there is no loss for whatever channel. Now we start to consider something more subtle: the advantage conditional on the prior information of the asymmetry property of noise of the physical channel.

5.3 Six-State Protocol With Two-Way Classical Communications

The symmetric channel noise for a six-state protocol is defined by $q_{x0} = q_{y0} = q_{z0}$, if this condition is broken, we regard it as an asymmetric channel for six-state protocols. In the standard six-state protocol [41,42,43], symmetrization is used, i.e., the key-bits are equally consisted by X-, Y-, Z-bits. When the channel noise itself is symmetric, i.e., $q_{x0} = q_{y0} = q_{z0}$, a six-state protocol can have a higher noise threshold than that of a four-state protocol. This is because in the six-state protocol, the σ_y type of channel error rate is also detected. In removing the bit flip error, σ_y error is also reduced, therefore the phase-flip error is partially removed. However, in a four-state protocol, σ_y error is never tested therefore we have to assume the worst situation that $q_{y0} = 0$ [42].

We shall show that one can have a higher tolerable channel errors if one modifies the existing protocols, given the asymmetric channel (i.e., the channel with its Y-bits flipping error being different from that of X-bits or Z-bits), for example, in the case that $q_{y0} = 0$ and $q_{x0} = q_{z0}$. The different types of error rates to the transmitted qubits are

$$q_x = q_{x0}, \quad q_y = 0, \quad q_z = q_{z0},$$

for those Z-bits;

$$q_x = q_{z0}, \quad q_y = 0, \quad q_z = q_{x0},$$

for those X-bits and

$$q_x = q_{z0}, \quad q_y = q_{x0}, \quad q_z = 0,$$

for Y-bits. The average error rates over all transmitted bits are:

$$\bar{q}_x = q, \quad \bar{q}_y = q/3, \quad \bar{q}_z = 2q/3, \quad q = q_{x0} = q_{z0}. \tag{68}$$

With such a fact, the threshold of total channel noise $q_{t0} = (q_{x0} + q_{y0} + q_{z0})$ for the protocol [43] is 41.4%, the same as the case with symmetric noise [43]. Actually, by our numerical calculation we find that the threshold of total channel noise for the Chau protocol [43] is almost unchanged with whatever value of q_{y0}. However, if *all* key bits were prepared in Y-basis (the basis of $\{|y\pm\rangle = \frac{1}{\sqrt{2}}(|0\rangle \pm i|1\rangle)\}$), there would be no σ_z-type of error, therefore one only needs to correct the bit-flip error. To see this we can regard a Y-qubit as a qubit treated in the following order: *prepared in Z basis, T transform, transmitted over the noisy channel, T^{-1} transform, measurement in Z basis.* Here $T = \frac{1}{\sqrt{2}} \begin{pmatrix} 1 & i \\ 1 & -i \end{pmatrix}$; it changes states $|0,1\rangle$ into $|y\pm\rangle$. The following facts

$$T\sigma_x T^{-1} = \sigma_y, \quad T\sigma_y T^{-1} = \sigma_z, \quad T\sigma_z T^{-1} = \sigma_x \qquad (69)$$

cause the consequence that a channel flip of the type $\sigma_x, \sigma_y, \sigma_z$ will cause an error to Y-bits in the type of $\sigma_y, \sigma_z, \sigma_x$, respectively. With this fact, if the channel error of p_{y0} is 0, the σ_z type of error to the Y-qubit is also 0. Using the iteration formula given in [43], once all bit-flip error is removed, all errors are removed. Therefore the error rate threshold is

$$Q_x = Q_z = 25\%, \qquad (70)$$

i.e., a total error rate of 50%. In practice, it is not likely that σ_y-type of channel flip is exactly 0. Numerical calculation (Fig. 3) shows that using Y-basis as the only basis for all key-bits always has an advantage provided that the channel flipping rate satisfies $q_{y0} < q_{x0} = q_{z0}$. For the purpose of improving the noise threshold, we propose the following protocol:

1: Alice creates a random binary string b with $(6+\delta)N$ bits.
2: Alice generates $(6+\delta)N$ quantum state according to each element of b. For each bit in b, if it is 0, she produces a quantum state of either $|0\rangle$, or $|+\rangle$ or $|y+\rangle$ with probability $1/4, 1/4, 1/2$, respectively; if it is 1, she creates a quantum state $|1\rangle$ or $|-\rangle$ or $|y-\rangle$ with probability $1/4, 1/4, 1/2$, respectively.
3: Alice sends all qubits to Bob.
4: Bob receives the $(6+\delta)N$ qubits, measuring each one in basis of either $\{|0\rangle, |1\rangle\}$ or $\{|\pm\rangle\}$ or $\{|y\pm\rangle\}$ randomly, with equal probability.
5: Alice announces basis information of each qubits.
6: Bob discards all those qubits he measured in a wrong basis. With high probability, there are at least $2N$ bits left (if not, abort the protocol). Alice randomly chooses $N/2$ Y bits for the distillation of final key later, and uses the remaining $3N/2$ bits as check bits. (Among all the check bits, approximately $N/2$ are X-bits, $N/2$ are Z-bits, and $N/2$ are Y-bits.)
7: Alice and Bob announce the values of their check bits. If too many of them are different, they abort the protocol.
8: Alice randomly groups the key bits with each group consisting of two bits. Alice and Bob compare the parity values on each side to each group. If the values agree, they discard one bit and keep the other one. If the parity values disagree, they discard both bits of that group. They repeatedly do this for a number of rounds until they believe they can find a certain integer k so that both bit-flip error and phase-flip error are less than 5% after the following step is done.

9: They randomly group the remaining key bits with each group consisting k bits. They use the parity value of each group as the new bits after this step.
10: Alice and Bob use classical CSS code to distill the final key.

Remark: The above protocol is unconditionally secure. This means, under whatever type of intercept-and-resend attack, Eve's information to the final key is exponentially small. The security proof can be done through the purification and reduction procedure given by [10]. The only thing that is a bit different here is that Alice and Bob will take measurement in Y basis to make the final key, after the distillation. The conditional advantage requires the users to first test the properties of the physical channel before doing QKD. And we assume that the physical channel itself is stable. Note that we do not require anything for Eve's channel. As we have discussed in the beginning, the physical channel is in general different from Eve's channel since Eve may take over the whole channel only at the time Alice and Bob do the QKD. However, if Eve wants to hide her presence, she must respect the expected results of the error test in the protocol. In our protocol, Eve's operation must not change the error rates of the physical channel, though she can change the error pattern. Since if these values are changed, Alice and Bob will find that their error test result is much different from the expected one, therefore Eve cannot hide her presence.

Besides the advantage of a higher tolerable error rate, there are also advantages in the key rate of our protocol with asymmetric channel noise. Obviously, when the error rate is higher than other protocols' threshold while lower than our protocol's threshold, our protocol always has an advantage in key rate. More interestingly, even in the case that the error rate is significantly lower than the threshold of Shor–Preskill's protocol, we may modify our protocol and the advantage in key rate may still hold. We modify our protocol in such a way: take one round bit-error rejection with two-way communication and then use CSS code to distill the final key. As it was shown in [43], the various flipping rates will change by the following formulas after the bit-flip error rejection:

$$\begin{cases} q_I &= \dfrac{p_{I0}^2 + p_{z0}^2}{(q_{I0} + q_{z0})^2 + (q_{x0} + q_{y0})^2}, \\ q_x &= \dfrac{q_{x0}^2 + q_{y0}^2}{(q_{I0} + q_{z0})^2 + (p_{x0} + p_{y0})^2}, \\ q_y &= \dfrac{2 q_{x0} q_{y0}}{(q_{I0} + q_{z0})^2 + (q_{x0} + q_{y0})^2}, \\ q_z &= \dfrac{2 q_{I0} q_{z0}}{(q_{I0} + q_{z0})^2 + (q_{x0} + p_{y0})^2}. \end{cases} \qquad (71)$$

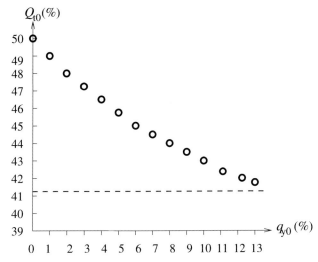

Fig. 3. Comparison of channel error threshold of different protocols. All values are in the unit of one percent. Q_{t0} is the threshold value of total channel noise given certain value of q_{y0}. In calculating Q_{t0}, we assume $q_{x0} = q_{z0}$. The *dashed line* is the threshold value of Chau protocol; the *circled curve* is the threshold of the six-state protocol given in this work

Also, it can be shown that the number of remaining pairs is

$$f = \frac{1}{2} \frac{1}{(q_{I0} + q_{z0})^2 + (q_{x0} + p_{y0})^2}. \tag{72}$$

The key rate of our protocol is given by

$$R = f \cdot (1 + q_x \log_2 q_x + q_y \log_2 q_y + q_z \log_2 q_z + q_I \log_2 q_I). \tag{73}$$

We shall compare this with the key rate of our modified six-state protocol where key bits are equally distributed over three different bases but we distill three batches of final key, i.e.,

$$r' = 1 + q_{x0} \log_2 q_{x0} + q_{y0} \log_2 q_{y0} + q_{z0} \log_2 q_{z0} + q_{I0} \log_2 q_{I0}. \tag{74}$$

We need not compare our results with the standard six-state protocol [41] since its key rate given by (66) is lower than that given by (67). The numerical results are given in Fig. 4.

6 Quantum Key Distribution With Encoded BB84 States

As we have seen in the previous section, for systems based on currently existing technologies, there are still some limitations to QKD in practical applications. The main limitation in practice is the channel loss. However, in some

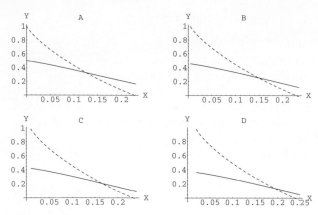

Fig. 4. Comparison of the key rate of our protocol (*solid line*) and six-state Shor–Preskill protocol (*dashed line*). The Y-axis is for the key rate, X-axis is $q_{x0} + q_{y0} + q_{z0}$; A, B, C, D are for the cases of $q_{y0} = 0, 0.5\%, 1\%, 2\%$, respectively. In all cases we have assumed $q_{x0} = q_{y0}$. We can see that, if q_{y0} is small, our protocol can have a higher efficiency than six-state Shor–Preskill protocol even the total error rate is significantly lower than the threshold value of the six-state Shor–Preskill protocol

cases, large channel noise can be the main barrier. We have to distill out a shorter final key for security, since the Eve may pretend her disturbance to be the noise from the physical channel. If the noise is too large, no final key can be obtained. To overcome this, one needs to design new fault-tolerant protocols or new physical realizations for quantum key distribution. There are two approaches to this problem: one is to find a new protocol which raises the threshold of channel noise unconditionally, such as the protocol with two-way classical communications [42,43]; the other way is first to study the noise pattern and then find a way to remove or decrease the noise itself, such as the various methods to cancel the collective errors [44,45,46]. Although so far it has not been systematically studied in which situations large error rates happen in practice, it is at least an interesting theoretical problem on how robust the QKD can actually be. A number of methods have been considered to remove the collective random unitary noise from the channel [44,45,46].

Here we propose another method to reduce the channel errors, or, equivalently, to raise the noise threshold. The main idea is to use encoded qubits in carrying out BB84 protocol. Since we shall only use a subspace of the two-qubit state, we can do error rejection to the raw bits: If the received state of an encoded qubit goes beyond the assumed subspace, we reject the code. In such a way, the expected error rate to all those accepted codes should be decreased. This method applies to both the collective channel noise [47] and independent channel noise [48]. We shall demonstrate our method in polarization space.

6.1 A Protocol For Collective Channel Noise

Our method does not require Bob to take any collective measurement here. Our method is based on the widely accepted assumptions that the flipping errors of polarization (mainly) come from the random rotation by the fiber or the molecules in the air, with the degree of the rotation fluctuating randomly. Also, if several qubits are transmitted simultaneously and they are spatially close to each other, the random unitaries to each of them must be identical, i.e., the error of the physical channel is *collective*.

Consider an arbitrary collective random unitary U which satisfies

$$U|0\rangle = \cos\theta|0\rangle + e^{i\phi}\sin\theta|1\rangle,$$
$$U|1\rangle = e^{i\Delta}(-e^{-i\phi}\sin\theta|0\rangle + \cos\theta|1\rangle). \tag{75}$$

Here $|0\rangle, |1\rangle$ represent horizontal and vertical polarization states, respectively. Note that the parameters Δ, ϕ and θ fluctuate with time, therefore one has no way to make unitary compensation to a single qubit. However, the channel unitary error is a type of collective error to all qubits sent simultaneously, therefore it is possible to send qubits robustly because the collective errors on different qubits may cancel each other. With such types of collective unitary errors, we shall take the QKD in the subspace of two-qubit state of

$$S = \{|01\rangle, |10\rangle\}. \tag{76}$$

In particular, we let Alice prepare and send Bob two-qubit states randomly chosen from $|01\rangle, |10\rangle, |\psi^\pm\rangle = \frac{1}{\sqrt{2}}(|01\rangle \pm |10\rangle)$. Although state $|\psi^-\rangle$ remains unchanged under the collective unitary errors [49], the other three states do not remain unchanged. However, in our protocol, we shall let Bob first take a parity check to the two-qubit state to see whether it belongs to subspace S. If it does, he accepts it; if it does not, he discards it. Although the two-qubit states could be distorted by the collective random unitary, most often the distortion will drive the codes out of subspace S, therefore the distorted codes will be discarded by the protocol itself. The error rates to those *accepted* codes are normally small, provided that the channel noise are mainly from the collective unitary and the averaged value θ is not too large. For example, our protocol gives a good key rate if the averaged value $|\sin\theta|$ is $1/2$. (The dispersion, ϕ value can be arbitrarily large.) Explicitly, any collective rotation cannot exchange states $|\psi^+\rangle$ and $|\psi^-\rangle$; it can only drive $|\psi^+\rangle$ out of the subspace S. However, any state outside of S will be rejected, as required by our protocol. Therefore the rate of flipping between $|\psi^+\rangle$ and $\psi^-\rangle$ (phase-flip rate) is zero. A collective rotation U will also take the following effects:

$$U^{\otimes 2}|01\rangle = U|0\rangle \otimes U|1\rangle$$
$$= \cos^2\theta|01\rangle - \sin\theta\cos\theta(e^{-i\phi}|00\rangle + e^{i\phi}|11\rangle) + \sin^2\theta|10\rangle. \tag{77}$$
$$U^{\otimes 2}|10\rangle = U|1\rangle \otimes U|0\rangle$$
$$= \cos^2\theta|10\rangle - \sin\theta\cos\theta(e^{-i\phi}|11\rangle + e^{i\phi}|00\rangle) + \sin^2\theta|01\rangle. \tag{78}$$

Since the states outside the subspace S will be discarded, the net flipping rate between $|01\rangle$ and $|10\rangle$ (bit-flip rate) is $r_b = \frac{\sin^4\theta}{\cos^4\theta + \sin^4\theta}$. Therefore, if the average rotating angle is small, the flipping rate r_b will be also small. (If we directly use the BB84 protocol, the bit-flip rate is $\sin^2\theta$, one order of magnitude larger than ours.) Moreover, in the ideal case that all flips come from the random rotation, since the phase-flip rate is zero, one can *always* distill some bits of final key provided that $r_b \neq 1/2$. The key rate is $1 + r_b \log_2 r_b + (1 - r_b) \log_2(1 - r_b)$. Note that if $r_b > 1/2$ one can simply reverse all bit values given by $|01\rangle$ and $|10\rangle$ and also distill some bits of final key. In practice, if θ does not change too fast, we can divide the data into many blocks, say, each block contains the data with several seconds. We invert the bit-values of those blocks with larger than $1/2$ error rate after the decoding. Note that we assume the phase-flip error to be always very small by our protocol.

6.1.1 Protocol 1 and Security Proof

For clarity, we now give a protocol with collective measurements first and then reduce it to a practically feasible protocol without any collective measurements.

Protocol 1. **1: Preparation of the encoded BB84 states.** Alice creates $3(1+\delta)n$ single-qubit states randomly chosen from $\{|0\rangle, |1\rangle\}$, $(1+\delta)n$ qubits randomly chosen from $\{|\pm\rangle\}$. She put down each one's preparation basis and bit value: state $|0\rangle, |+\rangle$ for bit value 0, the other two states are for 1. She randomly permutes them. She also prepares $4(1+\delta)n$ ancilla which are all in state $|0\rangle$. She then encodes each individual qubit with an ancilla into a two-qubit code through the following CNOT operation: $|00\rangle \longrightarrow |01\rangle$; $|10\rangle \longrightarrow |10\rangle$; $|11\rangle \longrightarrow |11\rangle$; $|01\rangle \longrightarrow |00\rangle$. The second digit in each state is for the ancilla. Such an encoding operation changes $(|0\rangle, |1\rangle)$ into $(|01\rangle, |10\rangle)$ and $|\pm\rangle$ into $|\psi^\pm\rangle$. **2: State transmission.** Alice sends those two-qubit codes to Bob. **3: Error-rejection and decoding.** Bob takes the same CNOT operation as used by Alice in encoding. He then measures the second qubit in Z basis: if it is $|1\rangle$, he discards both qubits and notifies Alice; if he obtains $|0\rangle$, he measures the first qubit in either X basis or Z basis and records the basis as his "measurement basis" in the QKD protocol. The bit-value of a code is determined by the measurement outcome of the first qubit after decoding, $|0\rangle, |+\rangle$ for bit value 0, $|1\rangle, |-\rangle$ correspond to 1. **4: Basis announcement.** Through public discussion, they discard all those decoded qubits with different measurement bases in two sides. There are at least $2k$ bits left among which about $k/2$ are prepared and measured in X basis (if not, abort the protocol). **5: Error test.** They announce the values of all X bits and the same number of randomly chosen Z bits. If too many values disagree, they abort the protocol. Otherwise they distill the remaining bits for the final key. **6: Final key distillation.** Alice and Bob distill the final key by the classical CSS code [10].

Quantum Key Distribution: Security, Feasibility and Robustness   221

The unconditional security here is equivalent to that of BB84 [7,10] with a lossy noisy channel: *Protocol 1* can be regarded as an encoded BB84 protocol with additional steps of encoding, error rejection and decoding. If Eve can attack *Protocol 1* successfully with operation $\hat{A}$ during the stage of codes transmission, she can also attack BB84 protocol successfully with

$$\hat{A}' = \hat{E} \longrightarrow \hat{A} \longrightarrow \hat{R} \longrightarrow \hat{D} \tag{79}$$

during the qubit transmission and then pass the decoded qubit to Bob, where $\hat{E}$, $\hat{R}$, $\hat{D}$ are encoding, error rejection and quantum decoding, respectively. (The operation of encoding, error rejection or decoding does not requires any information about the unknown state itself.) Obviously, BB84 protocol with attack $\hat{A}'$ is identical to *Protocol 1* with attack $\hat{A}$. To Alice and Bob, BB84 protocol with Eve's attack $\hat{A}'$ is just a BB84 with a lossy channel. (Eve must discard some codes in the error rejection step.) Therefore *Protocol 1* must be secure, since BB84 protocol is unconditionally secure even with a lossy channel.

6.1.2 Protocol 2

Though we have demonstrated the unconditional security of *Protocol 1*, we do not directly use *Protocol 1* in practice since it requires the local CNOT operation in encoding and decoding. We now reduce it to another protocol without any collective operations. First, since there are only four candidates in the set of BB84 states, instead of encoding from BB84 states, Alice may directly produce four random states of $|01\rangle, |10\rangle, |\psi^+\rangle, |\psi^-\rangle$. Note that except for Alice herself, no one else can see whether the two-qubit codes in transmission are directly produced or the encoding result from BB84 states. One may simply produce the states of those two-qubit codes by the spontaneous parametric down-conversion [50,51]. Second, in the decoding and error rejection step, Bob can carry out the task by *post-selection*. For all those codes originally in state $|01\rangle$ or $|10\rangle$, Bob can simply take local measurements in Z-basis to each qubit and then discard those outcomes of $|0\rangle \otimes |0\rangle$ or $|1\rangle \otimes |1\rangle$ and only accept the outcome $|0\rangle \otimes |1\rangle$, which is regarded as a bit value 0, and $|1\rangle \otimes |0\rangle$ which is regarded as bit value 1. The net flipping rate between $|01\rangle$ and $|10\rangle$ is regarded as bit-flip rate. The nontrivial point is the phase-flip rate, i.e., the net flipping rate between states $|\psi^\pm\rangle$. Note that all these codes only take the role of indicating the phase-flip rate; we do not have to know explicitly which one is flipped and which one is not flipped. Instead, we only need to know the average flipping rate between $|\psi^\pm\rangle$. To obtain such information, we actually do not have to really carry out the error rejection and decoding steps to each of these codes. What we need to do is simply to answer what the flipping rate *would* be if Bob really *took* the error rejection step and decoding step to each code of $|\psi^\pm\rangle$. One straightforward way is to let Bob take a Bell measurement to each code that was in state $|\psi^\pm\rangle$ originally. (*We shall* *call them ψ^+ codes or ψ^- codes hereafter.*) For example, consider ψ^- codes, after transmission, if the distribution over four Bell states $|\psi^+\rangle, |\psi^-\rangle, |\phi^+\rangle, |\phi^-\rangle$ are $p_{\psi^-}, p_{\psi^+}, p_{\phi^+}, p_{\phi^-}$, respectively, after the Bell measurements, we conclude that the channel flipping rate of $|\psi^+\rangle \longrightarrow |\psi^-\rangle$ is $p_{\psi^-}/(p_{\psi^+}+p_{\psi^-})$. This rate is equivalent to the flipping rate of $|+\rangle \longrightarrow |-\rangle$ in BB84 protocol. Note that the rate of $q_{\phi\pm}$ has been excluded here since their corresponding states are outside of the subspace S and should be discarded by our protocol.

Now we see how to get rid of the Bell measurements. Bell measurement is not the unique way to see the distribution over four Bell states for a set of states. We can also simply divide the set into three subsets and take collective measurements ZZ to subset 1, XX to subset 2, and YY to subset 3. We can then *deduce* the distribution over the four Bell states. Here ZZ, XX, YY are parity measurements to a two-qubit code in Z, X, Y basis, respectively. (Y: measurement basis of $\{|y\pm\rangle = \frac{1}{\sqrt{2}}(|0\rangle \pm i|1\rangle)\}$.)

Before going into the reduced protocol, we show the explicit relationship between the phase-flip rate and the local measurement results. Note that Bob has randomly divided all the received two-qubit codes into three subsets, and he will take local measurement $Z \otimes Z, X \otimes X, Y \otimes Y$ to each of the qubits of each code in subsets 1, 2, 3, respectively. Consider all ψ^- codes first. Denote $\epsilon_z, \epsilon_x, \epsilon_y$ for the rate of wrong outcome for ψ^- codes in subset 1, 2, 3, respectively, i.e., the rate of codes whose two qubit has the same bit values in basis Z, X, Y, respectively. Given values $\epsilon_{z,x,y}$ we immediately have

$$p_{\phi^+} + p_{\phi^-} = \epsilon_z, \quad p_{\psi^+} + p_{\phi^+} = \epsilon_x, \quad p_{\psi^+} + p_{\phi^-} = \epsilon_y. \tag{80}$$

Our aim is only to see the flipping rate from $|\psi^-\rangle$ to $|\psi^+\rangle$. Other types of errors are discarded since they have gone out of the given subspace S. The net flipping rate from $|\psi^-\rangle$ to $|\psi^+\rangle$ is

$$t_{\psi^- \to \psi^+} = \frac{p_{\psi^+}}{p_{\psi^-} + p_{\psi^+}} = \frac{\epsilon_x + \epsilon_y - \epsilon_z}{2(1 - \epsilon_z)}. \tag{81}$$

In a similar way we can also have the formula for the value of $t_{\psi^+ \to \psi^-}$, the flipping rate from $|\psi^+\rangle$ to $|\psi^-\rangle$:

$$t_{\psi^+ \to \psi^-} = \frac{\epsilon'_x + \epsilon'_y - \epsilon'_z}{2(1 - \epsilon'_z)}. \tag{82}$$

Here $\epsilon'_{x,y,z}$ are the rate of wrong outcome in local measurement basis $X \otimes X, Y \otimes Y, Z \otimes Z$, respectively, to all codes originally in $|\psi^+\rangle$. The total phase-flip error is

$$t_p = \frac{t_{\psi^- \to \psi^+} + t_{\psi^+ \to \psi^-}}{2}. \tag{83}$$

Protocol 1 is now replaced by the following practically feasible protocol without any collective operation:

Protocol 2. **1: Preparation of the encoded BB84 states.** Alice creates $3(1+\delta)n$ two-qubit states randomly chosen from $\{|01\rangle, |10\rangle\}$, and $(1+\delta)n$ two-qubit codes randomly chosen from $\{|\psi^\pm\rangle\}$. She randomly permutes all the $4(1+\delta)n$ quantum codes. For each two-qubit code, she puts down the "preparation basis" as "Z-basis" ($\{|0\rangle, |1\rangle\}$ basis) if it is in state $|01\rangle$ or $|10\rangle$, and the preparation basis as "X-basis" ($\{|\pm\rangle\}$) if it is in one of the states $\{\frac{1}{\sqrt{2}}(|01\rangle+|10\rangle), \frac{1}{\sqrt{2}}(|01\rangle-|10\rangle)\}$. For those code states of $|01\rangle$ or $\{\frac{1}{\sqrt{2}}(|01\rangle+|10\rangle)\}$, she denotes a bit value 0 ; for those code states of $|10\rangle$ or $\{\frac{1}{\sqrt{2}}(|01\rangle-|10\rangle)\}$, she denotes a bit value 1. **2: Transmission.** Alice sends all the two-qubit codes to Bob. **3: Measurement.** To each code, Bob measures the two qubits separately in a basis randomly chosen from Z, X, Y, i.e., random basis from $\{Z\otimes Z, X\otimes X, Y\otimes Y\}$. **4: Basis announcement.** Alice announces her "preparation basis" for each code. Bob announces his measurement basis to the qubits in each code. For those codes originally prepared in $|01\rangle$ or $|10\rangle$, they discard all results of which Bob has used a basis other than $Z\otimes Z$. They discard all codes outside the subspace S. **5: Error test.** To all the survived results, they announce all results from codes originally in $|\psi^+\rangle$ or $|\psi^-\rangle$. From the announced results they can calculate the phase-flip rate by (83). They also randomly choose $n/3$ results from those survived codes which are originally in $|01\rangle$ or $|10\rangle$ and announce them. From these announced results they can estimate the bit-flip rate. **6: Final key distillation.** Alice and Bob distill the final key by using the classical CSS code [10].

6.1.3 Physical Realization

Now we give a specific physical realization of *Protocol 2*. There are two parts in the realization. One is the source for the required four different two-qubit states at Alice's side. The other is the measurement device at Bob's side. Both of them can be realized with simple linear optical devices. The requested source states can be generated by SPDC process [50, 51], as shown in Fig. 5. The measurement with random basis at Bob's side can be done by a polarizing beam splitter (PBS) and a rotator driven electrically, as shown in Fig. 6.

6.1.4 Another Protocol For Robust QKD With Swinging Objects

In some cases, especially in free space, the dispersion can be small while the random rotation angle θ can be large. We consider the extreme case that ϕ in unitary U is 0, or otherwise can be compensated to almost 0, but θ is random and can be arbitrarily large. The swinging angle of an airplane can be very large in certain cases. We can exactly use the collective unitary model, with all elements in U being real if there is no dispersion. Then we have a better method. It is well known that both states $|\phi^+\rangle$ and $|\psi^-\rangle$ are invariant under whatever real rotation. Any linear superposed states of these two are also invariant. Therefore we use the following for states $\{|\bar{0}\rangle = |\phi^+\rangle, |\bar{1}\rangle =$

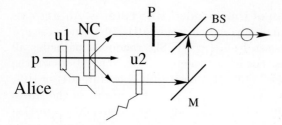

Fig. 5. The source of two-qubit state. *P:* $\pi/2$ rotator. *BS:* beam splitter, *M:* mirror, *NC:* nonlinear crystals used in SPDC process, *p:* pump light in horizontal polarization, *u1:* unitary rotator, *u2:* phase shifter. u1 takes the value of 0, $\pi/2$, $\pi/4$ to produce emission state $|01\rangle, |10\rangle, |\psi^+\rangle$, respectively; u2 can be either I or σ_z

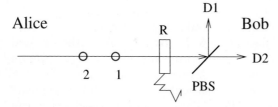

Fig. 6. Measurement device at Bob's side. R is a rotator driven electrically. R offers a random rotation to both qubits in the same code. The rotation is randomly from unity, $(|0\rangle, |1\rangle) \longrightarrow (|+\rangle, |-\rangle)$, $(|0\rangle, |1\rangle) \longrightarrow (|y+\rangle, |y-\rangle)$. The rotation helps to set the measurement basis randomly chosen from Z, X, Y. The event of two clicks on one detector (D1 or D2) show that the two qubits of the code have the same bit value; two clicks on different detectors show that the two qubits have different bit values

$|\psi^-\rangle; |+'\rangle = \frac{1}{\sqrt{2}}(|\bar{0}\rangle + |\bar{1}\rangle) = \frac{1}{\sqrt{2}}(|0\rangle|+\rangle - |1\rangle|-\rangle); |-'\rangle = \frac{1}{\sqrt{2}}(|0\rangle|-\rangle + |1\rangle|+\rangle)\}$. Bob need not take any collective measurement to determine the bit value. If he chooses "Z"-basis, he measure each of the two qubits in Z-basis, 00 or 11 for bit value 0 while 01 or 10 for bit value 1. If he chooses "X"-basis, he measures the first qubit in Z-basis and the second in X-basis, $|0\rangle|+\rangle$ or $|1\rangle|-\rangle$ for bit value 0 and $|0\rangle|-\rangle$ or $|1\rangle|+\rangle$ for bit value 1. There is no error-rejection step here because it is expected that there is no error after decoding, given the real rotarion channel. Even for the QKD with fixed object there is still a little bit of advantage: They do not need take any bases alignment with each other. Each of them only needs to make sure their local measurement bases are BB84-like, i.e., the inner product of two bases are $\frac{1}{\sqrt{2}}$.

6.1.5 Summary and Discussion

We have presented a robust QKD protocol in polarization space given that the collective random unitaries are dominant channel errors [47]. Our protocol can obviously be extended to the six-state-like protocol [41] if we add one more candidate state of $\frac{1}{\sqrt{2}}(|0\rangle \pm i|1\rangle)$ in the source. The advantage of our protocol has been experiment verified [52,53].

6.2 A Protocol With Independent Noise

In the case of independent channel noise, encoded states can also be more fault tolerant. Here we shall demonstrate this with two-bit quantum codes. The protocol is unconditionally secure under whatever type of intercept-and-resend attack. Given the symmetric and independent errors to the transmitted qubits, our scheme can tolerate a bit error rate up to 26% in four-state protocol and 30% in six-state protocol, respectively. These values are higher than all currently known threshold values for the prepare-and-measure protocols. Moreover, we shall also give a practically implementable linear optics realization for our scheme. The drawback is that the scheme requires Bob to make a collective measurement with a polarizing beamsplitter. Here we shall use two-way classical communication for final key distillation (2-EPP) [42,43].

6.2.1 The Method and the Main Idea

The linear optical realization is shown in Fig. 7. In our scheme, Alice shall send both qubits of the quantum codes to Bob, therefore they do not need any quantum storage. Bob will first check the parity of the two qubits by the polarizing beam splitter (PBS) and then decode the code with postselection. The two-bit code is produced by the spontaneous parametric down conversion (SPDC) process [51], see in Fig. 7.

We propose a revised 2-EPP scheme which is unconditionally secure and which can further increase the thresholds of error rates given the independent channel errors. We propose to let Alice send Bob the quantum states randomly chosen from $\{\frac{1}{\sqrt{2}}(|00\rangle + |11\rangle), \frac{1}{\sqrt{2}}(|00\rangle - |11\rangle), |00\rangle, |11\rangle\}$. As we shall see, these states are just the quantum phase-flip error-rejection (QPFER) code for the BB84 state $\{|0\rangle, |1\rangle, \frac{1}{\sqrt{2}}(|0\rangle + |1\rangle), \frac{1}{\sqrt{2}}(|0\rangle - |1\rangle)\}$.

In our four-state protocol, the tolerable channel bit-flip and phase-flip rates are raised to 26% for the symmetric channel with independent noise. (A symmetric channel is defined as the one with equal distribution of errors of $\sigma_x, \sigma_z, \sigma_y$.) Our six-state protocol can tolerate the channel flipping error rate up to 30%. Note that the theoretical upper bound of 25% [42] only holds for those four-state schemes where Alice and Bob only test the error rate *before* any error-removing steps. However, this is not true with the *delay* of error test. Considering the standard purification protocol [14] with symmetric channel, one may distill the maximally entangled states out of the raw pairs whose initial bit-flip error and phase-flip error are 33.3% (the threshold that the fidelity of raw pairs is larger than 50%). In our four-state protocol, since we do not assume any quantum storage, we are not able to delay the error test until the whole distillation is completed, but we can still delay the error test by one step of purification with two-bit QPFER code. This one-step delay significantly raises the tolerable channel flipping rates.

6.2.2 The QPFER Code

We shall use the following quantum phase flip error rejection (QPFER) code:

$$|0\rangle|0\rangle \longrightarrow (|00\rangle + |11\rangle)/\sqrt{2},$$
$$|1\rangle|0\rangle \longrightarrow (|00\rangle - |11\rangle)/\sqrt{2}. \tag{84}$$

Here the second qubit in the left side of the arrow is the ancilla for the encoding. This code is not assumed to reduce the errors in all cases. But in the case that the channel noise is uncorrelated or nearly uncorrelated, it works effectively. Consider an arbitrary state $\alpha|0\rangle_1 + \beta|1\rangle_1$ (qubit 1) and an ancilla state $|0\rangle_2$ (qubit 2). Taking unitary transformation of (1) we obtain the following un-normalized state:

$$\alpha(|0\rangle_1|0\rangle_2 + |1\rangle_1|1\rangle_2) + \beta(|0\rangle_1|0\rangle_2 - |1\rangle_1|1\rangle_2). \tag{85}$$

This can be regarded as the encoded state for $\alpha|0\rangle_1 + \beta|1\rangle_1$. Alice then sends both qubits to Bob. In receiving them, Bob first takes a parity check, i.e., he compares the bit value of the two qubits in Z-basis. Note that this *collective* measurement does not destroy the code state itself since they are the eigenstate of the measurement operator. Specifically, the parity check operation can be done by the PBS in Fig. 7: there states $|0\rangle, |1\rangle$ are for horizontal and vertical polarization photon states, respectively. Since a PBS transmits $|0\rangle$ and reflects $|1\rangle$, therefore if the bit values of two incident beams of the PBS are same, there must be one photon on each output beam; if the bit values of two incident beams are different, one output beam must be empty. After the parity check, if bit values are different, Bob discards the whole two-qubit code, if they are same, Bob decodes the code. In decoding, he measures qubit 1 in X-basis: if he obtains $|+\rangle$, he takes a Hadamard transformation $H = \frac{1}{\sqrt{2}}\begin{pmatrix} 1 & 1 \\ 1 & -1 \end{pmatrix}$ to qubit 2. If he obtains $|-\rangle$ for qubit 1, he takes the Hadamard transformation to qubit 2 and then flips qubit 2 in Z-basis. Suppose the original channel error rates of $\sigma_x, \sigma_y, \sigma_z$ types are p_{x0}, p_{y0}, p_{z0}, respectively. Let $p_{I0} = 1 - p_{x0} - p_{y0} - p_{z0}$. One may easily verify the probability distribution and error type for the survived and decoded states (qubit 2) in following table:

JCE	Probability	Decoded state	Error type
$I \otimes I$	p_{I0}^2	$\alpha\|0\rangle + \beta\|1\rangle$	I
$\{I \otimes \sigma_z\}$	$2p_{I0}p_{z0}$	$\alpha\|1\rangle + \beta\|0\rangle$	σ_x
$\sigma_z \otimes \sigma_z$	p_{z0}^2	$\alpha\|0\rangle + \beta\|1\rangle$	I
$\sigma_y \otimes \sigma_y$	p_{y0}^2	$\alpha\|0\rangle - \beta\|1\rangle$	σ_z
$\sigma_x \otimes \sigma_x$	p_{x0}^2	$\alpha\|0\rangle - \beta\|1\rangle$	σ_z
$\{\sigma_x \otimes \sigma_y\}$	$2p_{x0}p_{y0}$	$\alpha\|1\rangle - \beta\|0\rangle$	σ_y

The first column lists the various types of joint channel errors (JCE) before decoding. $\{\alpha \otimes \beta\}$ denotes both $\alpha \otimes \beta$ and $\beta \otimes \alpha$. According to this table, the error rate distribution for the survived raw pairs after decoding is:

$$\begin{cases} p_I = \dfrac{p_{I0}^2 + p_{z0}^2}{(p_{I0}+p_{z0})^2 + (p_{x0}+p_{y0})^2}, \\[2mm] p_z = \dfrac{p_{x0}^2 + p_{y0}^2}{(p_{I0}+p_{z0})^2 + (p_{x0}+p_{y0})^2}, \\[2mm] p_y = \dfrac{2p_{x0}p_{y0}}{(p_{I0}+p_{z0})^2 + (p_{x0}+p_{y0})^2}, \\[2mm] p_x = \dfrac{2p_{I0}p_{z0}}{(p_{I0}+p_{z0})^2 + (p_{x0}+p_{y0})^2}. \end{cases} \qquad (86)$$

With this formula, the phase flip error to the shared raw pairs is obviously reduced. Note that in the QKD task by Fig. 7, Alice and Bob will not test the error rate of all incident beams of the PBS. They will only consider those cases where each output beam of the PBS contains one photon (other cases are discarded). They then decode the survived codes by measuring the lower output beam in X-basis and taking Hadamard transform and conditional flipping to the upper output beam of the PBS. After decoding, the state of the upper output beam is expected to be the original state from Alice. They only check the error rate of such decoded states. If the tested error rate on decoded beam is significantly larger than the expected value from (86), they give up the protocol. (As we shall show later, they do not really need to take the Hadamard transform and the conditional flipping to the upper output beam, instead they simply measure it and take the conditional flipping to classical bit value only.)

6.2.3 The Protocol and Its Linear Optical Realization

In the BB84 protocol, there are only four different states. Therefore Alice may directly prepare random states from the set of $\{\frac{1}{\sqrt{2}}(|00\rangle + |11\rangle), \frac{1}{\sqrt{2}}(|00\rangle - |11\rangle), |00\rangle, |11\rangle\}$ and send them to Bob. This is equivalent to first preparing the BB84 states and then encoding them by (84). We propose the following four-state protocol:

1. Alice prepares N two-qubit quantum codes with $N/4$ of them being prepared in $|00\rangle$ or $|11\rangle$ with equal probability and $3N/4$ of them being prepared in $\frac{1}{\sqrt{2}}(|00\rangle \pm |11\rangle)$ with equal probability. All codes are put in random order. She records the "preparation basis" as X-basis for code $|00\rangle$ or $|11\rangle$; and as Z-basis for code $\frac{1}{\sqrt{2}}(|00\rangle \pm |11\rangle)$[1]. And she records the bit value of 0 for the code $|00\rangle$ or $\frac{1}{\sqrt{2}}(|00\rangle + |11\rangle)$; bit value 1 for the code $|11\rangle$ or $\frac{1}{\sqrt{2}}(|00\rangle - |11\rangle)$. She sends each two-qubit code to Bob. In Fig. 7, any of the above four states can be produced from the nonlinear crystal by appropriately setting the polarization of the pump light [51]. **2.** Bob checks the parity of each two-qubit code in Z-basis. He discards the codes whenever the two bits have different values, and he takes the following measurement if they have same values: He measures qubit 1 in X-basis and qubit 2 in either X-basis or Z-basis with equal probability. If Bob measure B2 in X- (or) Z-basis, he records the "measurement basis" as Z- (or) X-basis,[2] and we shall simply call the qubit as Z-bit (or X-bit) later on. If he obtains $|+\rangle|+\rangle$, $|+\rangle|0\rangle$, $|-\rangle|-\rangle$ or $|-\rangle|0\rangle$, he records bit value 0 for that code; if he obtains $|-\rangle|+\rangle$, $|-\rangle|1\rangle$, $|+\rangle|-\rangle$ or $|+\rangle|1\rangle$, he records bit value 1 for that code. In Fig. 1, parity check and decoding are done jointly by postselection: If both detectors are clicked, this means the two incident beams have the same polarization and he deduces the corresponding bit value by the above mentioned rule. If there is only one detector click or no detector clicks, the incident code is disregarded. **3.** Bob announces which codes have been discarded. Alice and Bob compare the "preparation basis" and "measurement basis" of each bit decoded from the survived codes by classical communication. They discard those bits whose "measurement basis" disagrees with the "preparation basis". Bob announces the bit value of all X-bits. He also randomly chooses the same number of Z-bits and announces their values. If too many of them disagree with Alice's record, they abort the protocol. **4.** Now they regard the tested error rates on Z-bits as the bit-flip rate and the tested error rate on X-bits as phase flip rate. They reduce the bit-flip rate in the following way: They randomly group all their unchecked bits with each group containing two bits. They compare the parity of each group. If the results are different, they discard both bits. If the results are the same, they discard one bit and keep the other. They repeatedly do so for a number of rounds until they believe that both bit-flip rate and phase-flip rate will be decreased to less than 5% after step 5. **5.** They then randomly group the remaining bits with each group containing r bits. They use the parities of each groups as the new bits. **6.** They use the classical CSS code [10] to distill the final key.

---

[1] The "preparation basis" here is defined by the basis of the original qubit which corresponds the two-qubit code by our encoding method.

[2] Here Bob has omitted the Hadamard transformation as appeared in the EPP, therefore the measurement basis to B2 and Bob's recorded " measurement basis" to the decoded qubit are different.

Remark: In this protocol, Alice and Bob first assumed the channel noise to be independent. However, Eve may do whatever in her attacking therefore (86) is not necessarily always true. Before the protocol, they *expect* that the verified error rates after decoding and error rejection are in agreement with (86). For security, Alice and Bob check the bit errors *after* decoding the two-bit quantum codes.

If the detected errors are significantly larger than the expected values calculated from (86), they will abort the protocol. That is to say, if (86) really works, they continue; if it does not work, they abort it. After any round of bit-flip error rejection in step 4, the error rate will be iterated by (71) [43].

Given p_x, p_y, p_z, if there exits a finite number k, after k rounds of bit-flip error rejection, we can find a r which satisfies

$$r(p_x + p_y) \leq 5\%, \quad \text{and} \quad e^{-2r(0.5-p_z-p_y)^2} \leq 5\%. \tag{87}$$

One can then obtain the unconditionally secure and faithful final key with a classical CSS code [10].

In the four-state protocol, we do not detect the σ_y error for the states decoded from the survived codes, therefore we have to assume $p_y = 0$ after the quantum parity check and decoding. But we do not have to use the assumption of $p_{y0} = 0$, actually. Alice and Bob never test any error rate before decoding. However, *if* the channel noise is symmetric and uncorrelated, after the quantum decoding, both σ_z error (p_z) and σ_y error (p_y) are reduced, i.e., the detectable phase error rate has been reduced in a rate as it should be in the case of symmetric noisy channel. We then start from the asymmetric error rate with assumption $p_y = 0$ and p_x, p_z being the detected bit-flip rate and phase-flip rate, respectively. After the calculation, we find that the tolerable error rate of bit flip or phase flip is 26% for the four-state protocol. Moreover, in the case that the channel error distribution itself is $p_{y0} = 0$, $p_{x0} = p_{z0}$, the tolerable channel error rate for our protocol is $p_{x0} = p_{z0} \leq 21.7\%$.

Our protocol can obviously be extended to the six-state protocol [41]. In doing so, Alice just changes the initially random codes by adding $N/4$ codes from $\{\frac{1}{2}[(|00\rangle+|11\rangle)\pm i(|00\rangle-|11\rangle)]\}$. This is equivalent to $\frac{1}{\sqrt{2}}(\{|00\rangle \mp i|11\rangle)\}$. She regards all this type of codes as Y-bits. In decoding the codes, Bob's "measurement basis" is randomly chosen from three bases, X, Y and Z. All decoded X-bits, Y-bits and the same number of randomly chosen decoded Z will be used as the check bits. Since the Hadamard transform switches the two eigenstates of σ_y, after decoding if Bob measures qubit 2 in Y-basis, he needs to flip the measurement outcome in the Y-basis so as to obtain everything the same as that in the 2-EPP with quantum storages. In such a way, if the channel is symmetric, Bob will find $p_y \neq 0$. And he will know p_x, p_y, p_z exactly instead of assuming $p_y = 0$. This will increase the tolerable error rate accordingly.

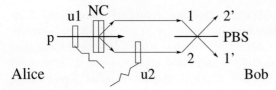

Fig. 7. QKD scheme with two-bit quantum codes. *PBS:* polarizing beam splitter. *NC:* nonlinear crystals used in SPDC process, *p:* pump light in horizontal polarization; *u1:* unitary rotator; *u2:* phase shifter. u1 takes the value of 0, $\pi/2$, $\pi/4$ to produce emission state $|11\rangle, |00\rangle, |\phi^+\rangle$, respectively. u2 can be either I or σ_z

6.2.4 Security Proof

The protocol above contains the encoded BB84 states and final key distillation with two-way classical communication. If we replace the encoded BB84 states with original BB84 states, the protocol must be secure because it is totally equivalent to the one based on entanglement purification [42]. Since the encoded BB84 states and the original BB84 states can be connected by a unitary transformation, the one using encoded BB84 states must be also secure because otherwise the one using original BB84 states is insecure with channel loss. This completes the security proof.

6.2.5 Subtlety of the "Conditional Advantage"

Although the advantage of a higher threshold is conditional, the security of our protocol is *unconditional*. That is to say, whenever our protocol produces any final key, Eve's information to that key must be exponentially close to zero, no matter whether Eve uses coherent attack or individual attack. *Our protocol is totally different from the almost useless protocol which is only secure with uncorrelated channel noise.* There are two conditions for the error threshold advantage: (1) The noise of physical channel should be the type where (86) holds. (2) Eve is not detected in the error test, i.e., the result of error test must be in agreement with the expected result given by (86).

Both conditions here are verifiable by the protocol itself. The second condition is a condition for *any* QKD protocol. The first condition is on the *known* physical channel rather than Eve's channel in QKD. In our protocol, Eve's attack must not affect the error rates detected on the decoded qubits if she wants to hide her presence. That is to say, if Eve hides her presence, all results about the final key of our protocol can be correctly estimated based on the known properties of the physical channel, no matter what type of attack she has used. *Given a physical channel with its noise being uncorrelated and symmetric and higher than the thresholds of all other prepare-and-measure protocols but lower than that of our protocol, our protocol is the only one that works.* In practice, one may simply separate the two qubits of the quantum code substantially to guarantee the uncorrelation of the physical channel noise. This is to say, the error threshold advantage of our protocol is actually unconditional in practice.

7 Summary and Concluding Remarks

Although many classical protocols have been widely believed to be secure for private communication, none of them has been proven to be secure, while some of them have been proven to be insecure by either a quantum algorithm or a classical algorithm. Quantum key distribution can, in principle, offer unconditional security for private communication. The foundation of the security for QKD is the principle of quantum mechanics. The BB84 protocol with a single-photon source is secure because it is equivalent to a protocol based on entanglement purification. The BB84 protocol with an imperfect source is also secure if the fraction of tagged bits is not too large. Given the pulses from a traditional laser device, we can make a tight estimation of the fraction of tagged bits by switching the intensity of each pulse randomly from three values. This is the so-called decoy-state method. With this method, unconditionally secure QKD over a distance larger than 100 km is immediately possible using the existing technology. The noise threshold for BB84 protocol is 11%. But we can raise both the noise threshold and key rate through two-way classical communication, asymmetric channel noise or encoded BB84 states.

References

[1] M. A. Nielsen, I. L. Chuang: *Quantum Computation and Quantum Information* (Cambridge Univ. Press, Cambridge 2000)
[2] P. Shor: in *Proc. 35th Ann. Symp. on Found. of Computer Science* (2004) p. 124
[3] X. Wang, H. Yu, Y. L. Yin: Efficient collision attacks on SHA-0, in *Crypto'05*
[4] X. Wang, Y. Yin, H. Yu: Finding collisions in the full SHA-1 collision search attacks on SHA1, in H. Yu, Y. Yin (Eds.): *Crypto'05* (2005)
[5] X. Wang: Collision for some hash functions MD4, MD5 HAVAL-128, primed, in *Crypto'04* (2004)
[6] N. Gisin, G. Ribordy, W. Tittel, H. Zbinden: Rev. Mod. Phys. **74**, 145 (2002)
[7] C. H. Bennett, G. Brassard: in *Proc. IEEE Int. Conf. on Computers, Systems and Signal Processing* (1984) p. 175
[8] D. Mayers: Advances in cryptology – proc. crypto'96, in N. Koblitz (Ed.):, LNCS **1109** (Springer, New York 2001) p. 343
[9] D. Mayers: J. Assoc. Comput. Mach. **48**, 351 (2001)
[10] P. W. Shor, J. Preskill: Phys. Rev. Lett. **85**, 441 (2000) and references therein
[11] P. W. Shor: Phys. Rev. A **52**, 2493(R) (1995)
[12] D. Gottesman, J. Preskill: Phys. Rev. A **63**, 022309 (2001)
[13] M. Hamada: J. Phys. A: Math. Gen. **37**, 8303 (2004)

[14] C. H. Bennett, D. P. DiVincenzo, J. A. Smolin, W. K. Wootters: Phys. Rev. A **54**, 3824 (1996)
[15] X.-B. Wang: arXiv:quant-ph/0409099
[16] A. K. Ekert: Phys. Rev. Lett. **67**, 661 (1991)
[17] D. Deutsch, A. Ekert, R. Jozsa, C. Macchiavello, S. Popescu, A. Sanpera: Phys. Rev. Lett. **77**, 2818 (1996) erratum: Phys. Rev. Lett. **80**, 2022
[18] A. Tomita, K. Nakamura: Opt. Lett. **27**, 1827 (2002)
[19] H. Kosaka, A. Tomita, Y. Nambu, T. Kimura, K. Nakamura: Electron. Lett. **39**, 1199 (2003)
[20] H. Zbinden, J. D. Gautier, N. Gisin, B. Huttner, A. Muller, W. Tittel: Electron. Lett. **33**, 586 (1997)
[21] C. Gobby, Z. L. Yuan, A. J. Shields: Appl. Phys. Lett. **84**, 3762 (2004)
[22] Z. L. Yuan, A. J. Shields: Opt. Express **13**, 660 (2005)
[23] P. D. Townsend, J. G. Rarity, P. R. . Tapster: Electron. Lett. **29**, 634 (1993)
[24] A. Muller, T. Herzog, B. Huttner, W. Tittel, H. Zbinden, N. Gisin: Appl. Phys. **70**, 793 (1997)
[25] G. Ribordy, J. D. Gautier, N. Gisin, O. Guinnard, H. Zbinden: Electron. Lett. **34**, 2116 (1998)
[26] H. Inamori, N. Lütkenhaus, D. Mayers: (2001) arXiv:quant-ph/0107017
[27] D. Gottesman, H. K. Lo, N. Lütkenhaus, J. Preskill: Quantum Information and Computation **4**, 325 (2004)
[28] B. Huttner: Phys. Rev. A **51**, 1863 (1995)
[29] G. Brassard, N. Lutkenhaus, T. Mor, B. C. Sanders: Phys. Rev. Lett. **85**, 1330 (2000)
[30] N. Lütkenhaus, M. Jahma: New J. Phys. **4**, 44 (2002)
[31] W.-Y. Hwang: Phys. Rev. Lett. **91**, 057901 (2003)
[32] X.-B. Wang: Phys. Rev. Lett. **94**, 230504 (2005)
[33] H. K. Lo: in *Proc. 2004 IEEE Int. Symp. on Inf. Theor.* (2004) p. 17
[34] X.-B. Wang: arXiv:quant-ph/0501143
[35] X.-B. Wang: Phys. Rev. A **72**, 012322 (2005)
[36] H. K. Lo, X. F. Ma, K. Chen: Phys. Rev. Lett. **94**, 230504 (2005)
[37] J. W. Harrington, J. M. Ettinger, R. J. Hughes, J. E. Nordholt: arXiv:quant-ph/0503002
[38] C. H. Bennett: Phys. Rev. Lett. **68**, 3121 (1992)
[39] M. Koashi: Phys. Rev. Lett. **93**, 120501
[40] V. Scarani, A. Acin, G. Ribordy, N. Gisin: Phys. Rev. Lett. **92**, 057901 (2004)
[41] D. Bruss: Phys. Rev. Lett. **81**, 3018 (1998)
[42] D. Gottesman, H. K. Lo: IEEE Trans. Inform. Theory **49**, 457 (2003)
[43] H. F. Chau: Phys. Rev. A **66**, 060302(R) (2002)
[44] G. M. Palma, K. A. Suominen, A. K. Ekert: Proc. R. Soc. Lon. Ser.-A **452**, 567 (1996)
[45] Z. D. Walton, et al.: Phys. Rev. Lett. **91**, 087901 (2003)
[46] J. C. Boileau, D. Gottesman, R. Laflamme, D. Poulin, R. W. Spekkens: Phys. Rev. Lett. **92**, 17901 (2004)
[47] X.-B. Wang: Phys. Rev. A **72(R)**, 057901 (2005)
[48] X.-B. Wang: Phys. Rev. Lett. **92**, 077902 (2004)
[49] P. G. Kwiat, A. J. Berglund, J. B. Altpeter, A. G. White: Science **290**, 498 (2000)

[50] P. G. Kwiat, E. Waks, A. G. White, I. Appelbaim, P. H. Eberhard: Phys. Rev. A **60**, R773 (1999)
[51] P. G. Kwiat, K. Mattle, H. Weinfurter, A. Zeilinger, A. V. Sergienko, Y. H. Shih: Phys. Rev. Lett. **75**, 4337 (1995)
[52] Y. K. Jiang, X. B. Wang, B. S. Shi, A. Tomita: Opt. Express **13**, 9415 (2005)
[53] Q. Zhang, J. Yin, T. Y. Chen, S. Lu, J. Zhang, X. Q. Li, T. Yang, X. B. Wang, J. W. Pan: Phys. Rev. A **73**, 020301 (2006)

Index

asymmetric channel noise, 210

BB84, 186, 218, 220, 221, 224, 227, 230
biCNOT, 193–195
bit-flip, 191, 194, 196, 211, 215, 220

CNOT, 193, 194, 220
coherent states, 196
collective channel noise, 218
collective measurement, 190, 225
counting rate, 201–204

decoy-state method, 200

entanglement purification, 188, 190
error correction, 187, 189, 195
error rejection, 216, 220, 225
error test, 189

final key, 195
four-state protocol, 214, 225

hashing, 188

imperfect source, 196
individual measurement, 190

key rate, 196, 198, 206, 209

multiphoton pulses, 196, 199

one-way function, 185

phase-flip, 191, 194, 196, 197, 211, 215, 220, 225
PNS attack, 199
prepare-and-measure, 187, 195
privacy amplification, 187, 195
public key, 185

quantum key distribution (QKD), 186

RSA, 186

secure private communication, 185
six-state protocol, 213, 217, 225
spontaneous parametric down conversion (SPDC), 223, 225, 230

tagged bits, 196, 199, 200, 205

unconditional security, 187, 206, 216, 230

Why Quantum Steganography Can Be Stronger Than Classical Steganography

Shin Natori[1]

Graduate School of Science, The University of Tokyo
natori@adm.s.u-tokyo.ac.jp

Abstract. Steganography is the art and science of hiding the existence of a message by embedding it into another message. In this paper we first define a quantum steganography model by extending the classical one. Next we show that quantum steganography can be stronger than classical steganography, by introducing a quantum steganography system that cannot be imitated by classical one.

1 Introduction

Steganography is the art and science of hiding data in innocent-looking cover data so that no one can detect the existence of the hidden data [1, 2]. It is different from cryptography, since the goal of steganography is undetectability, not secrecy only. For example, a ciphertext may contain peculiar words like "QJYZQDFLKJ," but a stego-text (data-embedded text file) should be read as an ordinary text file so as not to draw suspicion of a secret message. Speaking more precisely, a steganography system needs a priori existence of cover message, into which the steganography encoder hides secret message.

2 Definitions

2.1 General Model of Steganography System

Figure 1 is the commonly accepted model of information hiding [1]. If Alice wants to send some data (embedded data) to Bob secretly, she computes a message (stego-data) using a key, embedded data and an innocent looking cover data. (In some cases, cover data is omitted for computation.) Then Alice sends the stego-data to Bob. Bob computes the original embedded data from the key and the stego-data. If the stego-data looks like the cover data, it may be difficult for eavesdroppers to detect the existence of the secret message. Most of the steganography systems so far proposed take image or audio files as cover data.

We introduce here a more formal and general model of classical steganography systems. Without a steganography system, Alice normally sends an innocent-looking message (C, cover data) to Bob. (In this case, eavesdropper Eve sees C.) The cover data is computed from environmental data (V)

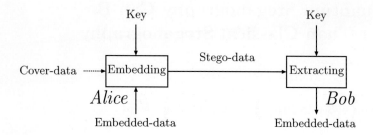

Fig. 1. Commonly accepted model of information hiding

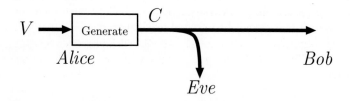

Fig. 2. Communication without steganography (classical model)

using cover generator algorithm ($\mathcal{G}$). (See Fig. 2. In most cases, $V = C$ and $\mathcal{G}(x) = x$ holds.)

A steganography system is a pair of an embedder ($\mathcal{E}$) and an extractor ($\mathcal{D}$). If Alice uses a steganography system ($\mathcal{E}, \mathcal{D}$), Alice first computes stego-data (S) from environmental data (V), key (K) and embedded data (E) using $\mathcal{E}$ (Fig. 3.) Then, Alice sends the stego-data to Bob. (In this case, Eve eavesdrops S.) Finally, Bob uses the extractor to compute the original embedded data from the stego-data and key. The embedder satisfies the following equation:

$$\mathcal{E}(V, E, K) = S,$$

and the extractor usually satisfies the following one:

$$\mathcal{D}(S, K) = E. \tag{1}$$

We allow here extraction error ($\mathcal{D}(S, K) \neq E$) so long as the channel capacity of the steganography channel is not 0.

Eve's task is to detect the usage of steganography by eavesdropping C or S. If C and S are indistinguishable, the steganography system is secure.

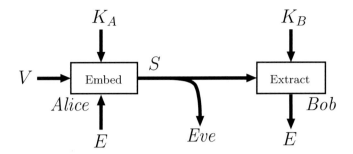

Fig. 3. Communication with steganography (classical model)

2.2 Classical Model of Steganography System

In classical steganography, C, E, K, V, S are all random variables over certain alphabets. If the probability distribution of S is equal to that of C, we call the system perfectly secure.

3 Related Works

Curty et al. proposed three steganography systems that exploit quantum information characteristics [3]. The first one hides one classical bit $E \in \{0, 1\}$ into one noise-looking qubit (such as least significant bits of quantized values), by replacing the qubit with $|+\rangle = 1/\sqrt{2}(|0\rangle + |1\rangle)$ (if $E = 1$)) or $|-\rangle = 1/\sqrt{2}(|0\rangle - |1\rangle)$ (if $E = 0$)). The second one hides two classical bits into one noise-looking qubit by replacing the qubit with dense-coding. The security of these systems depends on the similarity between the noise-looking qubit and true white noise (the density matrix of which is $\frac{1}{2}\mathbf{I}$). The third one sends a qubit over a classical steganography channel by using quantum teleportation. The security of this steganography system is equal to that of the underlying classical steganography system.

4 Quantum Steganography

4.1 Model of Quantum Steganography System

In this section we extend the classical steganography model to support quantum steganography (Figs. 4 and 5). In this model, all random variables are replaced by quantum registers. We allow Eve destructive measurement of either C or S. And since K cannot be cloned, we explicitly add an initialization step. The environmental input V is divided as $V = V_1 \otimes V_2$, and V_1 is used for key setup and V_2 is used for embedding.

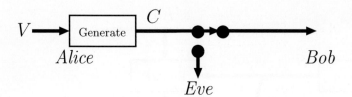

Fig. 4. Communication without steganography (quantum model)

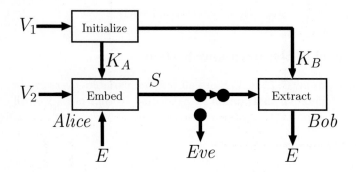

Fig. 5. Communication with steganography (quantum model)

In this model, Eve's task is to distinguish C and S by measuring them. Therefore, the perfect security condition is

$$\rho_c = \rho_s.$$

(ρ_c and ρ_s are density matrices of C and S respectively.)

4.2 Comparison Between Classical and Quantum Steganography

In this section, we show that quantum steganography can be strictly securer than classical one. In general, no classical steganography system can be perfectly secure if its cover data is the result of a measurement and the distribution of the cover data is unknown [4], since any modification of cover data may distort the original cover distribution. But under the same conditions, one can construct perfectly secure quantum steganography systems in certain situations. Here we give an example of such steganography systems. This steganography system, S_1, is a modified version of dense-coding system, and is practically useless since it is elaborately created to show the power of quantum steganography.

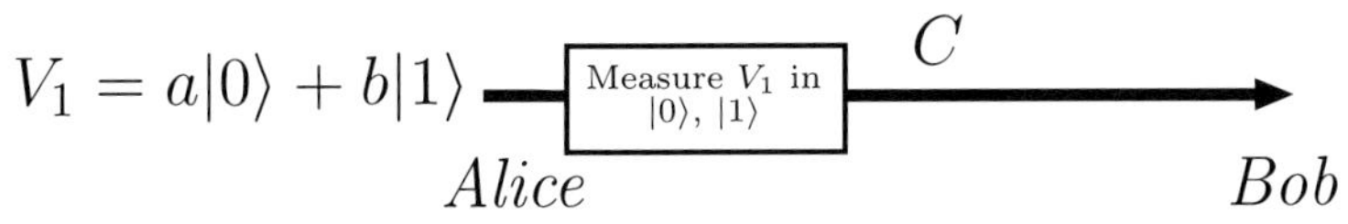

$V_1 = a|0\rangle + b|1\rangle$ — [Measure V_1 in $|0\rangle, |1\rangle$] — C →

Alice   Bob

Fig. 6. Communication without steganography ($\mathcal{S}_1$)

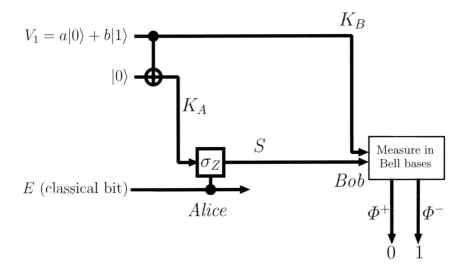

Fig. 7. Communication with steganography ($\mathcal{S}_1$)

In the example steganography system $\mathcal{S}_1$ (Figs. 6 and 7)

a, b : two unknown complex numbers that satisfy
$|a|^2 + |b|^2 = 1, |a + b|^2 > 1$
C : $\rho_C = |a|^2 |0\rangle\langle 0| + |b|^2 |1\rangle\langle 1|$
V_1 : $a|0\rangle + b|1\rangle$
$\mathcal{G}$: measurement over $\{|0\rangle, |1\rangle\}$ $(\mathcal{G}(V_1) = C)$
K_B : V_1
K_A : $|V_1 \oplus 0\rangle$ (output from CNOT$(V_1, 0)$)
E : 0 or 1 (classical bit)
$\mathcal{E}$: if $E = 1$ then output $\sigma_Z K_A$ else output K_A
$\mathcal{D}$: measure $|\psi\rangle_{K_B S}$ over Bell bases, and
output 0 if the result is $|\Phi^+\rangle = \frac{1}{\sqrt{2}}(|00\rangle + |11\rangle)$,
output 1 otherwise

Since

$$|\psi\rangle_{K_B S} = a|00\rangle + (-1)^E b|11\rangle$$

holds, it is easy to prove the following perfect security condition:
$$\rho_S = \mathbf{Tr}_{K_B}(|\psi\rangle_{K_BS}\langle\psi|_{K_BS}) = |a|^2 |0\rangle\langle 0| + |b|^2 |1\rangle\langle 1| = \rho_C.$$

The error rate of this steganography system is

$$\begin{aligned}
\Pr\{&E \neq \mathcal{D}(S, K_B)\} \\
&= \Pr\{E = 0 \wedge \mathcal{D}(S, K_B) = 1\} + \Pr\{E = 1 \wedge \mathcal{D}(S, K_B) = 0\}, \\
&= \Pr\{E = 0\}\left|\langle\Phi^-|\left(a|00\rangle + (-1)^0 b|11\rangle\right)\right|^2, \\
&\quad + \Pr\{E = 1\}\left|\langle\Phi^+|\left(a|00\rangle + (-1)^1 b|11\rangle\right)\right|^2, \\
&= (\Pr\{E = 0\} + \Pr\{E = 1\})\frac{|a-b|^2}{2} = 1 - \frac{|a+b|^2}{2} < \frac{1}{2}.
\end{aligned}$$

This ensures the positive capacity of the steganography channel.

5 Conclusions and Future Work

In this paper we extended the model of classical steganography systems and defined a model of a quantum steganography system. Based on this system, we showed that a quantum steganography system can be strictly securer than classical one. However, the steganography system we introduced in Sect. 4.2 is practically useless. Whether quantum steganography is superior to classical one or not in practical use is still an open question.

References

[1] First International Workshop IH'96 Proceedings, in R. Anderson (Ed.): *Information Hiding*, vol. 1174, LNCS (Springer, Berlin, Heidelberg, New York 1996)
[2] Second International Workshop IH'98 Proceedings, in D. Aucsmith (Ed.): *Information Hiding*, vol. 1525, LNCS (Springer, Berlin, Heidelberg, New York 1998)
[3] M. Curty, D. J. Santos: Quantum steganography, in *2nd Bielefeld Workshop on Quantum Information and Complexity* (Bielefeld, Germany 2000) pp. 12–14
[4] S. Natori: One-time hash steganography, in LNCS (1999) pp. 17–28

Index

cover data, 235

embedder, 236

extractor, 236

steganography, 235

stego-data, 235

Part IV

Realization of Quantum Information System

Photonic Realization of Quantum Information Systems

Akihisa Tomita and Bao-Sen Shi

ERATO Quantum Computation and Information Project, Japan Science and Technology Agency, Miyukigaoka 34, Tsukuba, Ibaraki 305-8501, Japan
tomita@qci.jst.go.jp

Abstract. This Chapter to introduces research to implement quantum information systems with photonics. Photonics provides strong tools to realize systems working on a single qubit. Quantum key distribution systems using improved photon detectors have been developed for commercial products in two directions: longer distances and high speeds. A photonic circuit has demonstrated the quantum Fourier transform operation over 1024 qubit. Entangled photon genaration in spontaneous parametric down conversion has been also improved.

1 Introduction

In designing actual systems, we first need to decide the medium to represent quantum information, or qubits. Among candidates for representing qubits, we focused on photon states. Photons have advantages to implement qubits as follows: SU(2) space is easily implemented by polarization, they are very weakly coupled to the environment, and single-photon measurement technique is available. In short, photons will provide "cleaner" physical realization and clearer correspondence to theories than materials. Moreover, in practice, we can utilize the fruits of the extensive research and development efforts of the optical communication industry. Currently, improved devices are commercially available with affordable costs in fiber optics. We can construct quantum circuits that consist of one-qubit operations (including classically controlled gates) with those devices. Fiber optics also resolves the mode matching problems in conventional optics, and provides mechanically stable optical circuits. It would be worth exploring the feasibility of quantum information processing based on photons, in particular, with fiber optics.

In general, research activities would be conducted in two directions: One is to demonstrate functions required in quantum information processing with the current technology; the other is to create novel devices. We believe that it is important to show how far we can go with our currently available equipment and what we really need to develop to go further. The following sections describe the achievements in both directions in the areas of cryptography, computation, and information. First, we show the experiments on quantum cryptography, which is thought to be close to commercial applications. Second, we introduce a fiber-optic implementation of quantum Fourier transform followed by measurement (MQFT). The circuit, constructed with commercially available fiber optics components, yields MQFT operation with up to 1024 qubit. Nevertheless, we need controlled unitary gates to complete remarkable quantum computation protocols, such as phase estimation. Finally, we describe the experiments to generate entangled photon pairs, which will play an indispensable role in advanced quantum information processing.

2 Cryptography

Quantum cryptography, quantum key distribution (QKD), in particular, is an important application of photonic quantum information technology, because it is closest to the practical (commercial) use. QKD allows two remote parties (Alice and Bob) to generate a secret key, with privacy guaranteed by quantum mechanics [1,2]. The security will never be threatened by any progress in computer hardware and software; even a quantum computer cannot break it. The secret key will provide unconditional security when it is used in the one-time-pad cryptosystem. Though QKD can be performed with only current technology, there still remain many things to be improved to satisfy specifications that would be necessary for practical systems. Transmission distance and transmission speed are the two most important issues. Then, how long should the QKD transmission cover? The longer, the better. However, the first customer would be government offices (military, foreign affairs, and so on), or finance (banks or stock traders, for example.) They usually stay in the centers of cities. The city center lies within a circle of about ten kilometers radius in most big cities. However, if we consider the use of networks, the fiber distance would reach several tens or a hundred kilometers. Our first goal would be several tens (say, forty) kilometers, then over a hundred. Practical use of the key will also require fast key transmission, or key generation. The rate can be slower than current data transmission, which exceeds Gbps even in a local network, because the amount of valuable information that requires unconditional security would be much smaller. Nevertheless, fast and secure random numbers (i.e., cryptographic key) generation will open wider applications of the QKD. We here concentrate ourselves on the recent progress in single-photon transmission for QKD. To guarantee security, the sender needs to transmit only one photon at a time to a remote receiver, keeping the photon states to show high visibility. A problem in practical systems is the trade-off between long-distance transmission and high key generation rate. Two directions may be explored: One is slow but long distance transmission, and the other is short but high rate transmission.

2.1 High-Sensitivity Photon Detector

2.1.1 Requirement for Single-Photon Detectors

One of the most important devices in QKD transmission is the single-photon detector (SPD), which limits both the transmission distance and the transmission rate. The SPDs should show high detection efficiency, low dark count, and short response time. The ratio of the detection efficiency η to the dark count probability P_d determines the error rate e_B, as

$$e_B = \frac{1}{2} \frac{S(1-v)\eta + P_d}{P_{DET}} = \frac{1}{2} \frac{S(1-v) + P_d/\eta}{S(1-P_d) + P_d/\eta}, \tag{1}$$

where v is the visibility of the interferometer, and P_{DET} represents the detection probability per pulse. P_{DET} is related to the probability S that at least one photon arrives at the detector by

$$P_{DET} = S\eta + P_d - S\eta P_d. \tag{2}$$

The probability S is a function of the loss in the transmission line and the receiver. The photon loss in a L km-long fiber is given by αL [dB]. When we assume the receiver loss is β [dB], S is given by

$$S = 10^{-(\alpha L + \beta)/10} \tag{3}$$

for a single photon source, and

$$S = 1 - \exp\left[-\mu 10^{-(\alpha L + \beta)/10}\right] \tag{4}$$

for a coherent photon source with the average photon number of μ. The error rate (1) is given by the half of the inverse of signal-to-noise ratio S/N (noise will give the error with the probability of 1/2). (1) shows that the ratio P_d/η is a figure of merit of a SPD that determines the error probability, because P_d and $1-v$ are small. The error rate should be kept lower than a threshold for secure QKD. The threshold varies according to the assumptions on the method of Eve's attack and error correction. Typical threshold value is around 11% [3, 4]. Then the ratio P_d/η should be smaller than 10^{-3} for 100 km fiber transmission in 1550 nm even with an ideal single photon source.

Clock frequency of the system is limited mainly by the afterpulse, false photon detections caused by residual carriers created by the previous detections. We cannot send a photon pulse during the period of large afterpulse probability. The afterpulse effect remains typically 1 μs after photon detection. This period may vary on devices and operating conditions. The afterpulse effect on error probability can be formulated as follows. We assume two detectors 1 and 2 to discriminate bit values 0 and 1, respectively. The probabilities $p_1(t_n)$ that detector 1 fires and $p_2(t_n)$ that detector 2 fires are given by the bit value $b(t_n) = \{0,1\}$ at the nth clock t_n as

$$p_1(t_n) = S\eta q(t_n) + P_d + \sum_{i=-\infty}^{n-1} f(t_n - t_i)p_1(t_i), \tag{5}$$

$$p_2(t_n) = S\eta(1 - q(t_n)) + P_d + \sum_{i=-\infty}^{n-1} f(t_n - t_i)p_2(t_i), \tag{6}$$

where the function

$$q(t_n) = v(1 - b(t_n)) + (1 - v)b(t_n) \tag{7}$$

defines the fraction that a photon enters the detector 1, and the memory function $f(t_n - t_i)$ represents the afterpulse effect. A reasonable form of f would be $f(t) = A\exp[-\gamma t]$, but here we assume

$$f(t) = \begin{cases} A & (0 \leq t \leq t_M), \\ 0 & (t < 0, t > t_M), \end{cases} \tag{8}$$

for simplicity. The afterpulse probability A remains constant during M periods of the clock in this model. Then (6) can be solved for the asymptotic values. The error probability is given by

$$\hat{e}_B = \frac{1 - v + P_d/S\eta}{1 + 2P_d/S\eta} + \frac{v - 1/2}{1 + 2P_d/S\eta}AM \approx e_B + \frac{1}{2}AM, \tag{9}$$

if we neglect the events that both detectors fire simultaneously. Equation (9) shows that the afterpulse effect increases the error probability by $AM/2$. For example, typical values $A = 10^{-3}$ and $M = 100$ increase the error probability by 5%.

2.1.2 Improved Single-Photon Detector for Fiber Transmission

The QKD experiments at 1550 nm have employed the SPDs using In-GaAs/InP avalanche photodiode (APD) [5, 6, 7] in Gaiger mode, where a reverse bias higher than the breakdown voltage is applied. The high bias increases the avalanche gain to enable single-photon detection. However, this also results in large dark count probability and afterpulse, which cause errors in the qubit discrimination. The dark count probability and the afterpulse can be reduced by using gated mode, where gate pulses combined with DC bias are applied to the APD. The reverse bias exceeds the breakdown voltage only in the short pulse duration synchronized to the photon arrival. Though this method works well, the short pulses produce strong spikes on the transient signals. High threshold in the discriminator is therefore necessary to avoid errors, at the cost of detection efficiency. High gate pulse voltage is

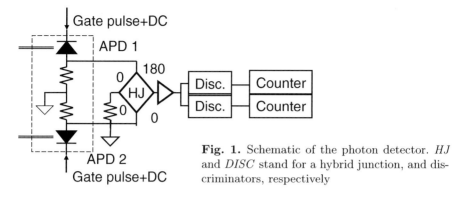

Fig. 1. Schematic of the photon detector. *HJ* and *DISC* stand for a hybrid junction, and discriminators, respectively

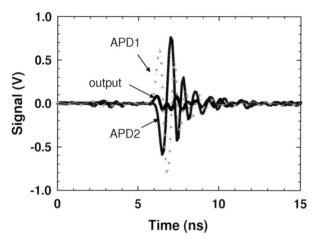

Fig. 2. Cancellation of the transient spike. *Dots*: APD 1, *Thin solid*: APD 2, *Thick solid*: the differential output of the APD 1 and the APD 2

also required to obtain large signal amplitude by increasing avalanche gain. Impedance matching helps to reduce the spikes to some extent [8]. *Bethune* and *Risk* introduced a coaxial cable reflection line to cancel the spikes [9]. We propose a much simpler method: canceling the spikes by taking the balanced output of the two APDs required for the qubit discrimination [10].

Figure 1 depicts the schematic of the SPD. Two APDs (Epitaxx EPM239BA) and load resisters were cooled to between $-133\,°C$ and $-60\,°C$ by an electric refrigerator. Short gate pulses of 2.5 V p-p and 750 ps duration were applied to the APDs after being combined with DC bias by bias-tees. The output signals from the APDs were subtracted by a 180° hybrid junction of 2 MHz to 2000 MHz bandwidth. The differential signal was amplified and discriminated by two discriminators. Since the spikes were the common mode input for the 180° hybrid junction, they would not appear at the output. APD 1 provided negative signal pulses at the output, while APD 2 provided positive pulses. We can determine which APD detects a photon from the sign of the output signals. Figure 2 shows the output signal of the amplifier without photon input. Almost identical I–V characteristics of the APDs enabled us to obtain a good suppression of the spikes. We observed the lowest dark count probability of 7×10^{-7} per pulse with detection efficiency of 11 % at $-96\,°\mathrm{C}$. The ratio P_d/η was as small as 6×10^{-6}, which corresponds to 270 km QKD transmission with an ideal photon source. The detection efficiency and the dark count probability are increasing functions of the bias. The maximum value of the detection efficiency is obtained when the DC bias is set to the breakdown voltage. We obtained larger values of the maximum detection efficiency at higher temperatures: the detection efficiency of 20 % at $-60\,°\mathrm{C}$ with the dark count probability of 3×10^{-5} per pulse. Afterpulse probability was measured by applying two successive gate pulses to the APDs. Afterpulse is prominent at low temperatures. We found that afterpulse probability remained about 10^{-4} for the 1 μs pulse interval at the temperatures higher than $-96\,°\mathrm{C}$. This corresponds to 10^{-5} error probability (per pulse) for 10 % detection efficiency. On the basis of the dark count probability and the afterpulse probability, we conclude that the optimal operation temperature for the present APDs is around $-96\,°\mathrm{C}$. The obtained afterpulse effect was shorter than the previous reports. We believe that this is due to the decrease of the gate pulse voltage. This is another advantage of our SPDs. Recently, we obtained the dark count probability of 2×10^{-7} per pulse at the detection efficiency of 10 % [11]. The S/N, or the ratio P_d/η is improved about 50 times (17 dB) as much as the values previously reported from other organizations.

2.2 Single-Photon Transmission Over 150 km in a Unidirectional System With Integrated Interferometers

Most of the successful QKD transmission experiments have been based on so called plug-and-play (P&P) system, which contains an autocompensation mechanism to achieve good interference performance with ease [7]. Although the P&P system works well for QKD systems using a weak pulse up to 100 km [11], extending the transmission distance will be difficult even if a lower noise SPD is developed. This is because backscattering noise in the fiber dominates the detector noise, which is intrinsic to the bidirectional autocompensating system. Although the use of storage line and burst photon trains would reduce the backscattering, this would also reduce the effective transmission rate by one third. Unidirectional systems are free from the above problem. The difficulty in the unidirectional system has been the stabilization of two remote interferometers to achieve high visibility. We propose a solution to this conflict between stability and transmission distance by showing a unidirectional system using integrated-optic interferometers based on planar lightwave circuit (PLC) technology [12]. Our system is also compatible with

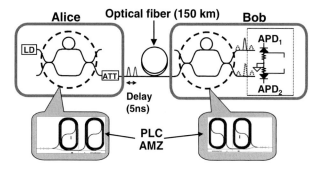

Fig. 3. Schematics of the integrated-optic interferometer system for quantum key distribution. LD: laser diode, ATT: attenuator, APD: avalanche photodiode, DS: discriminator, CT: counter, H: 180° hybrid junction

QKD systems using true single photon or quantum correlated photon pairs, which are believed to provide higher key rate after a long distance transmission. An asymmetric Mach-Zehnder interferometer (AMZs) with a 5-ns delay in one of the arms was fabricated on a silica-based PLC platform. Since the AMZs were fabricated using the photolithographic mask, they had the identical path length difference between the two arms. The optical loss was 2 dB (excluding the 3-dB intrinsic loss at the coupler). Polarization-dependent loss was negligible (0.32 dB). One of the couplers was made asymmetric to compensate for the difference in the optical loss between the two arms, so the device was effectively symmetric. A Peltier cooler attached to the back of the substrate enabled control of the device temperature with up to 0.01 °C precision. Polarization-maintaining fiber (PMF) pigtails aligned to the waveguide optic-axis were connected to the input and output of the AMZ.

Two AMZs were connected in series by optical fiber to produce a QKD interferometer system (Fig. 3). Optical pulses that were 200 ps long and linearly polarized along one of the two optical axes were introduced into the PMF pigtail of Alice's AMZ from a DFB laser at 1550 nm. The input pulse was divided into two coherent output pulses polarized along the optical axis of the output PMF, one passing through the short arm and the other through the long arm. The two optical pulses were attenuated to the average photon number of 0.2. The two weak pulses propagated along the optical fiber and experienced the same polarization transformation. This is because the polarization in fibers fluctuates much slower than the temporal separation between the two pulses. After traveling through Bob's AMZ, these pulses created three pulses in each of the two output ports. Among these three pulses, the middle presents the relative phase between the two pulses. The interfering signal at the middle pulses was discriminated by adjusting the applied gate pulse timing. The system repetition rate was 1 MHz to avoid APD afterpulsing.

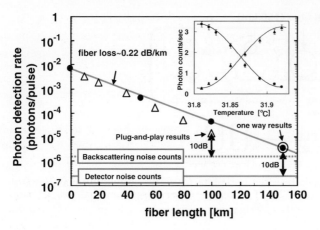

Fig. 4. Photon counting probability against transmission distance. *Open triangles* indicate the results in the P&P system. *Inset*: Fringe observed in photon count rate, obtained by changing the device temperature at 150 km

Precise control of the relative phase setting between the two AMZs and the birefringence in the two arms of Bob's AMZ is necessary to obtain high visibility. Control of both can be done by controlling the device temperatures. To set the phase, it is sufficient to control the path length difference within $\Delta L = \lambda/n$, where $n \sim 1.5$ is the refractive index of silica. The path length difference depends linearly on the device temperature with 5 μm/K, due to the thermal expansion of the Si substrate. The birefringence in the two arms can be balanced by controlling the relative phase shift between two polarization modes, because the two arms have the same well-defined optical axes on the substrate. If the path length difference is a multiple of the beat length $\Delta L_B = \lambda/\Delta n$, where Δn is the modal birefringence, the birefringence in the two arms is balanced and two pulses interfere at the output coupler of Bob's AMZ no matter what the input pulse polarization is. Since $\Delta n/n$ was the order of 0.01 for our device, the birefringence was much less sensitive to the device temperature than the relative phase. Therefore, we could easily manage both the phase setting and the birefringence balancing simultaneously.

We measured the photon counting probability given by the key generation rate divided by the system repetition rate and plotted it as a function of transmission distance (Fig. 4). The measured data fit well with the upper limit determined by the loss of the fiber used (0.22 dB/km). In Fig. 4, the base lines present the dark count probabilities. The interference fringe is shown in the inset. The visibility at 150 km was 82 % and 84 % for the two APDs [13], which corresponds to a quantum bit error rate (QBER) of 9 % and 8 %, respectively. These satisfy the rule of thumb for secure QKD. The interference was stable for over an hour, which is good enough for a QKD

2.3 Refinements Toward a Practical QKD System

2.3.1 Temperature-Insensitive Interferometer

We now turn our attention to the short distance system. A P&P system would be suitable for a short-distance QKD system, because of the simple optical control. In a practical systems, however, the system should be robust against the change in environment. Temperature in rack-mounted equipment may vary from $-5\,^\circ$C to $70\,^\circ$C. The temperature dependence of the rotation angle at the Faraday mirror (FM), the key device in a P&P system, causes errors, and thus the final secret key generation rate will vary. A temperature-insensitive system is indispensable for a practical installation. We propose a temperature-insensitive autocompensating device with a simple optical structure [14].

Before presenting our proposal, we summarize the role of the FM in P&P systems. Since the reflected light propagates the opposite direction, we need to be careful about the coordinates. In the following, we fix the direction of axes. The effect of FM in linear polarization basis leads σ_x rotation. The effect of transmission line (fiber) on the polarization can be expressed by the unitary transform:

$$U = e^{i\alpha} R_z(2\beta) R_y(2\gamma) R_z(2\delta) , \tag{10}$$

where R_y and R_z stand for the rotation on the y axis

$$R_y(2\gamma) = \begin{pmatrix} \cos\gamma & -\sin\gamma \\ \sin\gamma & \cos\gamma \end{pmatrix}, \tag{11}$$

and the rotation on z axis

$$R_z(2\delta) = \begin{pmatrix} e^{-i\delta} & 0 \\ 0 & e^{i\delta} \end{pmatrix}, \tag{12}$$

respectively. The above unitary transform (10) is general, as long as we can neglect depolarizing in the fiber. We can see that the total effect (not including the global phase) of going around the transmission line is just the transformation by the FM

$$R_z(2\delta) R_y(2\gamma) R_z(2\beta) \sigma_x R_z(2\beta) R_y(2\gamma) R_z(2\delta) = \sigma_x. \tag{13}$$

The outward and homeward polarizations are orthogonal, regardless of the disturbance at the transmission line and the initial polarization. This condition is essential for stable interference. However, the rotation angle of the FM depends on the temperature and the transformation by the FM deviates

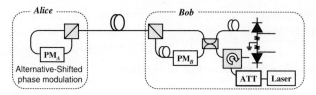

Fig. 5. Schematics of the proposed quantum key distribution system

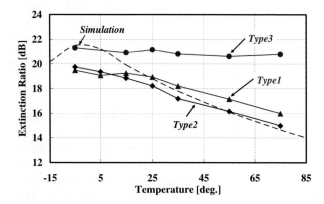

Fig. 6. Extinction ratios versus temperature change. *Type 1* and *type 2* represent the result of conventional systems. *Type 3* shows the temperature dependence in the proposed system

from σ_x as the temperature change. Autocompensation becomes no longer perfect.

We found that a loop mirror depicted in Fig. 5 can provide the same effect as a FM. Two input/output terminals of a polarization beam splitter (PBS) are connected by PMF to make a loop. The polarizations at the terminals are aligned to the slow axis of the PMF, so that one defined polarization runs in the fiber loop. Note that input horizontal polarization turns to the vertical polarization at the output, and vice versa. A phase modulator (PMA) is placed on an off-center position in the loop. In a P&P system, two pulses S and L enter Alice's loop mirror. The PBS divides the input photons by the polarization. Then, the four pulses travel in the loop: S_H, S_V, L_H, and L_V. PMA can apply the phase shift to the four pulses independently by the timing of the modulation. We put the following phase shifts to the four pulses: none to S_H, π to S_V, φ_A to L_H, and $\varphi_A + \pi$ to L_V (we named it "alternative-shifted phase modulation"). The four pulses are combined by PBS into two (S and L.) The two experience the transforms $S \to \sigma_x S$ and $L \to \exp[i\varphi_A]\sigma_x L$, respectively. Temperature dependence of phase modulators is smaller than that of FMs, and it can easily be adjusted by the pulse voltage to the PMA.

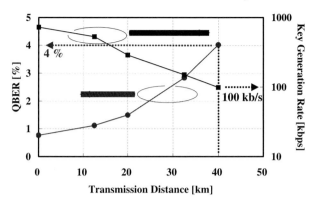

Fig. 7. QBER and raw key generation rate versus transmission distance

We examined the temperature dependence of the P&P QKD systems. Type 1 and type 2 used typical FM, whereas type 3 used the proposed loop mirror. The ambient temperature of Alice was changed from −5 to 75 °C. A polarization controller scrambled the polarization randomly with the four random digits generated with the "48-bit linear congruential method". Figure 6 shows the extinction ratios against the temperature. The solid line shows the simulated results based on the assumption that the depolarization was 0.8 %, and the rotation angle changed by −0.013 deg/K. The dashed and dotted lines show the measured results. We can see a great difference between Type 1, Type 2 and Type 3 in Fig. 6. As the temperature increased, the extinction ratio decreased in both Type 1 and Type 2 systems, whereas the extinction ratio remained high in Type 3 system. This demonstrated the advantages of our system over conventional P&P systems.

2.3.2 High-Speed Operation

As stated before, the clock rate of QKD systems is limited by the afterpulse effect. In a short-distance system, we can increase the clock rate by optimizing SPD for small afterpulse effects. The SPDs in the previous sections were optimized for low dark count probability to increase the S/N for long-distance transmission. In a short-distance system, the effect of the fiber loss is less serious, so that error probability can be kept below the security criteria with larger P_d/η ratio.

We implemented a high-speed secret key generation experiment using the Type 3 QKD system described above. At Bob, a 1550 nm directly modulated DFB-LD creates 500 ps-wide pulses with a repetition rate of 62.5 MHz. The sequence of optical pulses is split by a polarization maintained coupler (PMC). Transmission over a single mode fiber (SMF) was carried out. The optical power was adjusted so that the average photon number at Alice (μ) becomes 0.6 photon/pulse. We used NEC's APDs at higher temperature

($-40\,°C$) in the balanced SPD. The measured value of the dark count probability, the detection efficiency, and the afterpulse probability were about 1×10^{-4}, 7%, and 1×10^{-3}, respectively, for operation at 62.5 MHz. Figure 7 shows QBER and raw key generation rate against the transmission distance. The solid line shows the simulated results considering P_d, η, μ, and loss, but no reflection or scattering. The dashed and the dotted lines show the measured results. The good agreement between the calculated and the measured results shows little impairment was caused by reflection and scattering. We obtained a raw key generation rate of about 100 kHz and QBER of 4% over SMF 40 km transmission [14]. We sent the sequences of all "0"s and all "1"s in this experiment. Because the afterpulse probability was not negligibly small, (9) suggests that QBER would increase about a few percent for random bit sequences. Even if so, QBERs remain lower than the security criteria, and moreover, the afterpulse-induced impairment can be avoided by setting an adequate dead time at the APDs. We have used light pulses with slightly higher average photon numbers than the reported experiments. However, we believe the QKD system of this distance is still secure. As shown in a recent proposal, we can circumvent photon number splitting (PNS) attack by sending decoy state to detect the PNS attack. Readers can find a detailed discussion on the security against PNS attack in the Chapter by *Wang*. QKD is shown to be secure against other practical attacks [15].

So far, the transmission experiments were done in fairly stable environments. Equipment stays in an air-conditioned laboratory. Even in the transmission experiments with installed fibers, the fibers were buried under the lake or well-maintained as a test-bed. Long-term stable quantum key generation in an office environment, where temperatures are not constant and may vary from 0 to $40\,°C$, requires a temperature-independent, reliable system under such a wide temperature range. The alternative phase shift modulation in the interferometer is one of the techniques to make the equipment stable against temperature changes in offices. In commercial fiber networks, which contain many connections and reflecting points, the loss and the backscattering may differ from fiber to fiber. In order to avoid the scattering of light in the fiber and reflection of light from the connection point, burst-mode transmission technology should be installed in the system. Access links for end users sometimes use fibers installed in the open air. Such aerial fibers tend to experience mechanical vibration and temperature fluctuation. For example, a temperature rise of $10\,°C$ will cause a 20 ns delay in the photon arrival time after 16.3 km fiber transmission.

The system for quantum communications should be designed to keep stability against the fluctuation of the environment. The QKD systems should have a clock synchronization system, which can trace the shift of fiber length due to thermal expansion and keep the optimum timing. A watch-dog system is also necessary to monitor key generation rate and error rate, and the transmission system should automatically reset and calibrate itself on system errors. A QKD system fully equipped with the above functions has

Fig. 8. Prototype of a QKD system fully equipped with functions for stable operation (*left photograph*). A fortnight quantum key generation field trial over commercial access fibers was done with the system. The fiber cable was installed on the electric poles (*right photograph*)

been developed [16], where the all functions were installed in a 4U-height, 19-inch box commonly used for communication equipment, as shown in the picture (Fig. 8.) Bit synchronization, where the timing between the photon generation, the phase modulation, and the photon detection is kept optimum, was achieved with help of the clock pulses transmitted in the same fiber with the wavelength division technique. Frame synchronization is also required to synchronize the starting point of the bit sequences. We introduced fault detection and distinction by QBER monitoring. Two resynchronization mechanisms were installed to correct the bit synchronization errors and the frame synchronization errors. When the bit synchronization worsens, QBER slowly degrades, and when the frame synchronization is lost, QBER rapidly degrades to over 50 %. If the bit phase shift or theframe phase shift causes the QBER degradation, it should be improved after the resynchronization process. If not the case, the fault is classified as a fatal error, which may result from fetal system error or eavesdropping, and the system stops.

We carried out a fortnight quantum key generation field trial over commercial-access fibers offered by POWEWDCOM Inc. The fiber was an aerial fiber cable of a 16.3 km single-mode fiber installed on electric poles, and our prototype system was settled in an office room. We obtain the average quantum error rate of 7.5 % and the average final-key generation rate of 13.0 kbps. For this experiment, we set the discard rate in the privacy amplification (PA) at 0 %, where the CPU load was maximized and we could examine the computational effort required for the PA. A "hands-free" operation, and continuous final-key generation over two weeks was demonstrated in a real-world environment. We also confirmed a data transmission connecting IP phones to our prototype in the laboratory, and the voice of the IP data was encrypted by the final-keys using Vernam cryptography.

3 Quantum Computation

3.1 Measured Quantum Fourier Transform

3.1.1 Implementation and Experimental Results

A circuit constructed of commercially available fiber-optic devices has been built to perform Quantum Fourier Transform followed by measurement (MQFT) that is almost fault tolerantly up to 1024 qubits [17]. As is well known, quantum Fourier transform (QFT) plays an important role in quantum computation algorithms. We can find an example in phase estimation [18], where the heart of Shor's factorization [19] and its cousin algorithms lies. The phase estimation problem is given as follows: An eigenvalue of a unitary transform U defines a phase φ as $U\left|u\right\rangle = \exp\left[2\pi i\,\varphi\right]\left|u\right\rangle$. Our task is to determine the phase expressed in n bits by $\varphi = \varphi_1 2^{-1} + \cdots + \varphi_n 2^{-n} = 0.\varphi_1\cdots\varphi_n$. The task can be achieved by a quantum circuit of controlled-unitary operations (c-U's) and QFT on the control qubits.

Our implementation of the QFT is based on two facts. The first is that controlled-unitary operations commute with measurements when the control-qubits are measured in the computational basis. This implies that we can replace the controlled-unitary gates with the unitary gates controlled by the results of the measurements. Since the latter devices act on one qubit (target-qubit), they are much easier to obtain than the former. *Griffiths* and *Niu* [20] showed an alternative form of the QFT quantum circuit with Hadamard gate and rotation gates controlled by the measurement results of former bits. *Parker* and *Plenio* [21] found that the QFT for the phase estimation can be operated qubit by qubit with only one rotation at a time. We here refer it to phase estimation by serial QFT. *Beauregard* [22] used their observation to reduce the number of qubits to perform the Shor's algorithm. Figure 9a–c depicts quantum circuits for QFT. The classically controlled rotation R_k to the kth control qubit is defined by

$$R_k = \begin{pmatrix} 1 & 0 \\ 0 & \exp\left[-2\pi i \Phi_k\right] \end{pmatrix}, \tag{14}$$

$$\Phi_k = \sum_{j=1}^{k-1} \frac{1}{2^{j+1}} \varphi_{k-j}, \; \Phi_1 = 0\,.$$

We implemented a MQFT circuit with fiber-optic devices, as shown in Fig. 10. Qubits were represented by the polarization of a single photon, as the $\left|0\right\rangle$ state to be polarized horizontally and the $\left|1\right\rangle$ state to be polarized vertically. The input photons to the QFT circuit were generally elliptically polarized according to the phase between the basis states. The main part of the circuit is to apply the relative phase shift to the $\left|1\right\rangle$ state given by (14.) We employed a fiber loop structure, the so-called Sagnac interferometer, where

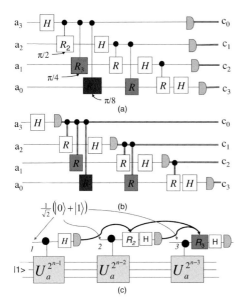

Fig. 9. Quantum circuits for measured quantum Fourier transform. (**a**): A quantum circuit with controlled-rotation gates. (**b**): A quantum circuit with classically controlled-rotation gates [20]. (**c**): A serial quantum circuit of phase estimation [21]

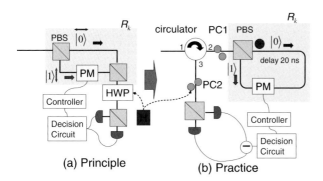

Fig. 10. Quantum circuit for measured quantum Fourier transform implemented by fiber optics. (**a**): The principle of the circuit. (**b**): Practical circuit. *PBS*, *PM*, *HWP*, and *PC* stand for polarization beam splitters, phase modulator, half-wave plate, and polarization controllers, respectively

the orthogonally polarized photons were propagated in the opposite directions through the same fiber. Therefore, the Sagnac interferometer guarantees the same additional phase fluctuation for the two basis states, in other words, the present QFT circuit is decoherence free. QFT operation was demonstrated by putting the single-photon pulse sequence elliptically polarized according

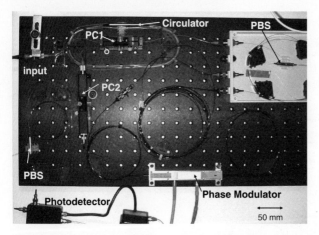

Fig. 11. Quantum circuit for measured quantum Fourier transform constructed on a breadboard

to a random number $j = j_1 \cdots j_n$ into the fiber-optic QFT circuit. The average photon number in the pulse was set to less than 1. We compared the measured bit values with the input bits to estimate the error probability. Figure 12a shows the distribution $E(n)$ of the first error qubit in 21 trials of 255-qubit-QFT, that is, the distribution of QFT trials done successfully up to the $(n-1)$th qubit. We truncated the calculation of the rotation angle at the fifth bit ($m = 5$). The estimated error probability was $p = 0.01$, corresponding to the expectation value of 100 qubits for the error-free QFT operations. Further statistical analysis showed that the error probability per qubit was in the range of $2.6 \times 10^{-3} \leq p \leq 1.6 \times 10^{-2}$ with the confidence level of 95%. A simple way to reduce error probability is to decide the bit value by majority (majority voting) of repeated measurements. QFT operations with 1024 qubits have been done to show the effect of decision by majority for $M = 10$. The rotation and measurement were done with the same input polarization states M times, and the bit value was determined by majority of the accumulated results. We set the error probability for one qubit to $p = 0.07$, intentionally higher than that for the single measurement. An estimation predicts that the decision by majority will decrease the error probability to $p_{10} = 3 \times 10^{-4}$. We have obtained 24 successful attempts out of 30 trials, as shown in Fig. 12b. The success probability was 80%, and the mean error probability was estimated to be 2.2×10^{-4} per qubit. The error probability per qubit lies in the range of $1.2 \times 10^{-4} \leq p \leq 4.3 \times 10^{-4}$ with the confidence level of 95%.

3.1.2 Effects of Imperfection

In the following, we consider effects of the imperfection of the experimental apparatus. Two types of failure may occur. One is failure in photon detection,

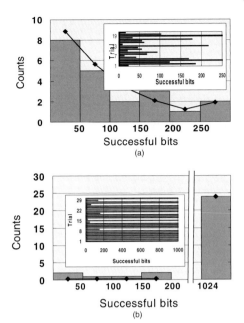

Fig. 12. Results of MQFT trial. **(a)** 21 trials of 255 qubits and **(b)** 30 trials of 1024 qubits. The bit values in (b) were determined by the decision by the majority of ten measurement outcomes. The inset shows the result of each trial. *Successful bits* refers to the number of bits for which a MQFT operation was done successfully

and the other is error in the measurement. If a pulse contains no photons, the operation will not affect the target qubits. We can thus continue the calculation by repeating the operation step. If a photon is lost in the MQFT circuit, it means that the c-U and measurement have been done without knowing the result. Therefore, the target states will be in a mixed state corresponding to the two possible outcomes. Even in this case, we can proceed with the calculation by repeating the same operation step. The repeated measurement will reduce the target qubit state into a pure state by selecting one of the possibilities.

Errors in the measurement originate from imperfection of the interferometer and from errors in the phase modulation. Dark counts are negligibly small in the photon detector [10]. The performance of the interferometer is characterized by visibility v, which defines the measurement on the control qubit as

$$M_0 = \sqrt{\frac{1+v}{2}} |0\rangle_c \langle 0| + \sqrt{\frac{1-v}{2}} |1\rangle_c \langle 1| ,$$
$$M_1 = \sqrt{\frac{1-v}{2}} |0\rangle_c \langle 0| + \sqrt{\frac{1+v}{2}} |1\rangle_c \langle 1| . \qquad (15)$$

The phase error, which shifts the rotation angle in (14) from $2\pi\Phi_k$ to $2\pi\Phi_k+\delta$, results in the control qubit after the rotation gate as $2^{-1/2}\left(|0\rangle_c + \exp[2\pi i\, 0.\varphi^s_{n-k+1} + i\delta]|1\rangle_c\right)$. The phase error originates from the approximation in the rotation angle Φ_k and from errors in converting the rotation angle into the drive voltage to the PM. The latter can be reduced by careful calibration, so that we only have to consider the effect of the truncation. The truncation at the mth bit results in the phase error

$$\delta = 2\pi \sum_{j=m}^{k-1} \frac{1}{2^{j+1}} \varphi_{n-k+j+1} \leq 2\pi \sum_{j=m}^{k-1} \frac{1}{2^{j+1}} < 2\pi \sum_{j=m}^{\infty} \frac{1}{2^{j+1}} = \frac{\pi}{2^{m-1}}. \quad (16)$$

The phase error should not be significant [23], because the contribution from the jth bit $(j > m)$ decreases with the factor of $2^{-(j+1)}$. The worst values of $\cos\delta = \pm 0.98$ obtained in the experiment correspond to a phase error of $\pi/16$, which agrees quite well with the prediction by (16) with $m=5$. The visibility and the phase error determine the error probability of the measurement by

$$p = \frac{1 - v\cos\delta}{2}. \quad (17)$$

The estimated error probabilities from (17), $p = 8.2\times 10^{-3}$ (by using $\langle\cos\delta\rangle = \pm 0.9936$) and $p = 1.5 \times 10^{-2}$ (by using $\cos\delta_{\max} = \pm 0.98$), agree well with the experiment.

3.1.3 Validity of Majority Voting

We consider the validity of majority voting. If the state of the target qubits is one of the eigenstates of the c-U, the unitary transform results in the same phase value to the control qubit. Every measurement will provide the result with the error probability given by (17). In general, however, the target qubit state is a superposition of the eigenstates. In this case, the measurement results will be probabilistic even in a perfect quantum circuit. The initial state at the kth operation step is given by

$$\rho = \frac{1}{2}\left(|0\rangle_c\langle 0| + |0\rangle_c\langle 1| + |1\rangle_c\langle 0| + |1\rangle_c\langle 1|\right)$$
$$\otimes\left(a_0 a_1 |0\rangle_t\langle 1| + a_0 a_1 |1\rangle_t\langle 0| + a_1^2 |1\rangle_t\langle 1|\right), \quad (18)$$

where the orthonormal bases $|0\rangle_t$ and $|1\rangle_t$ are defined by

$$|0\rangle_t = \frac{1}{a_0}|u_s^0\rangle, \quad |1\rangle_t = \frac{1}{a_1}|u_s^1\rangle.$$

The state vectors $|u_s^0\rangle$ and $|u_s^1\rangle$ belong to the kth bit values $\varphi_{n-k+j+1}^s = 0$ and $\varphi_{n-k+j+1}^s = 1$, respectively, as

$$|u_s^0\rangle = \sum_{s \in \{\varphi_{n-k+j+1}^s = 0\}} c_s |u_s\rangle ,$$

$$|u_s^1\rangle = \sum_{s \in \{\varphi_{n-k+j+1}^s = 1\}} c_s |u_s\rangle . \qquad (19)$$

The normalization constants are given by $a_0 = \sqrt{|\langle u_s^0 | u_s^0 \rangle|^2}$ and $a_1 = \sqrt{|\langle u_s^1 | u_s^1 \rangle|^2}$, where $a_0^2 + a_1^2 = 1$ is satisfied. The c-U gates, the rotation gate, and the Hadamard gate result in the transformation Q as

$$Q = \frac{1}{\sqrt{2}} \left(|0\rangle_c \langle 0| + e^{i\delta} |0\rangle_c \langle 1| + |1\rangle_c \langle 0| - e^{i\delta} |1\rangle_c \langle 1| \right) \otimes |0\rangle_t \langle 0|$$

$$+ \frac{1}{\sqrt{2}} \left(|0\rangle_c \langle 0| - e^{i\delta} |0\rangle_c \langle 1| + |1\rangle_c \langle 0| + e^{i\delta} |1\rangle_c \langle 1| \right) \otimes |1\rangle_t \langle 1| . \qquad (20)$$

Suppose measurement outcome is "0", then the target state collapses to

$$\rho^{(t)} = \frac{1}{p_0} tr_c \left[M_0 Q \rho (M_0 Q)^+ \right] ,$$

$$= \frac{1}{p_0} \left(\frac{1 + v \cos \delta}{2} a_0^2 |0\rangle_t \langle 0| + \frac{iv \sin \delta}{2} a_0 a_1 (|0\rangle_t \langle 1| - |1\rangle_t \langle 0|) \right.$$

$$\left. + \frac{1 - v \cos \delta}{2} a_1^2 |1\rangle_t \langle 1| \right) , \qquad (21)$$

where $p_0 = \left[1 + \left(2a_0^2 - 1 \right) v \cos \delta \right] / 2$ is the probability to obtain the outcome "0". Partial trace is taken, because we have no further information on the control bit state. A new control bit is supplied for the next measurement, and the total state is given by $\rho' = \frac{1}{2} \left(|0\rangle_c \langle 0| + |0\rangle_c \langle 1| + |1\rangle_c \langle 0| + |1\rangle_c \langle 1| \right) \otimes \rho^{(t)}$, in place of (18). As can be seen from (21), measurement outcome of "0" increases the matrix element $|0\rangle_t \langle 0|$, so that the target state is more like $|0\rangle_t$ and probability to obtain "0" in the following measurement is increased. It would be interesting to calculate the density matrix if the second measurement outcome is "1". (It can happen with a small probability.) The reduced density matrix for the target state is calculated using (15) and (20)

$$\rho^{(t)\prime} = \frac{1}{p_1'} tr_c \left[M_1 Q \rho' (M_1 Q)^+ \right] ,$$

$$= a_0^2 |0\rangle_t \langle 0| + \frac{v^2 \sin^2 \delta}{1 - v^2 \cos^2 \delta} a_0 a_1 |0\rangle_t \langle 1|$$

$$+ \frac{v^2 \sin^2 \delta}{1 - v^2 \cos^2 \delta} a_0 a_1 |1\rangle_t \langle 0| + a_1^2 |1\rangle_t \langle 1| , \qquad (22)$$

where we take a mean value of the probabilistic variable δ by assuming no correlation among δ's. The result (22) shows that the diagonal elements of the density matrix are recovered after successive measurements of different outcomes. The off-diagonal elements are decreased slightly. It will not affect the probability, because we take the trace. The probability to obtain the result "0" (or "1") in the following measurement is the same as the first measurement. Therefore, the error probability would be at most the value (17). Therefore, the accumulation of M measurement outcomes will reduce the error probability to

$$p_M = \sum_{j=0}^{[M/2]} \binom{M}{j} p^{M-j} (1-p)^j , \qquad (23)$$

which decreases rapidly for $p \ll 1$. For example, the error probability per operation $p = 0.07$ and accumulation $M = 10$ yields the error probability $p_{10} = 3 \times 10^{-4}$, which agrees with the value obtained from the experiment on 1024 qubits.

The above analysis can be applied only when the control qubits are separable. We need to handle the entangled qubits in a series. Majority voting may be possible on the final results. Reduction of the error probability remains open for further calculation.

3.1.4 Toward Quantum Computers

The main drawback of the MQFT is time, because it operates serially. The response time of the current circuit is limited by electronics. We may expect the response time to be reduced to several nanoseconds. If the operation time is decreased to 1 ns, the target qubits should remain coherent for at least 10 µs to complete controlled-unitary operations with a thousand control qubits (1 ns × 1000 qubits × 10 accumulations.) This coherence time would be possible with the help of atomic qubits, where the coherence time between the metastable states of an atom reaches 0.5 ms [24]. Here, we consider a quantum computer consists of photon control qubit and atom target qubits.

The remaining problems are in realizing controlled-unitary operation; the QFT by itself will not provide an exponential speed up in comparison with classical algorithms. One possibility is using atomic target qubits, where the unitary transformation is achieved by the interaction between atoms. The input photons induce a unitary transformation on the atom qubits. The controlled operation can be achieved by the polarization selection rules in the atomic transitions, for example. The photons are scattered by the atoms and obtain a phase shift through the kick-back effect. The phase shift will be analyzed by MQFT and provide a solution of the problem. This scheme shows an interesting analogy between the quantum computation and the spectroscopy; the change in the system (target qubits) state is probed by the change in the scattered photon state (control qubit.) If we introduce an interaction Hamiltonian

$$H_{int} = gS\left(|0\rangle_c \langle 0| - |1\rangle_c \langle 1|\right), \quad (24)$$

where S is a Hermitian operator on atoms, and assume the atom state $|s\rangle$ is an eigenstate of S, i.e., $S|s\rangle = \varphi_s |s\rangle$, the interaction between the atom and the photon results in

$$|s\rangle \left(|0\rangle_c + |1\rangle_c\right) \mapsto \exp\left(-ig\varphi_s t\right)|s\rangle \left(|0\rangle_c + \exp\left(2ig\varphi_s t\right)|1\rangle_c\right). \quad (25)$$

The simplest example of the Hamiltonian (24) is $S = S_z = \left(a_1^+ a_1 - a_2^+ a_2\right)/2$, which is known as a Hamiltonian for quantum nondemolition measurement [25, 26]. The Hamiltonian can be realized by a four-level atom of two ground states $\{|1\rangle, |2\rangle\}$ and two excited states $\{|3\rangle, |4\rangle\}$, interacting with off-resonant photon field, where the transition $|1\rangle \rightarrow |2\rangle$ is allowed by the photon $|0\rangle_c$, and the transition $|3\rangle \rightarrow |4\rangle$ is allowed by the photon $|1\rangle_c$, respectively [25, 26]. We can implement a more complicated operation by introducing many-atom operator for S. The construction of unitary transformations on the atomic qubits may be seen just moving the difficulty to the atoms. Nevertheless, we believe this scheme would make the problem simpler. It can exploit fairly large interaction between atomic qubits to obtain two qubit gates. Measurement on photonic control qubit will resolve the read-out problem. Another possibility is to combine with the linear-optics gates proposed by *Knill, Laflamme,* and *Milburn* [27] (KLM). Since the KLM scheme utilizes single photon states, it is well suited to the present MQFT circuit.

The present MQFT circuit may find its applications on such as finding eigenvalues and eigenvectors [28] and clock synchronization [29] in the near future. In these applications, fewer controlled-unitary operations are enough to achieve a meaningful task than those required in the factorization algorithm. The circuit would also be useful also for precise measurements. Suppose we want to measure birefringence of a medium. By transmitting photon pulses polarized linearly at $45°$ through the media of $2^{n-1}l, \ldots, l$ in length, and by applying the output pulses to the MQFT circuit, we can determine the value of the phase difference $2\pi\phi$ created by the medium in n-bit accuracy. Birefringence is then obtained by $\phi\lambda/l$. The required photon number scales as $O(n)$. Classically, we need to increase the photon number four times to reduce the error by half. The photon number scales with $O(n^2)$. Therefore, the MQFT method has an advantage of square root of n over the classical measurement. This can be applied to determination of any scattering potential. It might be interesting to compare this advantage with that of Grover's algorithm [30].

3.2 Control-Unitary Gates

As mentioned previously, we need to develop controlled-unitary gates to realize the exponential speed-up in quantum computers. A lot of implementation

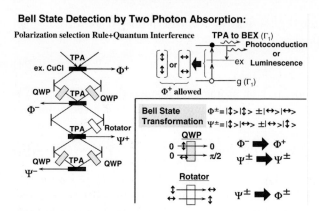

Fig. 13. A scheme for complete Bell state measurement using TPA and linear optics

schemes have been proposed. Yet, it is hard to tell what is most promising. We report basic studies toward controlled unitary gates by photon-exciton interaction.

3.2.1 Solid-State Bell State Measurement Devices by Two-Photon Absorption

Bell state measurement (BSM) is an indispensable element for quantum teleportation and teleportation-based quantum gates [31]. However, a realizable BSM method, which discriminates all the four Bell states, has been little known [32]. A controlled NOT gate transforms the Bell states into disentangled states to be easily discriminated, and provides a quantum circuit for the complete BSM. The problem is that a controlled NOT gate itself is what we want to construct by quantum teleportation.

We proposed a complete BSM by combining the discrimination of a Bell state and Bell state transform by linear optics [33]. Selection rules in two-photon absorption (TPA) enable the discrimination of a Bell state [34]. Unlike the quantum gates, coherency is not required in the output of the TPA detection scheme; the measurement is done in the TPA process. Therefore, the photon energy can be resonant to the atomic two-photon transition. This resonance may enhance the TPA to resolve the low efficiency problem of the nonlinear crystals.

One candidate for TPA is the optical transition between the lowest sublevels in a quantum dot. We assume the lowest hole states are the heavy hole states $|3/2, \pm 3/2\rangle_h$. Then the electron–hole pair states result from the optical transition are $|\uparrow\rangle = |1/2, -1/2\rangle_e |3/2, -3/2\rangle_h$ and $|\downarrow\rangle = |1/2, 1/2\rangle_e |3/2, 3/2\rangle_h$; the former is created by a right-handed circularly polarized photons $|\sigma^+\rangle$ and the latter by a left-handed circularly polarized photons $|\sigma^-\rangle$. Suppose one electron–hole pair exists in the quantum dot, and define Bell states of an electron–hole pair and a photon:

$$\left|\Phi^{(\pm)}\right\rangle = \frac{1}{\sqrt{2}}\left(|\uparrow\rangle\left|\sigma^+\right\rangle \pm |\downarrow\rangle\left|\sigma^-\right\rangle\right),\tag{26}$$

$$\left|\Psi^{(\pm)}\right\rangle = \frac{1}{\sqrt{2}}\left(|\uparrow\rangle\left|\sigma^-\right\rangle \pm |\downarrow\rangle\left|\sigma^+\right\rangle\right).$$

These states are created when the quantum dot absorbs one of the two photons in the Bell states. The state of two electron–hole pairs should be in the form $(1/\sqrt{2})(|\uparrow\rangle_1|\downarrow\rangle_2 + |\downarrow\rangle_1|\uparrow\rangle_2)$, so that only the $\Psi^{(+)}$ state in (26) is absorbed by the quantum dot. This controlled absorption results from the Pauli exclusion principle. Linear polarization elements transform the Bell states. A π-retarder, which transforms the $|\sigma^+\rangle$ polarization state to the $|\sigma^-\rangle$ state, interchanges the $\Phi^{(\pm)}$ states and the $\Psi^{(\pm)}$ states. A $\pi/2$-rotator, which provides relative phase (-1) between the $|\sigma^+\rangle$ polarization state and the $|\sigma^-\rangle$ state, exchange the signs as $\Phi^{(\pm)} \to \Phi^{(\mp)}$ and $\Psi^{(\pm)} \to \Psi^{(\mp)}$. The light beam should go through the quantum dot four times, because the electron stays in the excited quantum dot. The states are discriminated by the time of the photon detection event. Therefore, the electron–hole state should remain until the Bell state discrimination is completed. The time for the Bell discrimination would be determined by the time resolution of the photon detection. This requirement may limit the feasibility of the quantum dot BSM devices.

The Bell state detection requires a large TPA coefficient β. We need to combine highly nonlinear materials like quantum dots and a high-Q factor microcavity. The microcavity enhances the field strength of a photon and increases the interaction time. An estimation showed the Q value of the cavity to be larger than 8×10^3 is required, which can be satisfied by the current technology. The complete BSM by a solid state device is thus be realizable in the present TPA detection scheme.

4 Generation of Entangled Photon Pairs by SPDC

Among a variety of technologies required for quantum information, we here focus on the efficient generation of highly entangled photon pairs. As widely recognized, entanglement is one of the most important resources for quantum information processing. Entangled states of two or more particles make possible such phenomena as quantum teleportation [35], superdense coding [36], and quantum computation. It is very clear that the preparation of a maximally entangled state, or a Bell state, is a very important subject. We investigate an improved method for pulse-laser-pumped spontaneous parametric down conversion (SPDC). Use of photon pairs created by pulsed pump is indispensable to realize quantum teleportation and entanglement swapping, where different photons generated from different sources should interact. The time uncertainty of photon creation should be smaller than the coherence time of the photons. This condition can be satisfied with the SPDC photon pair generation by femtosecond laser pulses. Unfortunately, the femtosecond-pulse-pumped SPDC usually shows very poor quantum correlation compared to the continuous wave (cw) case due to large group velocity difference in two photon wave packets. We will show the efforts to improve SPDC in the following.

4.1 SPDC With Two-Crystal Geometry

SPDC is now widely used to generate entangled photon pairs. This method provides highly entangled states with a simple experimental setting. In particular, *Kwiat* et al. [37] have obtained a high flux of the photon pairs from a stack of two type-I phase-matched nonlinear crystals. As shown in Fig. 14, the nonlinear crystals (BBO), whose optical axes are set orthogonal to one another, are pumped by a pulsed UV light polarized in the 45-deg. direction to the optical axis of the crystals. One nonlinear crystal generates two photons polarized in the horizontal direction ($|HH\rangle$) from the vertical component of the pump light, and the other generates ones polarized in the vertical direction ($|VV\rangle$) from the horizontal component of the pump. If we use very thin (0.13 mm in our experiment) crystals, the directions of the photon waves are almost the same, so that one cannot distinguish which crystal generates a photon from the direction of the photons. Therefore, the two-photon state is given by a superposition:

$$\Phi(a,\phi) = a|HH\rangle + \sqrt{1-a^2}e^{i\phi}|VV\rangle. \tag{27}$$

The amplitude a and phase ϕ of the superposition are determined by the polarization state of the pump light. The 45-deg. polarized pump light will provide $a = 1/\sqrt{2}$ and $\phi = 0$. The two-photon state (27) then refers to the maximally entangled state $|\Phi^{(+)}\rangle = (1/\sqrt{2})(|HH\rangle + |VV\rangle)$.

A crucial condition to obtain a highly entangled state in the above scheme is to keep indistinguishability between the two SPDC processes. The group velocity dispersion and birefringence in the crystal may cause differences in the space–time position of the generated photons and make the two processes distinguishable [38]. For example, in the case of 266 nm pump light wavelength and 532 nm SPDC light wavelength, the horizontally polarized SPDC light travels through the first crystal earlier than the horizontally polarized pump light by 135 fs due to the group velocity dispersion and birefringence. The vertically polarized SPDC light generated in the second crystal takes 33 fs more than the horizontally polarized light to travel through the crystal. Therefore, the horizontally polarized SPDC light arrives at the detector 168 fs earlier than the vertically polarized light. This time delay is comparable to the inaccuracy of the SPDC generation equal to the pump pulse duration of 150 fs. The two SPDC processes can be distinguished. Fortunately, this

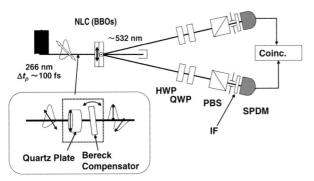

Fig. 14. Schematic of the entangled photon pair generation by spontaneous parametric down-conversion. Cascade of the nonlinear crystals (*NLC*) generates the photon pairs. Group velocity dispersion and birefringence in the NLCs are precompensated with quartz plates and a Bereck compensator. Two-photon states are analyzed with half-wave plates (*HWP*), quarter-wave plates (*QWP*), and polarization beam splitters (*PBS*). Interference filters (*IF*) are placed before the single photon counting modules (*SPCM*)

timing information can be erased by compensation; the horizontal component of the pump pulse should arrive at the nonlinear crystals earlier than the vertical component. The compensation can be done by putting a set of birefringence plates (quartz) and a variable wave-plate before the crystals. The two-photon states were analyzed by quantum state tomography and visibility of two-photon interference. Quantum state tomography provides 4×4 density matrix from the coincidence counts of the 16 combinations, $\{|H\rangle, |V\rangle, |D\rangle, |L\rangle\}_1 \otimes \{|H\rangle, |V\rangle, |D\rangle, |L\rangle\}_2$, where $|D\rangle$ and $|L\rangle$ stand for the linear polarized state at $45°$, and the circularly polarized state in the anti-clockwise direction, respectively. When the precompensation is optimal, the density matrix is close to that of the maximally entangled state, and the visibility is close to unity, as shown in Fig. 15b and Fig. 15e. It should be noted that only $HHHH, VVVV, VVHH, HHVV$ elements are dominant even in the density matrices for inadequate compensation [38], as seen in Fig. 15a and Fig. 15c, which implies that the density matrix can be approximately given by the classical mixture of the $|\Phi^{(+)}\rangle\langle\Phi^{(+)}|$ and $|\Phi^{(-)}\rangle\langle\Phi^{(-)}|$.

We proposed another method to produce a pulsed polarization entangled-photon pair in a two-crystal geometry [39]. In our geometry, two identical beta-barium-borate (BBO) crystals were stacked vertically, as seen in Fig. 16a. These two crystals, cut to satisfy the type-I phase matching condition, were oriented with their optical axes aligned in perpendicular directions. The crystals pumped simultaneously by a laser beam with diameter a and $45°$ polarization generate photon pairs. The upper crystal produces a horizontal polarization photon pair $|H\rangle$, and the lower produces a vertical polarization photon pair $|V\rangle$. These two processes are simultaneous, but separable in space. Therefore, if the spatial information is erased, the two possible down-

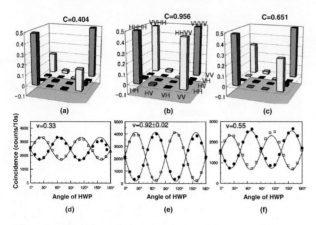

Fig. 15. Density matrices estimated by quantum tomography **(a)**–**(c)**, and the interference fringes **(d)**–**(f)** of the two photon states, without compensation (a), (d), optimal compensation (b), (e), and over compensation (c), (f)

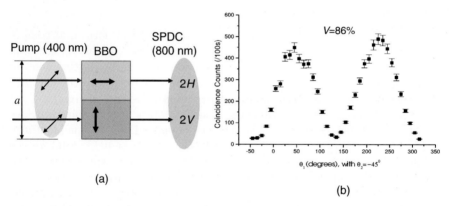

Fig. 16. **(a)** The two-crystal geometry. Two identical crystals are stacked, with their optical axes aligned in perpendicular directions. a is the diameter of the pump laser with 45° polarization. **(b)** Two-photon quantum interference for polarization variable. One polarizer is fixed at 45°, while the other polarizer is rotated with 10° steps

conversion processes are coherent and the photon pairs are entangled. The erasure of the spatial information was easily achieved by focusing the output photons into a single mode fiber.

We obtained more than 86 % of visibility in two-photon interference without any narrow-band filter nor time compensation, as shown in Fig. 16b. Such a high visibility is one evidence of quantum entanglement. The main advantage of this scheme is that we do not need to consider the suitable time compensation, nor the postselection on the spectrum by a narrow bandwidth

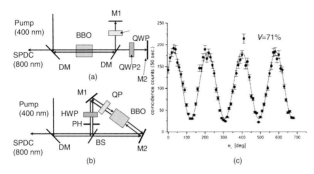

Fig. 17. (a) Michelson interferometer based SPDC. *QWP:* quarter-wave plate; QWP1 for 400 nm with 0°, QWP2 for 800 nm with 45°; *M1:* mirror for 400 nm; *M2:* mirror for 800 nm. (b) Sagnac interferometer based SPDC. *HWP:* half-wave plate for 800 nm, *BS:* beam splitter, *QP:* quartz plate, *PH:* pin hole. (c) Two-photon interference of SPDC photon pairs generated in a Sagnac interferometer. One polarizer is fixed at 45°; while the other polarizer is rotated

filter. Furthermore, visibility should be, in principle, insensitive to the thickness of crystal, so we can use the thicker crystal to increase the intensity of photon pair. The results show that our scheme will be a good scheme for generation of the polarization entangled state.

4.2 Interferometric Generation of Entangled Photon Pairs

An interferometric source of polarization-entangled photons has been proposed and demonstrated [40, 41], in which the outputs of two spatially separated SPDC processes are combined by an interferometer. It has been known that the interferometric technique could produce an entangled photon pair independent of wavelength and angle of emission. We followed the interferometric generation of a polarization entangled photons by coherently combining two collinear type-I SPDC processes via a Michelson interferometer constructed as Fig. 17a [42]. We have obtained a high visibility in the two-photon interference with a 10 nm interference filter by a femtosecond laser pump.

The main problem in the interferometric technique is to stabilize the interferometer against environmental disturbances for a long time. Therefore, one usually has to employ an active stabilization technique. We present a very simple solution of stabilization by using a Sagnac interferometer [43]. Sagnac interferometers, often used for optical sensors, particularly with fiber optics, show stable interference because the optical paths are common but different in direction. We obtained very stable generation of entangled photons in our experimental set-up exposed to air flows.

To generate a polarization-entangled photon pair, we placed a nonlinear crystal cut to satisfy the type-I phase matching condition to the Sagnac loop

(Fig. 17b). We also placed a half-wave plate (HWP) for the photons of frequency ω to rotate the angle of linear polarization by 90° for the photons of frequency ω, without any effects on the light of frequency 2ω. When 2ω pump light arrives at the Sagnac interferometer, half of the pump is transmitted through the BS to generate, for example, a horizontally polarized photon pair $|HH\rangle$. The HWP rotates the polarization of the generated photon pair to vertically polarized one $|VV\rangle$. The other half of the 2ω pump is reflected by the BS to pass through the HWP, where the polarization of the pump is unaffected, and to generate another horizontal photon pair $|HH\rangle$. These two possible processes are mixed at the BS and the information on which process generates a photon pair is erased. We obtain a polarization-entangled photon pair $|HH\rangle_{12} + |VV\rangle_{12}$, where subscripts 1 and 2 refer to the outputs of the BS. The erasure would be easy with a cw laser pump and a very thin BS. However, the process turns out to be more complicated, when it is pumped by femtosecond laser pulses. We cannot neglect the thickness of the BS and HWP, because they introduce different dispersions between the pump laser and the SPDC light, and make the two SPDC processes distinguishable. Therefore, dispersion compensation components are necessary to obtain highly entangled photons.

The experimental results are shown in Fig. 17c. The visibility was about 71%. The main reasons that visibility did not reach 100% were due to imperfections in the experiment, such as the incomplete dispersion compensation, and the difference in the transmittance between two directions of the Sagnac interferometer.

4.3 A New Material for SPDC: Periodically Poled KTP

Though SPDC generation of entangled photon pairs has been established, the photon pair production rate still needs to be increased in order to improve the signal-to-noise ratio in the measurement. The low production rate problem is serious in the experiments that require more than two entangled photon pairs. Recently, efficient generation of the photon pairs has been reported using periodically poled crystals in bulk or waveguide structure. The periodically-poled crystal can utilize the largest elements of a nonlinear optics tensor by the technique of quasi-phase matching (QPM). However, recent works are focused on SPDC from periodically-poled lithium niobate (PPLN) or periodically-poled potassium titanyl phosphate (PPKTP) by cw lasers. We have studied the generation of collinearly propagating pulsed photon pairs by pumping a type-0 phase matching bulk PPKTP crystal with an ultrashort pulse laser [44].

We measured the coincidence count of the SPDC photons to estimate the photon pair generation rate. The PPKTP crystal of 1.05 mm (height)×2.1 mm (width)×2.12 mm (length) with the grating period of 3.25 μm was antireflection coated for 800 nm and 400 nm on both facets. The crystal was placed at a temperature-stabilized (29°C) holder with a stability of 0.01°C. The peak wavelength of the pumping pulses (SHG of a mode locked Ti:S laser) was 399.44 nm with the spectral width of about 3.2 nm. A lens ($f = 20$ cm) focused the pumping pulses on to the crystal. The pumping power measured before PPKTP crystal was 1 mw. We used two red filters (color glass filter RG715) with transmission coefficient 90 % at 800 nm to cut the remaining pump pulses, then we coupled the SPDC photon pairs to a 2 m single-mode fiber (design wavelength 820 nm, operating wavelength range is typically 50 nm below and 200 nm above the design wavelength) with objective lens ($NA = 0.15$.) The output of the fiber was divided by a 50/50 fiber coupler and detected by single-photon detectors. The time window of coincidence counting was 4 ns. The measured coupling efficiency to the single mode was about 14.4 %. The loss of the coupling was about 50 %. Under these conditions, we obtained the coincidence counts about 3200 per second. If we consider the coupling efficiency and the 50 % loss by the fiber beam splitter, the coincidence counts should be estimated to be about 1.09×10^5 /s/mW. To be the best of our knowledge, this is the highest coincidence count in ultrashort pulse case.

The efficient photon pair generation from PPKTP enables us to explore many photon interference [45]. We used a PPKTP crystal of 1.05 mm (height)×2.1 mm (width)×5 mm (length) in the present experiment. Pump light of 400 nm was generated by SHG of mode locked Ti:S laser in another PPKTP crystal. The pump power was 5.3 mW at the SPDC crystal. We observed interference of SPDC photons by Michelson interferometer in the setup shown in Fig. 18. The output of the interferometer was coupled to the single-mode fiber and divided by a 50/50 fiber coupler. One output of the fiber coupler was forwarded to another fiber coupler. The outputs of the couplers were connected to photon detectors. Two- and four-photon interferences were observed by measuring the twofold and threefold coincidence, respectively. We took the interference fringe at several values (τ) of the coarse path difference, which were chosen to examine the effects of the coherence time of SPDC photons (Δt) and the coherence time of pump laser(ΔT).

Figure 19 shows the interference fringes for the single-photon counts and the threefold coincidence counts at small (≈ 0) coarse path difference. The single-photon counts were fitted well by $1 + \cos(2\pi x/x_1)$ with the period $x_1 = 783 \pm 192$ nm. The threefold coincidence counts were fitted by $3 + 4\cos(2\pi x/x_3) + \cos(2\pi x/x_3)$ with the period $x_3 = 795 \pm 195$ nm, which corresponds to the product of two photon interference. As we increased the coarse path difference, the single-photon interference disappeared. The htreefold coincidence counts still showed the oscillation at the coarse path difference of 400 μm, with period of 426 ± 105 nm. The oscillation disappeared in the tree coincidence counts at the coarse path difference 1.28 mm.

The above experiment results suggest the coherent length of the pump pulse was close to 1 mm, much longer than that of the SPDC photons (on the order of tens of nm). This estimation is reasonable, because the coherent length of the pump pulse is determined by the band width (0.09 nm) of the

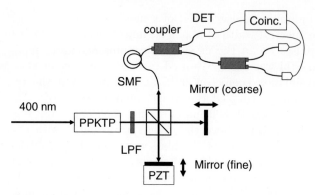

Fig. 18. Experimental setup for four-photon interference. SPDC photons from a PPKTP crystal were divided by two fiber couplers after passing thorough a Michelson interferometer composed of a beam splitter and two mirrors. The position of one mirror was tuned by PZT

PPKTP crystal that acts as a grating. The band width corresponds to the coherent length of 1.2 mm (i.e., the coherent time $\Delta t = 4$ ps), if we take account of the pulse shape (Gaussian). On the other hand, the coherent time of the SPDC photons is determined by the original pulse duration of the pump pulse ($\Delta T \approx 100 fs$).

From the analysis of the interference fringes, we conclude:

1. $\tau < \Delta t < \Delta T$: We observed all of the one-photon count, two-photon coincidence count, and three-photon coincidence count varied periodically with the change of τ. The period corresponded to 800 nm, the wavelength of SPDC photons.
2. $\Delta t < \tau < \Delta T$: We observed two-photon coincidence count and three-photon coincidence count varied periodically with the change of τ, whereas the one-photon count remained constant. The period corresponded to 400 nm. This implies the four photons in our experiment are not a four-photon state, but two independent photon pairs. If these four photons are in a true four-photon state, we should observe the period corresponding to 200 nm.
3. $\Delta t < \Delta T < \tau$: All the counts were independent of τ

The above observations agree well with the theory given by Riedmatten et al. [46]

5 Conclusion

So far, we have developed devices and systems to realize the promises made by quantum information theory. Some of the achievements are: demonstration of high-sensitivity photon detectors for optical communication wavelength,

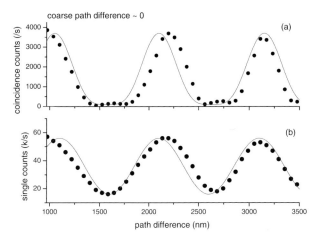

Fig. 19. The tree coincidence counts (**a**) and the single photon counts (**b**) at small (≈ 0) coarse path difference

proposal of Bell state measurement device, research on microscopic optical responses of semiconductor quantum dots, demonstration of quantum Fourier transform followed by measurement, and record breaking results on quantum key distribution systems. Remarkable progress in this field has been accomplished by the researchers in a number of institutes. Now the systems that rely only on a single photon (i.e., QKD system) cease being a proof-of-principle, and head for the market. Nevertheless, practical implementations of the many-qubit systems are still in their infancy. We need to develop reliable quantum memories and controlled unitary gates in order to realize a quantum computer that solves practically meaningful tasks. Even much shorter-term goals, such as quantum networks based on quantum repeaters, are beyond our current technology. Still, we believe the recent progress will turn out to be important steps to tackle the problem. Collaboration between the researchers of materials, devices, architecture, and theory will pave the way to construct the quantum information technology.

References

[1] C. H. Bennett, G. Brassard: in *Procedings of International Conference on Computer, System and Signal Processing* (Bangalore, India 1984) p. 175
[2] C. H. Bennett, F. Bessate, G. Brassard, L. Salvail, J. Smolin: J. Crypt. **5**, 3 (1992)
[3] N. Lütkenhaus: Phys. Rev. A **61**, 052304 (2000)
[4] M. Hamada: J. Phys. A: Math. Gen. **37**, 8303 (2004)
[5] P. D. Townsend, J. G. Rarity, P. R. Tapster: Electron. Lett. **29**, 634 (1993)
[6] R. Hughes, G. Morgan, C. Peterson: J. Mod. Opt **47**, 533 (2000)

[7] D. Stucki, N. Gisin, O. Guinnard, G. Ribordy, H. Zbinden: New J. Phys. **4**, 41 (2002)
[8] A. Yoshizawa, H. Tsuchida: Jpn. J. Appl. Phys. **40**, 200 (2001)
[9] D. S. Bethune, W. P. Risk: J. Quantum Electron., IEEE **36**, 340 (2000)
[10] A. Tomita, K. Nakamura: Opt. Lett. **27**, 1827 (2002)
[11] M. Dušek, M. Jahma, N. Lütkenhaus: Electron. Lett. **39**, 1199 (2003)
[12] Y. Nambu, T. Hatanaka, K. Nakamura: Jpn. J. Appl. Phys. **43**, L1109 (2004)
[13] T. Kimura, Y. Nambu, T. Hatanaka, A. Tomita, H. Kosaka, K. Nakamura: Jpn. J. Appl. Phys. **43**, L1217 (2004)
[14] A. Tanaka, A. Tomita, A. Tajima, T. Takeuchi, S. Takahashi, Y. Nambu: in *30th European Conference on Optical Communication (ECOC)* (2004)
[15] M. Dušek, M. Jahma, N. Lütkenhaus: Phys. Rev. A **62**, 022306 (2000)
[16] A. Tanaka, W. Maeda, A. Tajima, S. Takahashi: in *Proceedings of the 18th Annual Meeting of the IEEE Lasers and Electro-Optics Society* (Sidney, Australia 2005) p. 557
[17] A. Tomita, K. Nakamura: Int. J. Quantum Inf. **2**, 119 (2004)
[18] R. Cleve, A. Ekert, C. Macchiavello, M. Mosca: Proc. R. Soc. London A **454**, 339 (1998)
[19] P. W. Shor: SIAM J. Comp. **26**, 1484 (1997)
[20] R. B. Griffiths, C.-S. Niu: Phys. Rev. Lett. **76**, 3228 (1996)
[21] S. Parker, M. B. Plenio: Phys. Rev. Lett. **85**, 3049 (2000)
[22] S. Beauregard: Quantum Inf. and Comp. **3**, 175 (2003)
[23] A. Barenco, A. Ekert, K.-A. Suominen, P. Törmä: Phys. Rev. A **54**, 139 (1996)
[24] D. F. Phillips, A. Fleischhauer, A. Mair, R. L. Walsworth, M. D. Lukin: Phys. Rev. Lett. **86**, 783 (2001)
[25] A. Kuzmich, N. P. Bigelow, L. Mandel: Euro. Phys. Lett. **42**, 481 (1998)
[26] Y. Takahashi, K. Honda, K. Toyoda, K. Ishikawa, T. Yabuzaki: Phys. Rev. A **60**, 4974 (1999)
[27] E. Knill, R. Laflamme, G. J. Milburn: Nature **409**, 46 (2001)
[28] D. S. Abrams, S. Llyod: Phys. Rev. Lett. **83**, 5162 (1999)
[29] I. L. Chuang: Phys. Rev. Lett. **85**, 2006 (2000)
[30] L. K. Grover: Phys. Rev. Lett. **79**, 325 (1996)
[31] D. Gottesman, I. L. Chuang: Nature **402**, 390 (1999)
[32] N. Lütkenhaus, J. Calsamiglia, K.-A. Suominen: Phys. Rev. A **59**, 3295 (1999)
[33] A. Tomita: Phys. Lett. A **282**, 331 (2001)
[34] M. O. Scully, B.-G. Englert, C. J. Bednar: Phys. Rev. Lett. **83**, 4433 (1999)
[35] C. H. Bennett, G. Brassard, C. Crepeau, R. Jozsa, A. Peres, W. K. Wootters: Phys. Rev. Lett. **70**, 1895 (1993)
[36] C. H. Bennett, S. J. Wiesner: Phys. Rev. Lett. **69**, 2881 (1992)
[37] P. G. Kwiat, E. Waks, A. G. White, I. Appelbaum, P. H. Eberhard: Phys. Rev. A **60**, 773 (1999)
[38] Y. Nambu, K. Usami, Y. Tsuda, K. Matsumoto, K. Nakamura: Phys. Rev. A **66**, 033816 (2002)
[39] B.-S. Shi, A. Tomita: Phys. Rev. A **67**, 043804 (2003)
[40] Y.-H. Kim, M. V. Chekhova, S. P. Kulik, M. H. Rubin, Y. Shih: Phys. Rev. A **63**, 062301 (2001)
[41] Y. H. Kim, W. P. Grice: J. Mod. Opt. **49**, 2309 (2002)
[42] B.-S. Shi, A. Tomita: Opt. Commun. **235**, 247 (2004)
[43] B.-S. Shi, A. Tomita: Phys. Rev. A **69**, 247 (2004)

[44] B.-S. Shi, A. Tomita: J. Opt. Soc Am. **21**, 2081 (2004)
[45] B.-S. Shi, A. Tomita: J. Mod. Opt. **52**, 755 (2005)
[46] H. Riedmatten, V. Scarani, I. Marcikic, A. Acin, W. Tittel, H. Zbinden, N. Gisin: J. Mod. Opt. **51**, 1637 (2004)

Index

aerial fiber cable, 255
afterpulse, 245, 253
avalanche photodiode (APD), 246

Bell state, 264, 265
Bell state measurement (BSM), 264
birefringence, 250, 263, 266

coherence time, 271
coincidence, 270, 271
controlled-unitary operation, 262

dark count, 245
detection efficiency, 245
dispersion, 266

entangled photon, 265
entanglement, 265

Faraday mirror (FM), 251

Gaiger mode, 246
gated mode, 246

Hadamard gate, 256

interference fringe, 271
interferometer, Mach–Zehnder, 249
interferometer, Michelson, 269, 271
interferometer, Sagnac, 256, 269

kick-back effect, 262

loop mirror, 252

majority voting, 258, 260
maximally entangled state, 265

periodically-poled potassium titanyl phosphate (PPKTP), 270
phase estimation, 256
phase matching, 270
phase-matching, 266
planar lightwave circuit (PLC), 248
plug-and-play, 248, 251
PNS attack, 254

quantum bit error rate (QBER), 245, 250
quantum dot, 264
quantum Fourier transform followed by measurement (MQFT), 256
quantum key distribution (QKD), 244
quantum key generation field trial, 255
quantum state tomography, 267
quantum teleportation, 264, 265
quasi-phase matching (QPM), 270

separable, 262
single-photon detector, 245
spontaneous parametric down conversion (SPDC), 265

two-photon absorption (TPA), 264
two-photon interference, 268, 269

visibility, 245, 248, 250, 267, 268

Index

1-sensitive, 8
ϵ-biased oracle, 20
n-partite, 125, 129

aerial fiber cable, 255
afterpulse, 245, 253
Akaike's information criterion (AIC), 57
amplitude amplification, 23
asymmetric channel noise, 210
asymmetric quantum cloning machine, 105
avalanche photodiode (APD), 246

balanced matrix, 13
BB84, 186, 218, 220, 221, 224, 227, 230
beam splitter (BS), 50, 114
beam splitter entangler, 114
Bell state, 112, 120, 126, 264, 265
Bell state measurement (BSM), 128, 264
Bell type inequality, 123
Bernstein–Vazirani problem, 4
biCNOT, 193–195
bipartite, 129
bipartite,entanglement,infinite-dimensional space, 126
birefringence, 250, 263, 266
bit-flip, 191, 194, 196, 211, 215, 220
Bloch vector, 69
Boson–Fock space, 49, 114, 126

channel coding, 55
channel state, 139
cloning machine, 49, 65
CNOT, 193, 194, 220
coherence time, 271
coherent information, 119

coherent states, 196
coincidence, 270, 271
collective channel noise, 218
collective measurement, 190, 225
column flip, 8
compression, 55, 121
concurrence, 52, 121
controlled-unitary operation, 262
convex roof, 137
coset sampling method, 177, 179
counting rate, 201–204
covariant, 48
cover data, 235

dark count, 245
decoy-state method, 200
detection efficiency, 245
dispersion, 266

embedder, 236
entangled photon, 265
entangled state, 126, 128
entanglement, 52, 57, 114, 124, 265
entanglement concentration, 55, 112
entanglement cost, 137
entanglement distillation, 111, 119
entanglement of formation (EoF), 121, 137
entanglement purification, 188, 190
entropy of entanglement, 137
equivalence (EQ), 22
error correction, 187, 189, 195
error rejection, 216, 220, 225
error test, 189
exact learning, 15
extractor, 236

Faraday mirror (FM), 251
fidelity, 64, 118

final key, 195
four-state protocol, 214, 225

Gaiger mode, 246
gated mode, 246
Gaussian state, 49, 114
GHZ state, 126, 129
Goldreich–Levin (GL) problem, 21
graph automorphism problem, 177
Grover search, 4, 19

Hadamard gate, 256
halving algorithm, 11
hard-core predicate, 169
hashing, 188
hidden subgroup problem, 179
Holevo capacity, 134
hybrid argument, 5, 180
hybrid matrix, 13
hyperdeterminant, 125

imperfect oracle, 15
imperfect source, 196
individual measurement, 190
inner product (IP), 22
input oracle, 5
interference fringe, 271
interferometer, Mach–Zehnder, 249
interferometer, Michelson, 269, 271
interferometer, Sagnac, 256, 269

key rate, 196, 198, 206, 209
kick-back effect, 262

linear cost constraint, 140
LOCC (local operation and classical communication), 47, 54, 55, 112
loop mirror, 252

majority voting, 258, 260
maximally entangled state, 111, 265
minimal output entropy, 136
minimum output entropy, 135
multipartite, 124
multipartite,entangled state, 128
multiphoton pulses, 196, 199
multitarget Grover search, 9

no-cloning theorem, 63

one-way function, 168, 185
one-way permutation, 169–171
optimal average input, 153
oracle, 3
oracle computation, 3
oracle identification, 3
oracle identification problem, 5, 20

Peres–Horodecki criterion, 93
periodically-poled potassium titanyl phosphate (PPKTP), 270
phase covariant, 81
phase estimation, 256
phase matching, 270
phase-flip, 191, 194, 196, 197, 211, 215, 220, 225
phase-matching, 266
planar lightwave circuit (PLC), 248
plug-and-play, 248, 251
PNS attack, 199, 254
polynomial method, 5
POVM (positive operator-valued measure), 46, 54
prepare-and-measure, 187, 195
privacy amplification, 187, 195
probabilistic quantum cloning machine, 106
pseudo identity operator, 172
pseudo reflection operator, 172
pseudorandom generator, 169
public key, 185
public-key cryptosystem, 176

QECC (quantum error correcting code), 118–120
quantum adversary argument, 5
quantum adversary method, 5
quantum bit error rate (QBER), 245, 250
quantum capacity, 119
quantum computation, 5
quantum dot, 264
quantum Fourier transform followed by measurement (MQFT), 256
quantum key distribution (QKD), 120, 186, 244
quantum key generation field trial, 255
quantum learning theory, 11
quantum query complexity, 19

quantum state computational distinguishability, 176
quantum state tomography, 267
quantum teleportation, 264, 265
quantum adversary argument, 21
quasi-phase matching (QPM), 270
query complexity, 3, 5

reliability function, 118
remote state preparation (RSP), 128
robust, 20
robust quantum algorithm, 16
RSA, 186

Schwinger representation, 75
secure private communication, 185
separable, 134, 262
shrinking factor, 67
Simultaneous Schmidt decomposition, 123
single-photon detector, 245
six-state protocol, 213, 217, 225
SLOCC convertibility, 126
spontaneous parametric down conversion (SPDC), 56, 223, 225, 230, 265
squeezed state, 48

state entangled, 53
steganography, 235
stego-data, 235
stochastic local operation and classical communication (SLOCC), 124
Strong superadditivity, 137
strongly quantum one-way permutation, 172
symplectic code, 119

tagged bits, 196, 199, 200, 205
TP-CP (trace-preserving completely positive) map, 117
trapdoor property, 178
two-photon absorption (TPA), 264
two-photon interference, 268, 269

unconditional security, 187, 206, 216, 230
universal test, 169, 175
UQCM, 63, 65

visibility, 245, 248, 250, 267, 268

W state, 126
weakly quantum one-way permutation, 171

Topics in Applied Physics

85 **Solid–Liquid Interfaces**
    Macroscopic Phenomena – Microscopic Understanding
    By K. Wandelt and S. Thurgate (Eds.) 2003, 228 Figs. XVIII, 444 pages

86 **Infrared Holography for Optical Communications**
    Techniques, Materials, and Devices
    By P. Boffi, D. Piccinin, M. C. Ubaldi (Eds.) 2003, 90 Figs. XII, 182 pages

87 **Spin Dynamics in Confined Magnetic Structures II**
    By B. Hillebrands and K. Ounadjela (Eds.) 2003, 179 Figs. XVI, 321 pages

88 **Optical Nanotechnologies**
    The Manipulation of Surface and Local Plasmons
    By J. Tominaga and D. P. Tsai (Eds.) 2003, 168 Figs. XII, 212 pages

89 **Solid-State Mid-Infrared Laser Sources**
    By I. T. Sorokina and K. L. Vodopyanov (Eds.) 2003, 263 Figs. XVI, 557 pages

90 **Single Quantum Dots**
    Fundamentals, Applications, and New Concepts
    By P. Michler (Ed.) 2003, 181 Figs. XII, 352 pages

91 **Vortex Electronis and SQUIDs**
    By T. Kobayashi, H. Hayakawa, M. Tonouchi (Eds.) 2003, 259 Figs. XII, 302 pages

92 **Ultrafast Dynamical Processes in Semiconductors**
    By K.-T. Tsen (Ed.) 2004, 190 Figs. XI, 400 pages

93 **Ferroelectric Random Access Memories**
    Fundamentals and Applications
    By H. Ishiwara, M. Okuyama, Y. Arimoto (Eds.) 2004, Approx. 200 Figs. XIV, 288 pages

94 **Silicon Photonics**
    By L. Pavesi, D.J. Lockwood (Eds.) 2004, 262 Figs. XVI, 397 pages

95 **Few-Cycle Laser Pulse Generation and Its Applications**
    By Franz X. Kärtner (Ed.) 2004, 209 Figs. XIV, 448 pages

96 **Femtsosecond Technology for Technical and Medical Applications**
    By F. Dausinger, F. Lichtner, H. Lubatschowski (Eds.) 2004, 224 Figs. XIII 326 pages

97 **Terahertz Optoelectronics**
    By K. Sakai (Ed.) 2005, 270 Figs. XIII, 387 pages

98 **Ferroelectric Thin Films**
    Basic Properies and Device Physics for Memory Applications
    By M. Okuyama, Y. Ishibashi (Eds.) 2005, 172 Figs. XIII, 244 pages

99 **Cryogenic Particle Detection**
    By Ch. Enss (Ed.) 2005, 238 Figs. XVI, 509 pages

100 **Carbon**
    The Future Material for Advanced Technology Applications
    By G. Messina, S. Santangelo (Eds.) 2006, 245 Figs. XXII, 529 pages

101 **Spin Dynamics in Confined Magnetic Structures III**
    By B. Hillebrands, A. Thiaville (Eds.) 2006, approx. 165 Figs. XIV, 360 pages

102 **Quantum Computation and Information**
    From Theory to Experiment
    By H. Imai, M. Hayashi (Eds.) 2006, 49 Figs. XV, 281 pages

103 **Surface-Enhanced Raman Scattering**
    Physics and Applications
    By K. Kneipp, M. Moskovits, H. Kneipp (Eds.) 2006, 221 Figs. XVII, 471 pages

104 **Theory of Defects in Semiconductors**
    By D. A. Drabold, S. Estreicher (Eds.) 2006, approx. 60 Figs. XVI, 287 pages